Dietary ω3 and ω6 Fatty Acids

Biological Effects and Nutritional Essentiality

NATO ASI Series

Advanced Science Institutes Series

A series presenting the results of activities sponsored by the NATO Science Committee, which aims at the dissemination of advanced scientific and technological knowledge, with a view to strengthening links between scientific communities.

The series is published by an international board of publishers in conjunction with the NATO Scientific Affairs Division

| A | Life Sciences | Plenum Publishing Corporation |
| B | Physics | New York and London |

C	Mathematical and Physical Sciences	Kluwer Academic Publishers
D	Behavioral and Social Sciences	Dordrecht, Boston, and London
E	Applied Sciences	

F	Computer and Systems Sciences	Springer-Verlag
G	Ecological Sciences	Berlin, Heidelberg, New York, London,
H	Cell Biology	Paris, and Tokyo

Recent Volumes in this Series

Volume 166—Vascular Dynamics: Physiological Perspectives
edited by N. Westerhof and D. R. Gross

Volume 167—Human Apolipoprotein Mutants 2: From Gene Structure to
Phenotypic Expression
edited by C. R. Sirtori, G. Franceschini,
H. B. Brewer, Jr., and G. Assmann

Volume 168—Techniques and New Developments in Photosynthesis Research
edited by J. Barber and R. Malkin

Volume 169—Evolutionary Tinkering in Gene Expression
edited by Marianne Grunberg-Manago, Brian F. C. Clark,
and Hans G. Zachau

Volume 170—*ras* Oncogenes
edited by Demetrios Spandidos

Volume 171—Dietary $\omega 3$ and $\omega 6$ Fatty Acids: Biological Effects
and Nutritional Essentiality
edited by Claudio Galli and Artemis P. Simopoulos

Volume 172—Recent Trends in Regeneration Research
edited by V. Kiortsis, S. Koussoulakos, and H. Wallace

Series A: Life Sciences

Dietary ω3 and ω6 Fatty Acids

Biological Effects and Nutritional Essentiality

Edited by

Claudio Galli

Institute of Pharmacological Sciences
University of Milan
Milan, Italy

and

Artemis P. Simopoulos

ILSI Research Foundation
Washington, D.C.

Springer Science+Business Media, LLC

Proceedings of a NATO Advanced Research Workshop on
Dietary ω3 and ω6 Fatty Acids: Biological Effects
and Nutritional Essentiality,
held June 20–23, 1988,
in Belgirate, Italy

Library of Congress Cataloging in Publication Data

NATO Advanced Research Workshop on Dietary [Omega] 3 and [Omega] 6 Fatty
Acids: Biological Effects and Nutritional Essentiality (1988: Belgirate, Italy)
    Dietary [omega] 3 and [omega] 6 fatty acids.
    (NATO ASI series. Series A, Life sciences; v. 171)
    "Proceedings of a NATO Advanced Research Workshop on Dietary [Omega] 3
and [Omega] 6 Fatty Acids: Biological Effects and Nutritional Essentiality, held
June 20–23, 1988, Belgirate, Italy"—T.p. verso.
    "Published in cooperation with NATO Scientific Affairs Division."
    Includes bibliographies and index.
    1. Omega-3 fatty acids—Physiological effect—Congresses. 2. Omega-6 fatty
acids—Physiological effect—Congresses. 3. Unsaturated fatty acids in human
nutrition—Congresses. I. Galli, Claudio. II. Simopoulos, Artemis P., 1933-   .
III. North Atlantic Treaty Organization. Scientific Affairs Division. IV. Title. V.
Series. [DNLM: 1. Dietary Fats, Unsaturated—congresses. 2. Fatty Acids, Un-
saturated—congresses. 3. Fish Oils—congresses. QU 90 N2789d 1988]
QP752.F35N38   1988   612.3'97   89-8583
ISBN 978-1-4757-2045-7   ISBN 978-1-4757-2043-3 (eBook)
DOI 10.1007/978-1-4757-2043-3

PREFACE

On June 24-26, 1985, a major International Conference on the Health
Effects of Polyunsaturated Fatty Acids in Seafoods was held in Washington,
D.C. The conference had two objectives: (1) to review the research data
on the health effects of polyunsaturated fatty acids in seafoods in terms
of the impact of omega-3 fatty acids on eicosanoid formation, thrombosis
and inflammation, and the role of docosahexaenoic acid in membrane
function and metabolism, and (2) to develop a research agenda to determine
the spectrum of the health effects of polyunsaturated fatty acids of
seafood origin in the American diet.

The 1985 conference established the fact that omega-3 fatty acids of
marine origin - eicosapentaenoic acid (EPA) and docosahexaenoic acid
(DHA) - play important roles in prostaglandin metabolism, thrombosis and
atherosclerosis, immunology and inflammation, and membrane function.

In response to the conference recommendations, the Congress of the
United States provided special funding for the establishment of a "test
materials laboratory" within the US Department of Commerce to produce
under documented quality control the types and quantities of omega-3 test
materials required by biomedical researchers. The forms of test
materials to be produced include refined fish oil, polyunsaturated fatty
acid enhanced triglycerides, concentrates of esters of fatty acids,
purified omega-3 fatty acids, and omega-3 mono-, di- and triglyceride
mixtures. The proceedings of the conference were published (1) and in
response to the conference recommendations the National Institutes of
Health (NIH) developed a program announcement that led to expansion of
research on omega-3 fatty acids. This 1985 conference and other
conferences held in 1986, 1987 (2) and 1988 (3,4) focused on the role of
omega-3 fatty acids of marine origin. A NATO Advanced Research Workshop
was also held in 1986 devoted to "Advanced Technologies and Their
Nutritional Implications in the Production of Edible Fats" (5), in which
selected aspects concernings the biological effects of highly unsaturated
fatty acids of the omega-3 series were presented and discussed. Only
slight attention was given to the role of omega-3 fatty acids from
terrestrial sources such as alpha-linolenic acid and its metabolism to
its longer chain derivatives and its relationship to omega-6 fatty acids,
specifically linoleic acid and arachidonic acid, and most importantly the
question of the essentiality of omega-3 fatty acids. Research on these
issues has expanded over the past 5 years and the role of omega-3 fatty

acids in growth and development has been progressing in parallel to the research on the role of omega-3 fatty acids in thrombosis, atherosclerosis, inflammation, cancer, autoimmune disorders and aging.

The long chain members of the omega-3 fatty acid series EPA and DHA are found in relatively high concentrations in lipids from various tissues of marine animals, whereas in terrestrial animals, e.g. mammals, they appear to be concentrated in structural lipids of selected membranes, such as the retina and hte synaptic membranes of the central nervous system. In addition, in mammalian tissues mainly the very long chain members of the omega-3 series, DHA is present rather than the less unsaturated product EPA which is found in high concentrations in marine animals.

It would thus appear that in terrestrial animals and especially in the highest species, very complex regulatory processes are at work favoring the accumulation of selected highly unsaturated omega-3 fatty acids in specific sites. Also, the endogenous metabolic conversion of the short chain precursor of dietary origin, alpha-linolenic acid, to the long chain highly unsaturated derivatives, rather than the intake of preformed products in the diet appears to be the major process for the utilization of omega-3 fatty acids. Alpha-linolenic acid is very rapidly converted to the long chain members of the omega-3 series, but the incorporation of the formed products into cell phospholipids depends to a large extent on the selectivity of acyl transferase reactions for different substrates, in addition to those of the chain elongating and desaturating enzymes.

Finally the interactions of omega-6 and omega-3 fatty acids at various stages of their utilization (intake, desaturation and elongation, transport and uptake in cellular lipids) are certainly important features in the overall fates of long chain polyunsaturated fatty acids of both series.

Although the effects of the intake of omega-3 fatty acids on several physiological parameters show interesting applications for the dietary prevention of pathological states, it is felt that a more general and balanced evaluation of the biological roles of both omega-6 and omega-3 fatty acid series in different biological systems, of their relative distribution and stability in different food sources and of the comparative utilization of short chain precursors and long chain products of both series, is essential.

It was therefore thought timely and most appropriate to hold a NATO Advanced Research Workshop on Dietary w3 and w6 Fatty Acids: Biological Effects and Nutritional Essentiality.

The objectives of the workshop were to assess the comparative biological significance of omega-3 and omega-6 fatty acids in relation to the a) dietary availability, utilization and metabolism of different members of these fatty acid series; and b) roles of endogenous long chain members of the omega-3 (e.g. 20:5 and 22:6) and omega-6 (e.g. 20:3 and 20:4) series in different biological systems.

These aspects were extensively analyzed and discussed at the workshop. The workshop consisted of review papers on various aspects of omega-3 fatty acids and on selected topics on utilization, metabolism and interaction between the fatty acids of the two-series omega-3 and omega-6. There were extensive discussions following each presentation and a special round table discussion on the essentiality of omega-3 fatty acids.

Approximately 120 scientists from 15 countries participated. The program consisted of papers presented at plenary sessions and a poster session. There were seven sessions on the following topics: Dietary Sources of Omega-3 and Omega-6 Fatty Acids; Chemistry, Biosynthesis and Interactions of Omega-3 and Omega-6 Fatty Acids; The Role of Omega-3 and Omega-6 Fatty Acids in Development; Biological Effects of Omega-3 and Omega-6 Fatty Acids on Cell Activation Processes; two sessions on The Role of Omega-3 and Omega-6 Fatty Acids in Human Diseases; and a Round Table on the Essentiality of Omega-3 Fatty Acids.   Thirthy-two papers and an equal number of posters were presented.

The Executive Summary, at the end of the Proceedings, highlights the presentations, discussions, poster session papers and is followed by comprehensive "General Recommendations". We are indebted to the presenters at the workshop, and at the poster session for their excellent papers, and we are confident that the proceedings of this workshop will be an important milestone in the field. Many of the participants emphasized the importance of this workshop and recommended to hold such a workshop every 2 years in order to present new findings and exchange ideas and views in this very rapidly advancing field.

Finally, we would like to express our gratitude to the program and workshop participants and the financial sponsors of the workshop, mainly NATO, the Division of Nutritional Sciences-ILSI Research Foundation and the Nutrition Foundation of Italy, and the following cosponsors : BSN, Canola Council, DNA Plant Technology Corporation, Hoffmann-La Roche, ILSI-Europe, Nestle', Procter & Gamble Company, Quaker-Chiari & Forti S.p.A., Ross Laboratories, Star S.p.A., U.S. Department of Commerce-National Marine Fisheries Service, and Warner-Lambert Company.

As a last point in this preface, we would like to thank all those who contributed to the success of the workshop. Special thanks are addressed to Daniela Galli for all the burden of the local organization and of the preparation of this volume, to Brita Rolander-Chilo' for the administrative responsabilities, to Peggy Roberts for her help with the international liaisons, in addition to all the others who participated to the organization.

Claudio Galli

Artemis P. Simopoulos

REFERENCES

1. A.P. Simopoulos, R.R. Kifer and R.E. Martins,eds., "Health Effects of Polyunsaturated Fatty Acids in Seafoods", Academic Press , Orlando, Fl (1986)

2. The American Oil Chemists' Society (AOCS) Short Course on Polyunsaturated Fatty Acids and Eicosanoids, held May 14-17, 1987 in Biloxi, Mississippi

3. Health Effects of Fish and Fish Oils: An International Conference held July 30-August 2, 1988 in St. John's, Newfoundland

4. The Fifth Acta medica Scandinavica Symposium on N-3 Fatty Acids in health and Disease held August 11-13, 1988 in Trømso, Norway

5. C. Galli and E.Fedeli, eds.,"Fat Production and Consumption - Technologies and Nutritional Implications", Plenum Press, New York (1987).

CONTENTS

The role of w3 and w6 fatty acids
in development

Biological effects of w3 and w6 fatty acids
on cell activation processes

The role of w3 and w6 fatty acids
in human diseases

EARLY STUDIES ON THE BIOSYNTHESIS OF POLYUNSATURATED FATTY ACIDS

Konrad Bloch

Department of Chemistry
Harvard University
Cambridge, Massachusetts 02138

The synthesis of olefinic fatty acids, whether mono-, di-, or poly-unsaturated, begins in all eukaryotic cells with the aerobic desaturation of stearate to oleate and more rarely with the conversion of palmitate to palmitoleate. Andreasen and Stier, the first investigators to implicate molecular oxygen in this 9,10 hydrogen abstraction process observed in 1952 that oleate becomes an essential nutrient for yeast growing under strictly anaerobic conditions (1). The evidence available today points to a similar obligatory role of oxygen for all subsequent desaturation steps. The detailed reaction mechanisms responsible for the very earliest events in the desaturation chain may therefore be of some relevance to the steps that follow.

In 1959 we studied the stearate - oleate conversion catalyzed by a cell-free preparation of Saccharomyces (2). We found that stearoyl CoA was the active substrate, NADPH the electron donor and oxygen the mandatory electron acceptor, requirements which placed the desaturase into the category of mono- or mixed function oxygenases. Later on Strittmatter and collaborators purified and identified the electron transport components of the hepatic stearate desaturase as cytochrome b_5, cyt. b_5 reductase and the desaturase proper as a non-heme iron protein (3).

Turning to plant systems, we found here an interesting variant similar to but not identical with the particulate yeast and liver desaturases (4,5). First of all the plant enzyme, either from algae or higher plants, is freely soluble. Secondly, in these cases the ACP derivative rather than the CoA thioester is the desaturase substrate and thirdly, the electron transport chain to oxygen consists of or includes a NADPH reductase and ferredoxin. That the ACP derivatives are the intermediates in plant fatty acid synthesis, i.e. the chain elongation process, had been previously shown (6).

Proceeding from oleate to dienes, trienes and beyond, information on the properties of the respective enzymes becomes increasingly sparse. First of all the genetically based absence of oleate-linoleate desaturation in all animal species- the cause of essential fatty acid deficiency - restricts studies of the oleate - linoleate conversion to plants, fungi and some protozooans. Judging from our own early work with the obligate aerobic yeast Torulopsis utilis the formation of linoleate resembles the stearate - oleate conversion in some, but not in all respects (7). Notably,

1

as several laboratories have shown, linoleate formation occurs in two discreet steps (8,9,10). Oleyl thioester first enters into ester linkage with lysophospholipid, the resulting oleyl phospholipid, probably phosphatidylcholine, serving as the substrate for desaturation to a linoleate derivative. Incidentally these observations rule out the necessity for "high energy" thioester functions for facilitating desaturation at sites which in any event are quite distant from the "activated" fatty acid carboxyl groups.

Studies with the phytoflagellate Euglena gracilis, our preferred organism for investigating unsaturated fatty acid synthesis, revealed another version of the processes just described. As already mentioned, oleyl-ACP rather than oleyl-CoA is the product of stearate desaturation in plants. In cell-free extracts of Euglena, provided the cells are grown in the light, oleyl-ACP is incorporated into galactolipid, typical chloroplast constituents, rather than phospholipid (11). We have therefore raised the possibility that oleyl galactosyl glyceride may be the more direct substrate for linoleate and also linolenate formation. I am unaware, however, of any more recent evidence for a role of galactosyl glycerides as intermediates of polyunsaturated fatty acid synthesis in plants.

Of course the option of linoleic acid to enter diverging pathways to more highly unsaturated acids accounts for the existence of the omega-3 and omega-6 families. Presumably two distinct desaturases, Δ^6-and Δ^{15} - specific respectively, initiate the two branches. I have just alluded to the possible existence of two linoleyl derivatives, the galactosyl-glycerides which are plant-specific and the phospholipids which occur more universally. Conceivably therefore the ester linkage of linoleate, whether to the glycerol moiety of galactolipid or of phospholipid, might determine whether the third double bond is inserted into linoleate in the direction of the carboxyl group (omega-6) or the direction of the terminal fatty acid carbon (omega-3). This issue, the substrate specificity of the two desaturases which produce the isomeric linolenates and hence give rise to the two families of polyunsaturated fatty acids seems to me especially deserving of further investigation.

The further transformations of α-and γ-linolenates is achieved in both cases by alternating elongations and desaturations. Whether the two pathways share or compete for the respective enzymes is another issue that requires attention beyond the already existing evidence for the antagonistic effects of DHA and EPA on the arachidonic- prostaglandin conversions (12). It would also seem worthwhile to inquire whether members of the omega-3 family are precursors of substances with hormonal attributes in analogy to the hormonal metabolites of arachidonate. It should perhaps be stressed that DHA and EPA are at best very minor components if not absent entirely in plant lipids. It seems unlikely therefore that these rare plant acids play a significant role in the regulation of plant membrane fluidity. At any rate for the potential benefit of the human race, fish are certainly important food-chain intermediaries for the production of omega-3 acids provided they have access to α-linolenate.

In the remainder of my largely retrospective remarks I will describe some of our early experiments with Euglena gracilis (13-14), a versatile phytoflagellate that exhibits a phototrophic life style on mineral media in the light and grows equally well heterotrophically in the dark (15). Either state is reversibly convertible into the other. The bewildering variety of fatty acids the phytoflagellate synthesizes, in fact more than fifty, was first noted by Korn (16). Moreover their relative proportions depend critically on growth conditions, i.e. whether the cells grow in the light or dark. I will focus here on two fatty acids of interest, α-linolenic acid and 4,7,10,13-hexadecatetraenoic acid of which Euglena is an exceptionally rich source. The environmental conditions that affect the concentrations of these two omega-3 acids include light intensity, duration of illumination and the partial pressure of CO_2 all variables which modulate photosynthetic efficiency, i.e. the Hill reaction (17). Most striking is the difference in α-linolenate content between illuminated and etiolated cells, 35% compared to 1% or less. Some two thirds of α-linolenate is associated with monogalactosyl glyceride, a typical chloroplast lipid, while only a small fraction of this acid is found in phospholipid. Second only to α-linolenate in abundance, the C_{16} tetraenoic acid has received much less attention in the relevant literature. In parallel with α-linolenate this C_{16}-tetraenoic acid increases markedly as a function of light intensity and also at higher CO_2 pressures. As expected all these changes give rise to increased oxygen evolution and chlorophyll content of the cells. Finally, specific Hill reaction inhibitors such as CMU prevent the light-induced synthesis of both omega-3 acids. These correlations seemed to us sufficiently suggestive to indicate some specific role for omega-3 fatty acids in higher plant (type II) photosynthesis (17).

There are compelling arguments for regarding Euglena gracilis as the organism par excellence for studying the biosynthesis of omega-3 and omega-6 fatty acids, the enzymatic mechanisms as well as their control, largely because these cells harbor the genetic information for both plant-specific lipid biosyntheses and for the corresponding processes in animal tissues. Moreover numerous Euglena mutants are available or available potentially.

References

1. A. Andreasen, and T. J. B. Stier, J. Cell Comp. Physiol. 41:23 (1954).
2. D. Bloomfield, and K. Bloch, J. Biol. Chem. 235:337 (1960).
3. P. Strittmatter, M. J. Rogers, and L. Spatz, J. Biol. Chem. 247:7188 (1972).
4. J. Nagai, and K. Bloch, J. Biol. Chem. 241:1925 (1965).
5. J. Nagai, and K. Bloch, J. Biol. Chem. 242:357 (1967).
6. P. Overath, and P. Stumpf, J. Biol. Chem. 239:4103 (1964).
7. C. Yuan, and K. Bloch, J. Biol. Chem. 236:1277 (1961).
8. B. Talamo, N. Chang, and K. Bloch, J. Biol. Chem. 248:2738 (1973).
9. M. L. Gurr, M. P. Robinson, and A. T. James, Eur. J. Biochem. 9:70 (1969).
10. P. S. Sastry, and M. Kates, Can. J. Biochem. 44:459 (1966).
11. O. Renkonen, and K. Bloch, J. Biol. Chem. 244:4899 (1969).
12. M. Boudreau, P. Chanmugam, and D. H. Hwang, Fed. Proc. 46(4):1169 (1987).
13. J. Erwin, and K. Bloch, Science 143:1006 (1964).
14. D. Hulanicka, J. Erwin, and K. Bloch, J. Biol. Chem. 239:2778 (1964).
15. J. J. Wolken, Euglena, an Experimental Organism for Biochemical and Biophysical Studies, p. 5, Rutgers University Press (1961).
16. F. Davidoff, and E. D. Korn, J. Biol. Chem. 238:3199 (1963).
17. J. Erwin, and K. Bloch, Biochem. Ztschr. 338:495 (1963).

THE FOOD CHAIN FOR N-6 AND N-3 FATTY ACIDS

WITH SPECIAL REFERENCE TO ANIMAL PRODUCTS

M.A.Crawford, W. Doyle, P. Drury, K. Ghebremeskel, L. Harbige, J. Leyton, and G. Williams

Nuffield Laboratory of Comparative Medicine, The Institute of Zoology, The Zoological Society of London BR Regent's Park, London NW1 4RY, UK

1.0   INTRODUCTION AND COMPARATIVE PATHOLOGY

The present debate on nutrition and health, is concerned with the reasons for a high incidence of heart disease in certain countries and not in others. This approach assumes that those countries with a low incidence have developed a food structure which is "better" than those with a high incidence and it is common practice to make recommendations based on the food intakes of the low incidence countries.

Yet it can equally be argued that those countries with a low incidence of coronary heart disease (CHD), colon and breast cancer, have other nutrition related problems: eg. in Japan there is a high incidence of stroke and stomach cancer; around the North Western shore of Lake Victoria in East Africa, absence of CHD is offset by a high incidence of cardio-myopathy, primary liver carcinoma and volvulus of the sigmoid colon is the commonest surgical emergency (1).

Throughout 5 million years of evolution, man ate wild foods and it is worth asking what are the differences between those 'base-line' foods and the products of the contemporary or rather "new" food production systems. It appears likely that recent changes in our diet, relative to that to which our ancestors became accustomed through evolution, could account for some of this pathology. In discussing the food Europeans and Africans eat in relation to comparative pathology, it is usually assumed that both have been eating the same foods since their origin. What is missing is the concept of (i) a "base-line" and (ii) the time scales and rates of change: most contemporary foods are the product of an industrial revolution.

Major changes have occurred in the last few centuries, particularly since the industrial revolution allowing too short a time span for a physiological adaptation to them through natural selection. It is therefore not surprising that different communities using different modern foods have different nutritionally related disease patterns. The only true base-line which we can measure today, is that of wild-life. Consequently, we wish to compare some of the foods eaten today with wild foods which offered the nutrient balance to which man's physiology was exposed throughout his evolution.

In the current debate on diet and heart disease, dietary fats and in particular, a plausible causal link with saturated fats, occupies the central focus. We therefore wish to trace the fatty acids from the begining of life and photosynthetic systems.

2.0 EVOLUTION: THE ORIGIN OF LIFE AND N-3 FATTY ACIDS IN THE SEA

Evidence of the first life forms appears with reasonable certainty at 2.6 billion years before now, although some claim the fossil evidence records life at about 3.5 billion years ago. However, the important point is that life up until about 500 - 600 million years ago, was single celled and anaerobic with blue-green algae accounting for a preponderance of the fossil records so far described. The blue-green algae would have trapped the sun's energy, photosynthetically and a major end product of the photosynthetic system was oxygen. As Dr. Bloch described at the opening of this workshop, photosynthesis is associated with the synthesis of n-3 fatty acids.

After the appearance of life on the planet, some 1.6 - 2.6 billion years elapsed before the next major event coincided with the oxygen concentration reaching the Pasteur point; that is, the the oxygen tension sufficient to permit aerobic, catabolic reactions.

The blue-green algae were then replaced by aerobic systems: eukaryotes appeared and were rapidly followed by multicellular organisms. Then the basic invertebrate phylla, including the echinoderms, ancestral to the hemichordate-chordate and vertebrata lines became established. The first fish, the placoderms, appeared in the fossil record 340 Million years ago, adaptive radiation giving rise to the cartilaginous, jawed fish of which the sharks, skates and rays (elasmobranchii) are surviving examples. The bony fish came later.

As photosynthesis by most algae and phytoplankton, produces n-3 fatty acids, it seems most likely that n-3 fatty acids were major components of the lipid chemistry of the environment and the cells, since the beginning of life in the sea. The marine food chain is carnivorous and from today's analysis of algae, phytoplankton and fish, it is predominantly based on n-3 fatty acids (2). Although larger proportions of n-6 fatty acids are found in marine species at the beginning rather than the top of the food chain (table 1) and in warm waters (3), it is most likely that the n-3 dominance has been consistent throughout animal evolution in the sea. In view of the current interest on diet and heart disease it is worth commenting that there would be relatively little saturated fatty acids in the marine environment, especially in the colder waters where high degrees of polyunsaturation are achieved in cell membrane lipids. The ratio of n-3/n-6 in the structural lipids of cod muscle is between 40 and 50 to 1 and docosahexaenoic acid is the major fatty acid in membrane lipids (eg herring table 1).

Table 1   MARINE FOOD CHAIN FATTY ACIDS

FATTY ACIDS % OF TOTAL	PHYTO-PLANKTON*	ALGAE	MOLLUSC	SQUID	HERRING (MUSCLE EPG~)
LINOLEIC	0.8	1.2	0.9	0.5	0.2
ALPHA-LINOLENIC	0.3	2.6	1.3	0.8	0.1
20:4,n-6	0.7	12.4	2.3	5.8	0.6
20:5,n-3	21.9	15.8	12.6	14.9	4.6
22:5	0.1	2.5	1.1	2.3	0.9
22:6	10.5	0.8	22.0	21.3	62.8
n-6/n-3	0.02	0.65	0.09	0.16	0.01

*Monochrysis lutheri (2). Sea-weeds vary considerably, some have high amounts of 18 polyenoates & also 18:4,n-3 (2). ~Ethanolamine phosphoglyceride: an inner membrane lipid (24). Note: the plankton & algae have more eicosapentaenoic than docosahexaenoic whereas the herring has more docosahexaenoate.

3.0   THE N-3 PHOTORECEPTOR AND THE FIRST NERVOUS SYSTEMS

One interesting facet of this long history of n-3 dominance is that the evolution of successful animal systems depended on a photoreceptor and an interpretive nervous system which accepted visual information and told the rest of the animal what to do about it: i.e. where and how to catch it's food. It is interesting that photoreceptor systems of the Crustacea , Insecta , Gastropoda and Vertebrata are essentially similar in physio-chemical terms but they differ structurally in the design of the eyes e.g simple and compound systems. The structural component seems to represent multiple solutions to the same basic problem, woven around the same theme.

Contemporary evidence (4) from the laboratories of Bazan and Connors presented at this conference, illustrates the remarkable use made of n-3 in the photoreceptor even today. It is interesting that early studies of the retina identified the presence of docosahexaenoic acid packed in high concentration in the rods, not only of higher animal species like ourselves but also in fairly primitive species (eg. the frog), illustrating again the concept of biochemical conservatism.

Perhaps the next most important evolutionary development from a nutritional view was the colonisation of the land.

Plants came first and these were then followed by terrestial arthropods, amphibians and reptiles. The plants which dominated the land mass were giants and included the Lepidophyta (eg Lepidodendron) and Gymnosperms (eg Pteridosperms, Cycadophyta and Ginkyophyta) more commonly known as, ginkos, cycads, giant ferns and their allies. These plants were major participants in this epoch and so there would have been a similarity between the base of the land and marine food chains: i.e. both started with photosynthesis and both would have produced an n-3 dominated fatty acid profile (see table 1 & 2.)

Table 2   EXAMPLES OF THE FATTY ACID CONTENT OF GREEN LEAVES

% of TOTAL FATTY ACIDS IN:	18:2,n-6	18:3,n-3
SPINACH (UK)	16	51*
ACACIA (E. Africa)	14	49
OAK (UK)	12	57
BEECH (UK)	16	52
RYE GRASS (UK)	14	34
GINKO (UK)	4.8	45

* NOTE: Alpha-linolenic acid occupies 8-15% of the digestible energy. It is associated with a high proportion of cellulose. See 23 for further data.

4.0   THE ORIGIN OF FLOWERS, SEEDS AND N-6 FATTY ACIDS

The next most important change in evolution was the disappearance of the dinosaurs. With them went the giant ginkos, ferns and other large plants typical of that epoch. Whether the collapse was due to a population explosion, loss of soil fertility or a cataclysmic event is uncertain but some land animals, ginkos and a number of other plants belonging to that epoch survived; today however, they are only small plants.

Alongside the disappearance of the giant ginkos ferns and dinosaurs, another, major evolutionary event occured: namely, the evolution of the flowering plants (the Angiosperms) and the mammals. The flowering plants themselves introduced seeds with specialised coats and nutrients for the protection and nourishment of the sporophyte. This added to the food chain what are essentially concentrated packets of linoleic acid: ie the parent of the n-6 family of essential fatty acids.

The contemporary evidence indicates that the fish require n-3 fatty acids for reproduction (5,6). By contrast, the mammals require n-6 fatty acids (7,8). It is either a remarkable coincidence or alternatively, there is a logical biochemical connection between the simultaneous appearance of concentrated packets of n-6 fatty acids in the food chain and a new group of animals which required them for reproduction.

Hence early land-based animal life would have faced a preponderance of n-3 fatty acids in the diet up until the end of the Cretaceous period. Whilst the n-3 fatty acids had been available in abundance throughout animal evolution up to this time, the evolution of protected seeds added n-6 fatty acids to the land food chain: from the compositional data on land mammals (15) it seems that the n-6/n-3 ratio favours the n-6 fatty acids by 3 or 5 to 1.

Table 3 EXAMPLES OF ESSENTIAL FATTY ACIDS CONTENTS OF SOME SEEDS

% OF TOTAL FATTY ACIDS IN:	18:2, n-6	18:3, n-3
GRASS SEED	49	0.5-3.0
ACACIA (cyclops)	67	0.8
OAK ~	53	12
BEECH	54	2
MILLET	69	0.3
MARJORAM	20	55

~NOTE: 1) Linoleic acid occupies 40% of the digestible energy. Although there are exceptions, eg flax & clover, the dominance of linoleic acid is the general rule. In many seeds, the oil can account for 30-80% of the calorific value of the whole seed. 2) Although the leaf is mainly n-3 and it might be thought that seed material is less abundant, none-the-less, wild mammals incorporate into cell membranes more n-6 than n-3 (see tables 6 and 7 and ref. 15 & 26).

5.0 MAMMALS, HOMO SAPIENS, WILD PLANTS AND ANIMALS

Mammalian evolution may have started about 100 million years ago with homo sapiens apparently separating from the other apes about 5 million years ago. In nutritional terms, human physiology evolved in the context of wild plants and animals and from contemporary practice and history, we suspect that man would have made use of both aquatic and terrestrial foods. Obviously, human physiology evolved in the context of wild foods. Tables 4 and 5 provide details of parent essential fatty acid contents examples of terrestial wild foods still used by man today and which would have been eaten by other primates.

From the examples of the seeds and leaves given in tables 3, 4 and 5, it is evident that plant foods would have, like their marine equivalents, contained a high proportion of polyunsaturated fatty acids. The converse is that they also contained low proportions of saturated fatty acids. The difference between the land and marine systems today is the dominant use of the n-3 in the marine animals and the use of both n-6 anjd n-3 by terrestial mammals, particularly in the vascular system and the brain (15).

Table 4 AFRICAN WILD LAND PLANTS TRADITIONALLY USED AS FOOD

	LIPID*	18:2,n-6	18:3,n-3	P/S
MYRIANTHUS serratus seed (as seed)	32%	74	0.3	12
BOERHAVIA diffusa fruit (as fruit)	5.8%	29	0.2	1.0
HEISTERIA parvifolia nut (oil source)	72%	8.9	3.4	3.2
RICINODENDRON Mull. Arg. nut (powdered on meat and fish)	46%	10	10	1.0
ALTERNANTHERA Forsk. leaves (as spinach)	4.3%	16	50	3.3
CENCHRUS BIFLORUS Roxb Grain - leaves also used	9.3%	42.5	2.0	1.8

* Lipid is expressed as a proportion of the dry weight & fatty acids as a proportion of the total fatty acid content; the data illustrate the provision of EFA by wild food lipids (9).

Tubers generally contain little lipid but in the lipid of those few that have been analysed, the proportion of parent essential fatty acids is high and that of the saturated fatty acids, low. Mean values of wild foods eaten by contemporary Africans is presented in table 5 as an illustration of their general nature.

What is striking about the chemistry of wild plants is that with one or two notable exceptions, it is difficult to identify plants in which saturated fatty acids would be considered as major components. The converse is the case which implies that throughout evolution, plants would have supplied essential fatty acids but relatively little saturated fatty acids.

Table 5   BASE-LINE FOODS USED BY HOMO SAPIENS: PLANT FOODS

MEAN LINOLEIC ACID % IN SEED/NUT FOODS USED BY HADZA

AND BY WEST AFRICANS 40%

MEAN ALPHA-LINOLENIC ACID IN GREEN LEAFY FOOD

LIPIDS USED BY HADZA & IN WEST AFRICA .. 45%

MEAN P/S RATIO OF WHOLE DIET 2.0

Note: Means from 68 different plant species used by the
Hadza (hunters & gatherers) & West Africans (9 see also 23)

6.0   P/S RATIOS IN WILD ANIMAL PRODUCTS

The present health concern in Western countries, focuses on
the saturated fat intake and the quality of fat, is often
referred to as the P/S ratio. This is the ratio between
polyunsaturated and saturated fatty acids in the diet but in
practice is the ratio of linoleic acid to the saturated fatty
acids. Western food intakes, have over the last few decades,
provided ratios in the region of 0.2-0.3. It will be apparent
from the previous discussion that these low P/S ratios seem
inconsistent with natural plant foods and are largely due to the
consumption of animal fats from domestic species together with
cooking fats and fat spreads modelled on their composition.
Hence, it is necessary to ask if animals replace the essential
polyunsaturated fatty acids of plant life when they build their
own cell membranes? The answer for marine based animals is
already known: they do the opposite: they increase the degree of
polyunsaturation as the food chain is ascended. Saturated fatty
acids play a relatively minor role in sea foods so what is the
situation in land-based animals?

Data presented in table 6 of analyses of meat from a wild
pig living in Africa (from 10), answers this question with
regard to a species free to select its own food in its natural
habitat. Its muscle tissue presents an image of a
polyunsaturated rather than saturated fat and opposite to the
sea-foods, has higher proportion of n-6 than n-3 fatty acids
except in the adipose fat where the ratio is 1:1.

The wild pig is a mono-gastric animal but man also ate
ruminants and it is known that the bio-hydrogenation process in
the rumin, hydrogenates the polyunsaturated fats. It was
thought that the fat from meat in domestic ruminants are
predominantly saturated as a result. However, if these animals
are to build cell membranes they would be expected to require
essential polyunsaturated fatty acids for that purpose and
therefore their meat should be rich in polyunsaturated fatty
acids.

To test this hypothesis we analysed the meat from wild, ruminant antelopes and bovids and found that the meat was indeed rich in essential, polyunsaturated fatty acids as is shown in table 7. It is only the adipose fats in which saturated fatty acids dominate. Wild land herbivores and birds, lay down enough adipose fat for their requirement for energy expenditure and seasonal variations in the food supply. However, this is relatively small in amount compared to what is found in contemporary fatstock.

Table 6   BASE-LINE FOODS USED BY HOMO SAPIENS:
WILD˜ PIG MEAT LIPID CONTENT AND FATTY ACIDS

	MUSCLE	ADIPOSE FAT
TOTAL FAT CONTENT	2.4%	94%
% OF TOTAL FATTY ACIDS:		
LINOLEIC	32	17
ALPHA-LINOLENIC	5.0	17
LONG CHAIN 20:3	1.0	0.1
n-6 20:4	8.5	0.2
22:4	0.7	-
LONG CHAIN 20:5	2.2	0.1
n-3 22:5	3.0	0.1
22:6	0.6	-
P/S	1.8	1.0
n-6/n-3	3.98	1.0

˜ Phacochoerus aethiopicus

Table 7   BASE-LINE FOODS USED BY HOMO SAPIENS
ANTELOPE MEAT LIPID CONTENT AND FATTY ACIDS

	MUSCLE	ADIPOSE FAT
TOTAL FAT CONTENT	2.8%	95%
FATTY ACIDS % OF TOTAL:		
LINOLEIC	21.0	4.5
ALPHA-LINOLENIC	3.6	4.0
LONG CHAIN 20:3	0.3	-
n-6 20:4	7.4	-
22:4	0.5	-
LONG CHAIN 20:5	1.9	-
n-3 22:5	5.2	-
22:6	0.8	-
P/S	0.99	0.15
n-6/n-3	2.5	1.17

˜ Taurotragus oryx: we have published similar data for
   wild bovids (buffalo), other antelope and deer species.
   - = below 0.2%.

From what is known about the compartmentalisation of fatty acids, it is clear that these are selectively directed to tissues for specific purposes. The contrast in the chemistry of muscle cell membrane lipids and adipose fat suggests that the non-essential and saturated fats are directed to adipose fat where they may be used as fuel. In the ruminant species, it is to be assumed that the bio-hydrogenation process leaves relatively little essential fatty acid for cell membranes and these are selectively used in these. The mechanism of selective uptake is well documented and consistent with the quite different compsitions associated with different cell functions (eg brain, retina, liver etc). The very low levels of essential fatty acid in ruminant adipose fat will simply be a refection of the relatively small amounts left over after membrane lipids have exerted their priority demand.

Hence it seems that in the ruminant, the essential, polyunsaturated fatty acids are conserved for cell membranes whilst the non-essential fats are deposited in the energy stores.

Since the domestication of animals man has used ruminant milks which would havwe provided a source of saturated fatty acids. However, it is interesting that the Masai, who are often quoted in this respect, mix milk and blood. Even in ruminant species, blood contains high proportions of parent and long chain polyunsaturated fatty acids and of course, the rest of the Masia's food is of the wild type described here.

7.0 ESKIMOS AND JAPANESE

There are few remaining communities still dependent on wild foods; those that have been studied have a low incidence of CHD, breast and colon cancer. Fish and sea-foods are wild foods. The Japanese have the highest consumption of fish per capita and several authors have linked this with raised levels of eicosapentaenoic and docosahexanoic acids from the fish oils offering a protection against cardiovascular disease and thrombosis (11,12,13). The Japanese total fat intake used to be about 10 of the calories.

The Eskimos are also of interest because they eat meat from marine mammals (13,14) and there have been several reports on the n-3 polyunsaturated nature of their adipose fat (14). Marine mammals once lived on land and then migrated into the sea. Their muscle contains significant amounts of arachidonic acid showing that despite their new allegiance to the marine environment, they still maintain a significant link with n-6 fatty acids. This may well be obtained through biochemical selectivity (conserving n-6 for cell membranes) and by making use of foods near the base of the food chain as, for example, the use made of the cephalopods by the dolphins.

What is apparent from these data is that

1. Wild meats reflect the polyunsaturated charcteristic of the plants which the animals eat although there are wide species differences.

2. There are also tissue differences in membrane lipids
   which may reflect compartmentalisation to meet specific
   structural or messenger functions. In particular,
   membrane rich systems such as endothelium, kidney,
   brain, muscle and liver, make use of the essential,
   polyunsaturated fatty acids, whereas adipose tissue is
   quite different, in keeping with its function as an
   energy store.

8.0   MODERN DOMESTIC LIVESTOCK AND SATURATED FATS

   The data on wild life show that as opposed to saturating
fats for use in cell membranes, animals in their natural
habitats amplify the unsaturation. They extend the chain length
and degree of unsaturation: instead of 18 carbon chain length

Table 8   BASE-LINE FOODS USED BY HOMO SAPIENS
             MARINE ANIMALS: DOLPHINS AND SEALS

	DOLPHIN (EPG)	HARP SEAL (CPG)	SEA-LION (CPG)
FATTY ACIDS % OF TOTAL			
LINOLEIC	0.9	0.8	0.4
ALPHA-LINOLENIC	0.2	0.6	0.2
20:4,n-6	16.8	7.0	8.5
20:5,n-3	3.67	0.84	1.69
22:5	3.55	1.0	1.33
22:6	11.2	4.2	7.66
n-6/n-3	1.5:1	1.3:1	0.9:1

~ Dolphin muscle   Seal's liver (data from single animals)

fatty acids with 2 or 3 double bonds we find fatty acids with 4,
5 and 6 double bonds in chain lengths of 20 and 22 carbons.
Furthermore, these fatty acids are somehow directed to special
sites. If saturated fats are not a feature of the wild marine
or land based food chains, from where do the contemporary high
intakes of saturated fats come?

   Traditionally domestic livestock have been thought of as
producing saturated fats. The domestication of animals started
about 10,000 years ago but they were initially reared
extensively and wild life still continued to play an important
part in man's food supply until recently. Although our domestic
livestock are considered to produce meat rich in saturated fats,
the contemporary, nomadic, domestic species, which we have

studied in Africa, have little adipose fat and the fat
extractable from their meat is polyunsaturated in a similar
manner to the wild species. This fact is simply explained by
the large amounts of adipose fat (energy store as opposed to
structural lipid) which infiltrates the muscle tissue of
intensively reared animals fed high energy, fattening foods
without exercise.

Hence the idea that animal fats are saturated could only
have arisen as a result of exclusive studies on domesticated
animals, fed intensively to the point where adipose fat as
opposed to membrane lipid, is laid down in excessive amounts
(16).

The fattening of animals for market, intensive grazing
systems, high energy feeds, selection for rapid weight gain and
stall feeding date back to the industrial revolution. Weight
gain is most readily achieved by laying down fat so this process
resulted in the production of adipose tissue which was then
copied just over 100 years ago, to make butter analogues (hard
margarines) and cooking fats, so adding an additional fat which
had not previously been a component of human diets. It is
interesting that more recently, the margarine industry has been
using seed oils to make "polyunsaturated" margarines.

9.0   PROTEIN VS FAT PRODUCTION

If any doubt is left regarding the magnitude of shift in
food structure, it is possible to use another test of the
validity of the approach by refering fat quantity and quality to
protein. The generally accepted view is that animal production
provides us with protein. This view assumes that there is more
protein than anything else coming from animal production. In
wild plants and animals, protein synthesis goes more or less
hand in hand with membrane lipid synthesis, hence with essential
fatty acid synthesis and incorporation.

If protein is used as a reference, there would be expected
to be a balance between protein and essential fatty acids as
both are used in cell membranes. For example, approximately the
same amount of energy in human milk is presented as protein and
essential fatty acids (20). If we compare the energy invested
or derived from contemporary animal production with that which
is derived from animal fat, then the test again shows that there
is a greater investment in fat (largely adipose fat of the
saturated type) than protein. The extent of the discrepancy in
contemporary meat production in the UK by this process is
illustrated in table 9.

The Eland was chosen for table 9 for comparison with Beef
first because it is of similar size being a typical large meat
species (body weight over 1,000 lbs at 36 months) from the
locality in the world which many consider to be the cradle of
Homo sapiens. Secondly, there are good data on its carcass
composition (18) and thirdly, there are data on the essential,
polyunsaturated fatty acid content of its meat (19). Finally,
the Russians have a herd which is used for milk and meat in
Askania Nova. The partridge data are from our own data-base
analysis.

Table 9 illustrates that the ratio between fat and protein is in the order of a ten fold difference if man-made animals are compared with their free-living counterparts. The dilution of the protein value by the fat in the contemporary carcass also involves the same dilution by fat of iron, trace elements, vitamins and essential fatty acids.

The beef carcass fat calculation is based on recent UK targets for a lean carcass of 25% fat. EEC intervention subsidies came into operation at 25% carcass fat and above so the tendency in practice was to overshoot.

Table 9   RATIOS OF FAT TO PROTEIN (per Kg carcass weight)

	PROTEIN g/Kg	FAT g/Kg	ENERGY (CALORIES) PROTEIN	FAT	RATIO FAT:PROTEIN (as energy)
BROILER CHICKEN (17)					
Males 56 days	97	211	388	1,899	4.9
Females 56 days	94.3	233	377	2,097	5.6
BRITISH BEEF fat:	100	250	400	2,250	5.6
lean:	150	200	600	1,800	3.0
WILD:					
PARTRIDGE	120	30	479	270	0.56
ELAND	158	42	632	378	0.59

Data from 16, 17, 18 & 25 (Fat:= 25% carcass fat & 50% lean). In the domestic environment, the energy investment in fat production is greater than in protein. In nature, the investment in protein is twice that of fat. The mean carcass fat found in a study of 257 animals animals from 16 wild species was found by Ledger to be 3.66% associated with a mean lean carcass content of 79.5% in the same animals (18).

Note:- The protein aspect of meat also contains the trace elements, iron, vitamins & essential fatty acids.

The fallacy of this position was recognised by the joint FAO/WHO Expert Committee (20) which recommended over ten years ago, that animal production methods should be re-orientated to protein and nutrient rather than fat production. Recomendations were also made on the need to address this issue as one of the components of the attack on coronary heart disease by the Centre for Agricultural Strategy in the UK (21).

Yet despite this evidence, relatively little progress has been made in reducing the fat intake in Western countries. Indeed, the Japanese intake of modern Western types of fats is increasing at a surprising rate.

The health implication of the change in the nature of food has been discussed at recent international committees (22) including the Surgeon General's report, of the USA in mid 1988.

2.6 BILLION YEARS AGO -> TODAY

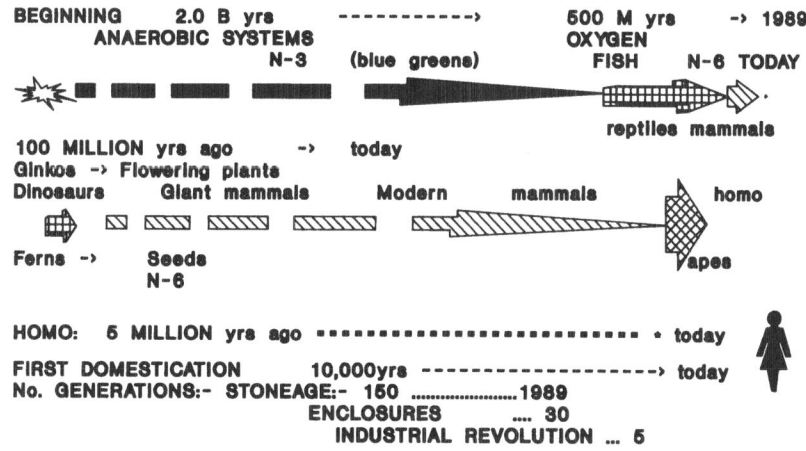

Figure 1.

Several other conferences have additionally drawn attention to the what are called the health benefits of fish oils and the concern on the decline in the consumption of fish and meat from marine mammals. Much of the data on fish oils are to be rediscussed at this conference, in Newfoundland and later in Tromso. An n-3 fish oil is now available in the UK, approved by the Department of Health, on doctors prescription, ironically as a lipid lowering and anti-thrombotic agent to protect against cardio-vascular disease considered to be the result of excessive fat and particularly saturated fats in the modern diet. The pointer which emerges from this recent interest in fish-oils and sea foods is that this resource is still wild. We are still using this wild resource today, as hunters and gatherers.

References

1. M.A. Crawford, Nutrition: heart disease and cancer. Are different diets necessary? Symposia of the Swedish Society of Medicine, IVth International Berzelius Symposium, Stockholm (Ed. Hallgren B. and Rossner, E.) Raven Press, New York, 149-158 (1986).

2. R.G Ackman, Fatty acid composition of fish oils, in "Nutritional evaluation of long-chain fatty acids in fish oil". (Barlow S.M. & Stansby, M.E. eds.) Academic Press, London, 25 (1982).

3. A.J. Sinclair, K. O'Dea and J.M. Naughton, Elevated levels of Arachidonic acid in fish from Northern Australian coastal waters. Lipids, 18(12): 877 (1983).

4. M. Neuringer and W.E. Connor, The importance of dietary n-3 fatty acids in the development of the retina and nervous system. Proc. AOCS ed. W.E.M. Lands, 301 (1987).

5. C.B. Cowey, Use of synthetic diets and biochemical criteria in the assessment of nutrient requirements of fish. J. Fish. Res. Board. Can. 33, 1040 (1976).

6. Yu, T.C. and Sinnhuber, R.O. (1979) Effect of dietary w3 and w6 fatty acids on growth and feed conversion efficiency of Coho salmon (oncorhynchus kisutch), Aquaculture, 16, 31-38.

7. G. O. Burr and M.M. Burr, On the nature and role of the fatty acids essential in nutrition. J Biol Chem. 86:587 (1930).

8. R.T. Holman, Biological Activities of and Requirements for Polyunsaturated Fatty Acids, Progr. Chem. Fats. & Other Lipids 9:611,-682.

9. M.A. Crawford, Fatty acid ratios in free-living and domestic animals, Lancet (i): 1329 (1968).

10. M.A. Crawford, M.M. Gale, and M.H. Woodford, M.H. Muscle and adipose tissue lipids of the Warthog (Phacochoerus aethiopicus). Int. J. Biochem. 1: 654 (1970).

11. Y. Kagawa, M. Nishizawa, M. Suzuki, T. Miyatake, T. Hamato, K. Goto, E. Montonaga, H. Izumikawa, H. Hirata and A. Ebihara, Eicosapolyenoic of serum lipids of Japanese islanders with low incidence of cardio-vascular disease. J. Nutr. Sci. Vitamol. 28:441 (1982).

12. A. Hirai, T. Hamazaki, T. Terano, T. Nishikawa, Y. Tamura, A. Kamugai and J. Sajiki. Eicosapentaenoic acid and platelet function, blood visocisity and red cell deformability in healthy human subjects. Atherosclerosis 46:321 (1983).

13. H.M. Sinclair, Nutrition and Atherosclerosis. Symp. Zool. Soc. Lond. 21: 275 (1968)

14. H.O. Bang, J. Dyerberg and H.M. Sinclair, The composition of Eskimo food in North Western Greenland. Amer. J. Clin. Nutr. 33:2657 (1980)

15. M.A. Crawford, N.M. Casperd, A.J. Sinclair, The long chain metabolites of linoleic and linolenic acids in liver and brain in herbivores and carnivores. Comp. Biochem. Physiol. 54B: 395 (1976).

16. M.A. Crawford, A re-evaluation of the nutrient role of animal products. Plenary lecture in Proc. Third World Conf. Animl. Product. Melbourne (1973), Reid, R.L. (Ed.). Sydney University Press: 21 (1975).

17. S. Leeson and J.D. Summers, Poultry Science 59: 786 (1980).

18. H.P. Ledger, Body composition as a basis for a comparative study of some East African mammals, Symp. zool Soc. 21: 289 (1968)

19. M.A. Crawford, M.M. Gale, M.H., Woodford, and N.M. Casperd, Comparative studies on fatty acid composition of wild and domestic meats. Int. J. Biochem. 1: 295 (1970).

20. FAO/WHO Conjoint Expert Consultation on 'The Role of Dietary Fats and Oils in Human Nutrition', Nutrition Report No.3 Rome, (1978).

21. Food, Health and Farming: report of panels on the implications for UK agriculture (1978) ed. Robins C.J. CAS paper 7, Centre for Agricultural Strategy, Reading UK (1978).

22. Prevention of Coronary Heart Disease, Report of a World Health Organsation Committee, Technical Report Series 678, Geneva, (1982).

23. T.O. Hilditch and P.N. Williams, The chemical composition of natural fats. Chapman & Hall (4th ed.) London (1964).

24. G. Williams, B.C. Davidson, P. Stevens and M.A. Crawford, Compoarative fatty acids of the dolphin and the herring. JAOCS 54:328 (1977).

25. A.J. Kempster, G.L. Cook and M. Grantley-Smith, National estimates of the body compositions of British cattle, sheep and pigs with special refernce to trends in fatness. A review. Meat Sci. 17:107 (1986).

26. M. A. Crawford, M.A. The balance between alpha-linolenic and linoleic acid. In: The Role Of Fats in Human Nutrition Brun, J.P., Podmore, J. & Nichols, B.W. (Eds). Chichester: Ellis Horwood 62 (1985).

SOME ASPECTS OF OMEGA-3 FATTY ACIDS FROM DIFFERENT FOODS

Joyce L. Beare-Rogers

Bureau of Nutritional Sciences, Food Directorate
Ottawa, (Canada)

INTRODUCTION

The most widely available omega-3 fatty acid in common diets comes from plants. Omega-3 linolenic acid has to be converted to docosahexaenoic acid to exert its essential role as a constituent of neural membranes. This paper will attempt to describe some dietary sources of omega-3 linolenic acid, its likelihood of conversion and an example of metabolic control with linolenic acid contrasted with marine omega-3 fatty acids.

DISTRIBUTION OF OMEGA-3 LINOLENIC ACID

Foods that can contribute to meeting a requirement for omega-3 fatty acid are widely available. As examples of concentrated sources, two common vegetable oils that contain both omega-6 linoleic acid and omega-3 linolenic acid have the generalized composition shown in Table 1. Canola oil is low erucic rapeseed oil and is characterized principally by a high level of monounsaturates (M) and a low level of saturates (S). Another important feature is the presence of both families of essential fatty acids, although the ratio of n-6 to n-3 is lower in canola oil than in soybean oil. In practical terms, this ratio is meaningful only for a total diet and not for an individual food, unless the diet contains only one source of fat.

Table 1. Vegetable oils containing omega-3 and
omega-6 fatty acids (% fatty acids)

	Canola	Soybean
S	6	15
M	62	24
n-6	22	54
n-3	10	7
n-6/n-3	2.2	7.7

Table 2. Essential Fatty Acids in Vegetables

Leafy Vegetables

	Month	18:2 n-6 % FA[a]	18:3 n-3 % FA
Romaine lettuce			
	Apr	18.7	55.2
	June	18.8	53.9
	Aug	21.5	46.5
	Mean	19.6	51.9
Iceberg lettuce			
	Apr	38.8	30.0
	June	39.6	30.9
	Aug	46.3	18.8
	Mean	41.6	26.6
Spinach			
	Apr	10.5	51.0
	June	10.6	49.9
	Aug	11.1	48.3
	Mean	10.7	49.7
Chicory			
	Apr	17.9	54.4
	June	16.7	56.0
	Aug	20.2	53.7
	Mean	18.2	54.7

Cruciferae Vegetables

	Month	18:2 n-6	18:3 n-3
Broccoli	Apr	17.4	46.3
	June	17.3	41.5
	Aug	14.4	47.4
	Mean	16.4	45.1
Brussels sprouts	Apr	22.0	49.0
	June	19.3	46.1
	Aug	22.9	45.9
	Mean	21.4	47.0
Cabbage	Apr	23.9	36.0
	June	26.6	40.6
	Aug	19.4	35.0
	Mean	23.3	37.2

Very low-fat foods can contribute to the intake of omega-3 linolenic acid, especially if they are rich in chloroplasts. Romaine lettuce contained more of this fatty acid than did iceberg lettuce (Table 2). Generally, the values were similar

Table 2 cont'd

Essential Fatty Acids in Vegetables

Legumes

Peas	Apr	54.2	9.4
	June	50.4	10.6
	Aug	48.6	10.4
	Mean	51.1	10.1
Green beans	Apr	28.6	36.7
	June	29.2	37.2
	Aug	14.8	52.4
	Mean	24.2	42.1
Wax beans	Apr	28.7	37.6
	June	25.7	39.4
	Aug	25.5	39.1
	Mean	26.6	38.7

[a] % total fatty acids

for April and June, but the values for August were higher for linoleic acid and lower for linolenic acid. In vegetable greens such as spinach and chicory, about half the total fatty acids were omega-3 linolenic acid. Also important sources, the cruciferae vegetables contained about twice as much of the omega-3 fatty acid as omega-6 fatty acids.

Most grains have more linoleic than linolenic acid, but products such as bread still contain a not insignificant amount of the omega-3 fatty acid (Table 3).

These sources of essential fatty acids are important for vegetarians. Non-vegetarians consume the longer-chain omega-3 fatty acids found in seafood and organ meats (Table 4). An unusually high level of eicosapentaenoic acid (30%), confirmed by mass spectrometry, was found in mussels that also contained domoic acid (Fig. 1). The value for omega-3 fatty acids in mussels is generally lower, but may be highly variable.

Table 3. Essential fatty acids in breads

Bread	Total FA	18:2	18:3
	mg/g	%	%
Rye	14.5	21.4	1.4
Whole wheat	35.8	19.3	1.3
Bran	17.2	29.5	2.7
Pumpernickel	11.8	50.3	5.2
White	21.9	22.8	1.3
	20.2	28.7	2.4

Table 4. Omega-3 Fatty Acids (FA) in
Cooked Foods (g/100g edible food)[a]

	Total Fat	(n-3) FA
Legumes		
Cowpeas	0.5	0.08
French beans	0.8	0.29
Kidney beans	0.8	0.28
Seafood		
Cod	0.9	0.16
Halibut	11.6	1.18
Mackerel	17.8	0.81
Salmon (Coho)	7.5	0.89
Tuna	4.3	0.68
Lobster	0.6	0.03
Mussels	4.5	0.55
Organ Meats		
Beef brain	11.1	0.97
Beef liver	4.9	0.25
Pork brain	9.5	0.88
Pork liver	3.1	0.04

[a] United States Department of Agriculture,
Agriculture Handbook Number 8, 1986.

Phytoplankton and zooplankton which are rich sources of
C20 and C22 omega-3 fatty acids for fish have not made inroads
into the human diet.

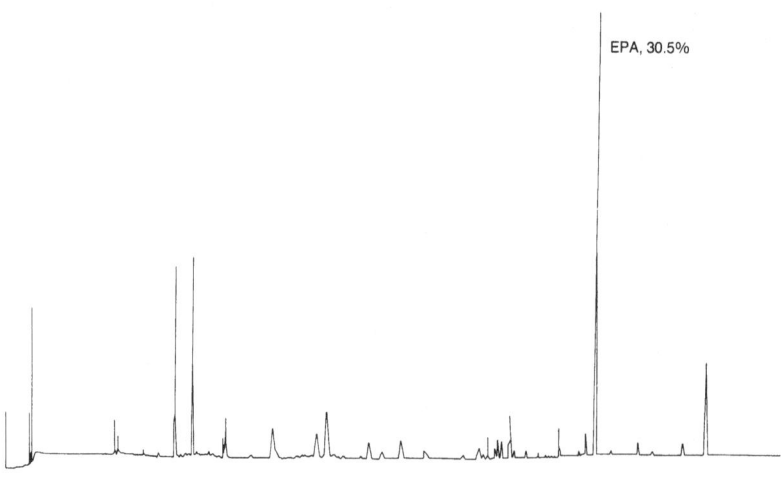

Fig. 1. Chromatogram of mussel fatty acids

The most critical food of all is the one fed to infants who are not given the luxury of variety. The essential fatty acids in formula are highly variable (Table 5). Powdered formulas, in particular, have frequently had extremely low levels of omega-3 fatty acids that can be overwhelmed by the omega-6 fatty acids.

CONVERSION OF OMEGA-3 LINOLENIC ACID

As has been demonstrated with a monkey infant formula containing soybean oil, dietary linolenic acid resulted in an enhanced level of its C20 and C22 derivatives (Neuringer et al., 1984). The infant rhesus monkey, delivered at 160 days of gestation, five days before the average full-term period, either had or could develop the enzymic capability to desaturate and elongate omega-3 linolenic acid.

Vegetarians who are dependent upon only plant sources of omega-3 linolenic acid produce the long derivatives, although in lower concentration than meat-eating controls (Sanders et al., 1978). A high proportion of linoleic acid to omega-3 linolenic acid was thought to be responsible for the suppressed conversion of the latter (Sander and Younger, 1981).

The dietary level of omega-3 linolenic acid can affect the relative concentrations of eicosapentaenoic acid and docosahexaenoic acid in rat liver (Mohrhauer and Holman, 1963). At low levels, representing less than 2% of food energy, the principal product was docosahexaenoic acid, the metabolite required in neural membranes; at high levels of intake, as achieved when feeding appreciable amounts of linseed oil, the principal product was eicosapentaenoic acid (Budowski, 1981). It was also observed that the highest concentration of eicosapentaenoic acid attained was in cholesteryl esters, not in glycerophospholipids that could be incorporated into membranes. In humans given linseed oil, the eicosapentaenoic acid increased more than the docosahexaenoic acid in serum glycerophospholipids (Mest et al., 1983). Organisms appear to have better mechanisms to transport and store excess eicosapentaenoic acid than excess docosahexaenoic acid.

Table 5. Classes of Fatty Acids in Infant Formula (% fatty acids)

	Powders	Liquids
No. of Products	9	16
Fatty Acids		
S	19.0 - 85.6	16.7 - 88.2
M	6.2 - 32.4	2.6 - 26.4
n-6	8.6 - 49.8	4.3 - 51.3
n-3	0.2 - 5.5	0.2 - 5.9
n-6/n-3	7.8 - 108	6.4 - 18.0

It has been stated, on the basis of the composition of the fatty acids found in human cells, that the omega-6 family must be preferred for desaturation and elongation (Dyerberg, 1986). Desaturases, however, favour a substrate with more double bonds (Brenner and Peluffo, 1966), but acyltransferase reactions determine the final selection of fatty acids. Arachidonic acid is more readily incorporated into phosphoglyceride than are eicosapentaenoic acid or docosahexaenoic acid (Iritani et al., 1984). The omega-3 linolenic acid would therefore have the advantage for desaturation, but its derivative, eicosapentaenoic acid, even at high concentration, could not prevent some incorporation of arachidonic acid into cellular membranes.

Human Milk

The requirements and role of essential fatty acids during brain development were recently reviewed by Agradi and Galli (1988) who stressed that quantitative aspects have yet to be determined and that the best reference is human milk. The concentration of omega-6 linoleic acid in milk exceeded that of omega-3 linolenic acid by more than 25-fold (Carlson et al., 1986), but would have depended upon the diet of the mother. The ratio of the longer omega-6 fatty acids to longer omega-3 fatty acids was 2 in the U.S.A. (1.8% vs 0.9% of total fatty acids). For five other countries, a corresponding value of 1.4 to 1 was calculated (Crawford, 1980).

It is interesting to note that in human milk, docosahexaenoic acid is much more prominent than eicosapentaenoic acid (Table 6). This is perhaps not surprising since docosahexaenoic acid is an important component of neural membranes (Tinoco et al., 1979) and the brain undergoes its greatest growth during early life (Crawford and Sinclair, 1972). If the principal role of

Table 6.   Fatty Acids in Human Milk (% fatty acids)

Fatty Acid	5 Countries[a]	U.K.[b] Vegans	U.K.[b] Control	U.S.[c]
S<18	45.3	27.3	35.9	36.6
18:0	6.4	5.2	10.8	5.9
M	33.9	32.5	38.9	39.2
18:2 n-6	9.1	31.7	6.9	16.0
18:3 n-3	0.8	1.5	0.8	0.6
20:2 n-6		0.5	0.2	0.6
20:3 n-6		0.3	0.2	0.4
20:4 n-6	1.0	0.7	0.5	0.6
22:4 n-6		0.1	0.1	0.2
22:5 n-6		0.1	0.2	0.2
20:5 n-3		0.04	0.2	0.03
22:5 n-3	1.4	0.3	0.5	0.1
22:6 n-3		0.2	0.6	0.2

a Crawford, 1980; b Sanders et al., 1978; c Carlson et al., 1986.

eicosapentaenoic acid is to modulate eicosanoid metabolism, it does not seem to have much usefulness in the early stages of development.

ERYTHROCYTE FATTY ACIDS OF CYNOMOLGUS MONKEY

Many animal experiments with omega-3 fatty acids have involved a single source of dietary fat, and sometimes a deficiency of linoleic acid. To avoid a study on one essential nutrient in the absence of another essential nutrient, linoleic acid was provided with omega-3 fatty acid from linseed oil or menhaden oil, as shown in Table 7 (Carman and Beare-Rogers, 1988). The diets had a similar distribution of saturated, monounsaturated and polyunsaturated fatty acids. An agar solution was mixed with each diet to produce dough balls and to reduce the surface exposed to air. These were made weekly, divided into daily portions and refrigerated until fed to the cynomolgus monkeys. The erythrocyte fatty acids responded to the dietary treatment with omega-6 fatty acids increasing at the expense of omega-3 fatty acids when the latter were missing from the diet. When the diet contained approximately equal proportions of linoleic and omega-3 linolenic acid, there was little change, but great increases in long-chain omega-3 fatty acids were seen when menhaden oil was provided. The ratios of the total omega-6 to omega-3 fatty acids in erythrocytes for each group of monkeys are shown in Figure 2. When the diet contained similar amounts of linoleic and omega-3 linolenic acid, the ratio of omega-6 to omega-3 fatty acids remained constant at approximately 2.4. This indicates a close regulation of the conversion of essential fatty acids to their long chain derivatives and their incorporation into the cellular membranes.

A comparison of the individual omega-3 fatty acids in erythrocytes from monkeys fed the different sources of omega-3 fatty acids showed that docosahexaenoic acid predominated when linseed oil was fed and eicosapentaenoic predominated when fish oil was fed. Sanders and Younger (1981) had found that a

Table 7. Characteristics of Monkey Diets

Diet	1	2	3
Oils[a]	L:CO	L:LSO	L:CO:MO
S	39.7	39.7	43.9
M	35.2	35.5	32.8
P	25.2	24.8	21.7
18:2 n-6	24.7	11.8	10.6
18:3 n-3	0.4	13.0	0.9
20:5 n-3	-	-	4.7
20:6 n-3	-	-	2.8

[a] L, lard; CO, corn oil; LSO, linseed oil; MO, menhaden oil

supplement of linseed oil in humans enhanced the
concentration of the long-chain derivatives in plasma and
platelet phosphoglycerides but to a lesser extent than did a
supplement of fish oil in which the predominant acid was
eicosapentaenoic acid. The enhanced incorporation of the
fish-type rather than the plant-type omega-3 fatty acids has
been observed particularly in cardiac tissue (Roshanai and
Sanders, 1985; Holmer and Beare-Rogers, 1985).

Animal experiments tend to deal with extreme situations
to find possible effects and to restrict the number of test
groups. This monkey experiment too deals with an extreme
situation. The omega-3 fatty acid is either linolenic acid or
fish fatty acids. A mixture should be tested since both plant

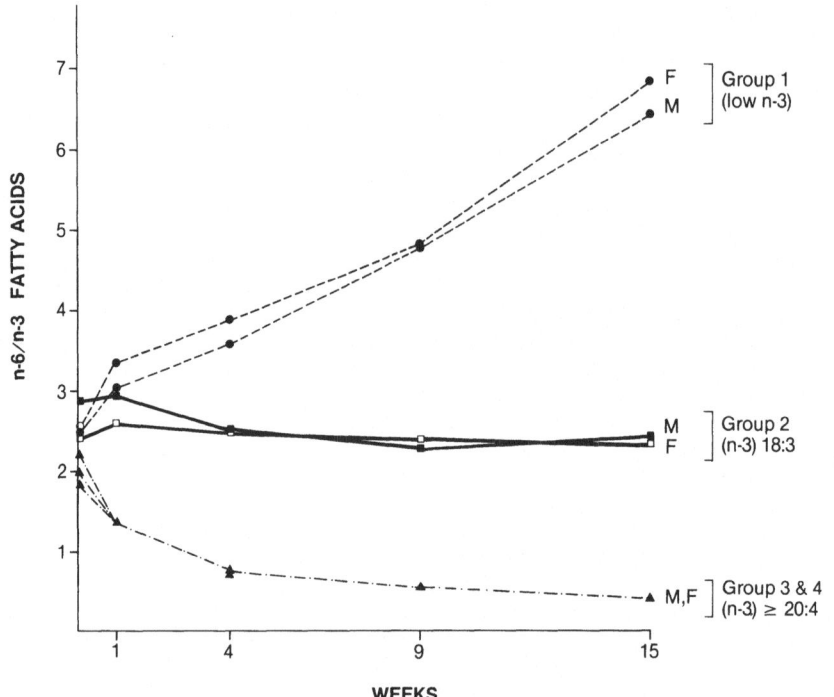

WEEKS

Fig. 2.   Ratio of omega-6 to omega-3 fatty acids in monkeys
          fed a diet low in omega-3 fatty acids, (group 1)
          linseed oil (group 2) or menhaden oil (group 3).

and animal sources of fatty acids are consumed by humans.
Also, an increased consumption of vegetables is being
encouraged because of their content of micro-nutrients and
fibre. Within the context of such nutrition recommendations,
omega-3 fatty acids of different chain lengths need to be
studied.

CONSIDERATIONS FOR INGESTING OMEGA-3 FATTY ACIDS

There is ample evidence that omega-3 linolenic acid can
be converted in man to its C20 and C22 derivatives. When
these substances are provided directly in the diet, more may
be incorporated into cell membranes, and if they are in high
enough concentration, pharmacological effects may be obtained.

Apart from the many desirable results obtained when some of the arachidonic acid in membranes is replaced by eicosapentaenoic acid (Leaf and Weber, 1988), there are concerns about elevated levels of blood glucose in diabetes-prone individuals. Non-insulin dependent diabetics treated with a fish oil concentrate showed a 51% decrease in VLDL and a 33% decrease in triacylglycerol, but a 29% increase in fasting glucose levels (Ensinck, 1988). The adverse effect on glucose homeostasis may be related to the dose of long-chain omega-3 fatty acids ingested. When pigs received 0.6 mg of eicosapentaenoic acid per day for 8 weeks, the hyperglycemic effect was evident, but not when the pigs received half the dose for twice the time (Hartog, 1987 a,b).

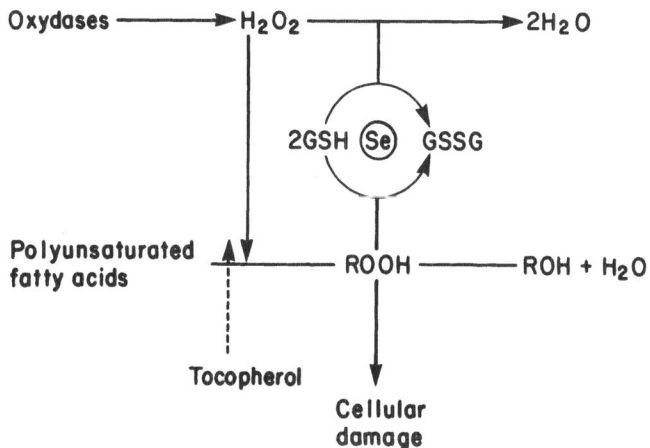

Fig. 3. Interrelationship of nutrients in oxidation of polyunsaturated fatty acids.

One of the advantages of consuming the long omega-3 fatty acids from fish rather than from extracted oils is that fish tissue contains other nutrients to assist in protecting cells from the effects of peroxidation. As illustrated in a modified scheme from Hockstra (1975), selenium is essential for the activity of glutathione peroxidase. The ingestion of polyunsaturated fatty acids also increases the requirement for tocopherol and perhaps other nutrients such as sulfur-containing amino acids, not shown in the diagram. The nutritional factors involved in the protection and metabolism of omega-3 fatty acids require further investigation. Such research would be particularly relevant to the design of food for the young.

REFERENCES

Agradi, E., and Galli, C. 1988, Requirement and role of essential fatty acids during brain development. Contr. Infusion Ther. Clin. Nutr., 19:128.

Brenner, R.R. and Peluffo, R., 1966, Effect of saturated and unsaturated fatty acids on the desaturation in vitro of palmitic, stearic, linoleic and linolenic acids, J. Biol. Chem., 241:5213.

Budowski, P., 1981, Review: nutritional effects of omega-3 polyunsaturated fatty acids. Isr. J. Med. Sc., 17:223.

Carlson, S.E., Rhodes, P.G., and Ferguson, M.G., 1986, Docosahexaenoic acid status of preterm infants at birth and following feeding with human milk or formula, Am. J. Clin. Nutr., 44:798.

Carman, M.A., and Beare-Rogers, J.L., 1988, Influence of diet on (n-3) and (n-6) fatty acids in monkey erythrocytes, Lipids, 23:501.

Crawford, M.A., 1980, Estimation of essential fatty acid requirements in pregnancy and lactation, Prog. Fd. Nutr. Sci., 4:75.

Crawford, M.A., and Sinclair, A.J., 1972, Nutritional influences in the evolution of mammalian brain, in "Lipids", pp. 267, Malnutrition and the developing brain, CIBA Foundation Symposium, Amsterdam, Elsevier.

Dyerberg, J., 1986, Linolenate metabolism, reply, Nutr. Res., 44:316.

Ensinck, J.W., 1988, Dietary omega-3 fatty acid supplementation in type II diabetic:diverse effects on glucose and lipoprotein metabolism. J. Am. Oil Chem., 65:509.

Hartog, J.M., Lamers, J.M.J., Montfoort, A., Becker, A.E., Klompe, M., Morse, H., ten Cate, F.J., van der Werf, L., Hulsmann, W.C., Hugenholtz, P.G., and Verdouw, P.D., 1987 a, Comparison of mackerel-oil and lard-fat enriched diets on plasma lipids, cardiac membrane phospholipids, cardiovascular performance, and morphology in young pigs, Am. J. Clin. Nutr., 46:258.

Hartog, J.M., Verdouw, P.D., Klompe,M., and Lamers, Jos, M.J., 1987 b, Dietary mackerel oil in pigs:effect on plasma lipids, cardiac sarcolemmal phospholipids and cardiovascular parameters, J. Nutr., 117:1371.

Hockstra, W.G., 1975, Biochemical function of selenium and its relation to vitamin E, Fed. Proc., 34:2083.

Holmer, G., and Beare-Rogers, J.L., 1985, Linseed oil and marine oil as sources of (n-3) fatty acids in rat heart, Nutr. Res., 5:1011.

Iritani, N., Ikeda, Y., and Kajitani, H., 1984, Selectivities of 1-acylglycerophosphorylcholine acyltransferase and acyl-CoA synthetase for n-3 polyunsaturated fatty acids in platelets and liver microsomes, Biochim. Biophys. Acta, 793:416.

Leaf, A., and Weber, P.C., 1988, Cardiovascular effects of n-3 fatty acids, New. Eng. J. Med. 318:549.

Mest, H.J., Bertz, J., Heinroth, I., Blick, H.U., and Forster, W., 1983, The influence of linseed oil diet on fatty acid pattern in phospholipids and thromboxane formation in platelets in man, <u>Klin. Wochenschr.</u>, 61:187.

Mohrhauer, H., and Holman, R.T., 1963, The effect of dose level of essential fatty acids upon fatty acid composition of the rat liver, <u>J. Lipid Res.</u>, 4:151.

Neuringer, M., Connor, W.E., Van Petten, C., and Barstad, L., 1984, Dietary omega-3 fatty acid deficiency and visual loss in infant rhesus monkeys, <u>J. Clin. Invest.</u>, 73:272.

Roshanai, F., and Sanders, T.A.B., 1985, Influence of different supplements of n-3 polyunsaturated fatty acids on blood and tissue lipids in rats receiving high intakes of linoleic acid, <u>Ann. Nutr.</u>, 29:189.

Sanders, T.A.B., and Younger, K.M., 1981, The effect of dietary supplements of omega-3 polyunsaturated fatty acids on the fatty acid composition of platelets and plasma choline phosphoglycerides, <u>Br. J. Nutr.</u>, 45:613.

Sanders, T.A.B., Ellis, F.R., and Dickerson, J.W.T., 1978, Studies of vegans:the fatty acid composition of plasma choline phosphoglycerides, erythrocytes, adipose tissue, and breast milk, and some indicators of susceptibility to ischemic heart disease in vegans and omnivore controls, <u>Am. J. Clin. Nutr.</u>, 31:805.

Tinoco, J., Babcock, R., Hincenbergs, I., Medwadowski, B., Miljanick, P., and Williams, M.A., 1979, Linolenic acid deficiency, <u>Lipids</u>, 14:166.

LINOLEIC AND LINOLENIC ACIDS INTAKE

O. Adam

Medical Policlinic
University of Munich
Pettenkoferstrasse 8a
D-8000 Munich 2

INTRODUCTION

Dietary fat supply is only one of several factors,[1,2] which determines fatty acid composition of body lipids. Research work of the last years showed that the incorporation of different fatty acids in plasma[3] and membrane lipids is a slow and controlled process[4,5], in which preferences for certain fatty acids can be found[5]. Moreover it has become clear that a competition exists between certain polyunsaturated fatty acids for incorporation into body lipids[6]. Thus the amount of fatty acid supply determines not solely the composition of body fat.

The essential fatty acids intake depends on the availability of foods, the individual taste, nutritional habits and on the prices of the food items, and only can be determined by an individual dietary recall. To evaluate an average essential fatty acids supply, it seems reasonable to calculate it from the amounts of fatty acids available for consumption. These figures can be evaluated by calculating the known precentage of polyunsaturates in these fats and oils.

WORLD PRODUCTION RATES OF FATS AND OILS

In 1984 the world production rates of fats and oils was 41 mmt plant oils, making up 63 % of the total fat production[7,8]. Animal fats accounted for about 35 %, while marine oils reached only 2 % of the worlds production of fats, fish oils making up 97 % of total marine oils.

The greatest amount of plant oils derived from soy bean, 14.0 mmt were produced in 1984. Sunflower-, palm-, rape seed-, cotton seed-, peanut- and coconut oil accounted for about 4 % each. Olive oil made up 1.6 mmt, while of the other oils, including safflor oil, less than 1 mmt per year was produced (Table 1).

Table1.   WORLD PRODUCTION OF PLANT OILS IN 1984

	mmt/y
Soybean oil	14.0
Sunflower oil	5.7
Palm oil	4.4
Rape seed oil	3.3
Cotton seed oil	3.1
Peanut oil	3.0
Coconut oil	2.9
Olive oil	1.6
Corn oil	0.6
Palm kernel oil	0.6
Safflower oil	0.3
Others	1.5
Total	41.0

If the average composition of the different oils is taken to compute the amount of linoleic and linolenic acid produced in the world, it becomes clear that nearly half of the available linoleic acid derives from soy bean, about 7.2 mmt per year, followed by 3.4 mmt per year from sunflower oil. The other oils contribute according to their content of linoleic acid to the supply of linoleic acid. Even with a production rate of only 3.1 mmt per year, cotton seed oil provides 1.5 mmt of linoleic acid per year. The other fats are less abundant in linoleic acid, so their contribution to the supply is smaller. The total world production of linoleic acid is, according to these figures, 16.6 mmt per year (Table 2).

Table 2. WORLD PRODUCTION OF LINOLEIC AND LINOLENIC ACIDS
FROM
PLANT OILS IN 1984

Source	18:2,n-6	18:3,n-3
	mmt / year	
Soy bean oil	7.2	1.1
Sunflower oil	3.4	0.03
Palm oil	0.5	0.02
Rape seed oil	0.6	0.3
Cotton seed oil	1.5	0.03
Peanut oil	0.7	–
Coconut oil	0.04	–
Olive oil	0.13	0.01
Corn oil	0.3	0.01
Palm kernel oil	0.01	–
Safflower oil	0.22	0.01
Total	16.6	1.5

For linolenic acid, also, soy bean oil is the most abundant source, followed by rape seed oil bringing 0.3 mmt every year to the market. Compared to these figures the linoleic acid supply with other oils is small. In total 1.5 mmt of linolenic acid are annually produced.

Compared to plant oils, animal fats contribute less to the worlds linoleic acid supply. From tallow, butter, lard and marine oils emerges 0.82 mmt linoleic acid, which is only little more than we get from rape seed oil. 0.16 mmt of linolenic acid are available for human nutrition from animal fats, which is a little more than we get from rape seed oil. With animal fats also 0.115 mmt of arachidonic acid are produced. From marine food derive 0.15 mmt of eicosapentaenoic acid and 0.11 mmt of docosahexaenoic acid (Table 3).

Table 3. WORLD PRODUCTION OF POLYUNSATURATED FATTY ACIDS FROM ANIMALS FATS IN 1984

Source	18:2,n-6	18:3,n-3	20:4,n-6	20:5,n-3	22:6,n-3
			mmt / year		
Tallow	0.3	0.03	0.015		
Butter	0.1	0.07			
Lard	0.4	0.05	0.09		
Marine oils	0.02	0.01	0.01	0.15	0.11
Total	0.82	0.16	0.115	0.15	0.11

In total the world production of PUFA with animal and plant fats is 17.42 mmt for linoleic acid and 1.83 mmt for linolenic acid. The supply of arachidonic acid is additionally enriched by meat and meat products, so the world production is about 0.23 mmt per year. In 1984 world population was 4.5 billion people and from this figure an average supply of PUFA can be calculated. With the actual production rate 11 g of linoleic acid are available for every person in the world. On a per capita basis 1.1 g/d of linolenic, 0.14 g of arachidonic, 0.1 g of eicosapentaenoic and 0.07 g of docosahexaenoic acids are available (Table 4).

Table 4. WORLD PRODUCTION OF POLYUNSATURATED FATTY ACIDS IN 1984

	mmt/y	g/d/person
Linoleic acid (18:2,n-6)	17.42	11
Linolenic acid (18:3,n-3)	1.83	1.1
Arachidonic acid (20:5,n-6)	0.23	0.14
Eicosapentaenoic acid (20:5,n-3)	0.16	0.1
Docosahexaenoic acid (22:6,n-3)	0.11	0.07

On a percental basis the supply of linoleic acid to linolenic acid is 10 : 1 and for eicosapentaenoic only 10 : 0.09. From these figures it is evident, that the wanted ratio of n-6 to n-3 fatty acids in our diet cannot be met by the actual fatty acid production in the world (Table 5).

Table 5. WORLD PRODUCTION OF POLYUNSATURATED FATTY ACIDS IN 1984

	ratio
Linoleic acid (18:2,n-6)	10
Linolenic acid (18:3,n-3)	1
Arachidonic acid (20:4,n-6)	0.13
Eicosapentaenoic acid (20:5,n-3)	0.09
Docosapentaenoic acid (22:6,n-3)	0.06

In fact n-3 fatty acids are underrepresented in western communities, where the consumption of fat is high. In Germany[8] the per capita consumption of fat was 55.3 kg per year and every person consumed on average 6.8 kg of polyenoic acids. This figure indicates that the ratio of PUFA to monounsaturated or saturated fatty acids is 1:2:6 and not 1:1:1 as it should be. This unfavourable relation is referrable to the high intake of animal fat, rich in saturated fatty acids, which accounts for 29 kg/person and makes up more than half on the total fat intake. Inspite of the high supply with 15 g/d linoleic acid in this community, the P/S-ratio is low, because of the high total fat intake (Table 6).

Table 6. FAT CONSUMPTION 1982 IN WESTERN GERMANY

	Polyenoic fatty acids content %of total	Total Fat consumption (kg/person)	Polyenoic fatty acids intake (kg/person)
visible fats			
cooking fats	5	1.8	0.09
cooking oils	50	5.4	2.7
animal fats	10	6.4	0.64
butter	3	6.0	0.18
margarines	26	6.7	1.74
invisible fats			
from meat and meat products, milk ect.	5	29.0	1.45
fat consumption per year		55.3	6.80

In greenland eskimos total fat consumption is comparable to western communities. But with their maritime food greenlanders ingest far less saturated fatty acids than danes. So the P/S ratio in greenlanders is 0.84, while it is only 0.24 in the danes. Furthermore in an eskimo-diet n-3 fatty acids are prevalent, while with western diets much more n-6 fatty acids are ingested.

It must be kept in mind that the actual supply of PUFA is higher. The given figures comprise only visible fats, which are produced and brought to the market as cooking fats or oils, butter or margarines. Not taken into account are invisible fats originating from meat products, milk and milk products. Those invisible fats contain only 5 % of PUFA, but[8] with the animal fat consumption in western societies about 4 g/d of linoleic acid are ingested.

With vegetables and grains addidionally PUFAs are entering the nutrition chain[9]. Plants rich in linoleic acid are black salsify, carrots, turnip-rooted parsley and florence fennel. The amount of fat in those plants is 0.1 %, but more than 50 % of it is linoleic acid. Plants rich in linolenic acid are given in Table 7. In these plants about 50 % of the 0.1 % fat content consists of linolenic acid. With a vegetarian diet about 0.4 g/d of linolenic acid are ingested with vegetables.

Table 7. PLANTS LIPIDS RICH IN LINOLENIC ACID (18:3,n-3)

	% of total fatty acids
Chinese Cabbage	59.5
White Cabbage	57.0
Savoy	46.4
Kale	50.0
Red Cabbage	39.8
Brussel Sprouts	58.7
Cauli flower	48.9
Kohlrabi	41.9
Swede	50.8
Cabbage lectuce	54.6
Parsley	41.7

Fats are often processed for special purposes by the industry[10]. To meet the need for high melting fats for the production of margarines or cooking fats polyunsaturated liquid oils are hardened, which means that double bonds are lost. Inspite of efforts to preserve essential fatty acids in this process, a loss of PUFA occurs in most cases. A further loss of PUFA may occure during the preparation of food in the industry as well as in households. With cooking or stewing temperatures do not exceed $100^{\circ}C$ and there is absolutely no deterioration of PUFAs. With frying or bakeing temperatures reach $250^{\circ}C$, but a decomposition of PUFAs only occurs at the surface, because temperatures inside the product are much lower. With barbecuing extremely high temperatures are reached, but most of the oxygenated fatty acids are pooring off the product and are

not ingested. With frying temperatures are between 160-180
^{O}C, but a deterioration of the fat may occur, because the
fat is boiled for days and air oxygen leads to oxidation of
PUFAs. Eder et al.[11] investigated the decay of linoleic
acid in different oils used for the commercial production
of pommes frites. The fat was analyzed before and after 40
hours of use, which is an average time in commercial
production of pommes frites. The decay of linoleic acid was
between 30 and 45 % of the initial value (Table 8).

Table 8. DECAY OF LINOLEIC ACID IN DIFFERENT OILS USED
FOR COMMERCIAL PRODUCTION OF POMMES FRITES

| | Linoleic acid | | |
	in new oil (mass%)	in used oil (mass%)	decay in %
Sunflower oil	64.2	40.2	37.4
Soy bean oil	52.3	33.6	35.8
hardened soy bean oil	26.2	18.3	30.2
Palm oil	10.6	5.8	45.3

Ingestion of linoleic and linoleic acids results in an
accumulation of both fatty acids in body lipids. But the
increase of linolenic acid is substantially less than that
of linoleic acid. We elaborated[12] this finding with liquid
formula diets, containing a constant amount of linoleic and
linolenic acid for two weeks each. With the dietary fat a
constant amount of 4 energy% of linoleic acid - the first
index of the diets - and an alpha linolenic acid intake of
0, 4, 8, 12 or 16 energy% - the second index of the diets -
was given for two weeks each. Measurements were done at the
end of each two weeks period. Linoleic acid incorporation
in plasma cholesterol esters was not depressed by linolenic
acid intake. Linolenic acid incorporation in plasma
cholesterol was low, even with a dietary supply of 16
energy%. Taken into account that linolenic acid ingestion
was fourfold that of linoleic acid, and that linolenic
reached only one seventh of the percentage of linleic acid,
the accumulation of linolenic acid was thirtyfold less than
of linoleic acid. Arachidonic acid in cholesterol esters
only decreased with a linolenic acid supply of more than 12
energy%. These results indicate a preferential
incorporation of linoleic into plasma lipids.

This finding was confirmed by another study[6], in which
linolenic acid was given in a constant amount of 8 energy%
with liquid formula diets, while the intake of linoleic
acid was increased from 0 to 1.7 and 4 energy%. For
comparison purposes a PUFA-deficient diet and a diet
providing an linoleic acid intake of 4 energy% was given.
At the end of each diet linoleic acid was incorporated
according to the dietary supply, but linolenic acid
decreased with augmented linoleic acid intake. This
indicates a preferential incorporation of linoleic aicd in
plasma cholesterol esters in man. In these experiments no
replacement of linoleic acid by linolenic acid was found,
indicating that no competition for incorporation into body

lipids exists for these fatty acids. On the contrary a competition has been established between arachidonic acid and eicosapentaenoic acid.

Animal experiments, done by Chow et al.[13] revealed that in rats the presence of other unsaturated fatty acids influences the absorption of 840 um of linoleic acid. Oleic, linolenic and arachidonic acid lessened the absorption of linoleic acid to one half of the control value. In rats the metabolism of PUFAs is different from humans, but it may be speculated that linoleic acid ingestion may decrease the resorption of linolenic acid in man. Jones et al.[14] compared the intestinal absorption of [13]-labeled stearic, oleic and linoleic acid in six healthy men. A bolus of each 1-^{13}C-labeled fatty acid was ingested in random order at 72-hours intervalls with the breakfast meal. The pooled 9-days stools were analyzed for the fatty acids. The absorption efficiency for stearic, oleic and linoleic acid was 78 %, 97 % and 99.9 % (Table 9). These data indicate that fatty acids are not necessarily absorbed, in the relation we ingest them.

Table 9. INFLUENCE OF OTHER UNSATURATED FATTY ACIDS ON 84 uM LINOLEIC ACID ABSORPTION IN RATS

Fatty acid added	No. of Animals	Absorption nmol/min/10cm[a]	P value[b]
None	5	48.5 $\pm$ 5.5	0.01
Oleic	4	28.4 $\pm$ 0.7	0.01
Linolenic	4	27.2 $\pm$ 1.3	0.01
Arachidonic	4	25.8 $\pm$ 0.7	0.01

[a]Mean $\pm$ S.E.
[b]statistical analysis was made by comparing the absorption rate of linoleic acid in the presence of other fatty acids to its absorption rate in their absence
SL Chow, D. Hollander:Lipids 14, 378-385,1978

Furthermore the storage of linoleic acid seems to be dependent on the nutritional status of the subject. Becker and Bruce[15] investigated the retention of linoleic acid in rats given different essential fatty acids levels with the diet. Rats on a low linoleic acid level in the diet retained half of the ingested linoleic acid. With higher linoleic acid intake the deposition of linoleic acid increased, but the percentage of retained linoleic acid was only 10 to 15 % (Table 10).

Table. 10. DEPOSITION OF N-6 FATTY ACIDS IN CARCASS
LIPIDS OF THE RAT

n-6 fatty acids

EFA level in diet	Consumed (g)	Deposited (g)	Retention (%)
Low	0.25 ± 0.02	0.12 ± 0.02	48.9 ± 8.6[**]
Normal	2.59 ± 0.33	0.27 ± 0.11	10.3 ± 3.9
High	5.97 ± 0.73	0.96 ± 0.27	15.9 ± 4.0

Values are means and SD
** p 0.01
W.Becker, A. Bruce: Lipids 21, 121-127, 1986

From these data it is evident that polyunsaturated fatty
acids intake in western societies is inadequate, because of
the unfavourable ratio of the individual fatty acids
ingested[6]. The consumption of saturated fatty acids with
meat products is much too high and the intake of n-3 fatty
acids is too low. Dietary recommendations for the
prevention of atherosclerosis should at first advice to
reduce the animal fat intake. We should became more aware
of the fact that absorption and the metabolism of the
individual fatty acids are different and furthermore
dependent on the composition of the food as well as on the
nutritional status of the individuum.

References

1. O. Adam, Ernährungsphysiologische Untersuchungen mit
   Formeldiäten: Der Stoffwechsel mehrfach
   ungesättigter Fettsäuren und die Prostaglandinbiosynthese beim Menschen, Klin. Wschr. 63:731
   (1985).
2. W. E.M. Lands, Biochemical observations on dietary long
   chain
   fatty acids from fish oil and their effect on
   prostaglandin synthesis in animals and man,
   in:"Nutritional factors: modulating effects on
   metabolic processes", R.F. Beers Jr and E.G. Bassett
   eds. Raven Press, New York (1981).
3. O. Adam, Polyenoic fatty acids metabolism and effects
   on prostaglandin biosynthesis in adults and aged
   persons, in: "Polyunsaturated fatty acids and
   eicosanoids", W E M Lands, ed., Am. Oil Chem. Soc.,
   Champaign, Ill (1987).
4. O. Adam, G. Wolfram and N. Zöllner, Effect of different
   linoleic acid intake on prostaglandin biosynthesis
   and kidney function in man, Am. J. Clin.Nutr.,40:763
   (1984).
5. O. Adam, G. Wolfram and N. Zöllner, Effect of alpha-
   linoleic acid in the human diet on linoleic acid
   metabolism ad prostaglandin biosynthesis, J. Lipid
   Res. 27:421 (1986).
6. O. Adam, Polyenfettsäuren und Prostaglandinbiosynthese,
   Verh.dt. Ges. Inn. Med. 92:600 (1986).

7. FAO/WHO, Dietary fats and oils in human nutrition,in: FAO food and nutrition paper 3 (1984).

8. Deutsche Gesellschaft für Ernährung (DGE), "Ernährungsbericht", Frankfurt a. M. (1984)

9. B. Nasirullah, G. Werner and A. Seher, Fatty acids composition of lipids from edible parts and seeds of vegetables, FSA 86:264 (1984).

10. G. Billek, Einfluß der industriellen Verarbeitung und der haushaltsmäßigen Zubereitung auf die Nahrungsfette, Biblitheca. Nutr. Dieta 34:82 (1985).

11. S. R.Eder and G. Guhr, Die Abnahme des Linolsäuregehaltes von Fetten und Ölen beim Braten und Fritieren, FSA 81:566 (1979).

12. O. Adam, G. Wolfram and N. Zöllner, Vergleich der Wirkung von alpha-Linolensäure und Eicosapentaen- säure auf die Prostaglandinbiosynthese und die Thrombozytenaggregation beim Menschen, Klin. Wschr. 64:274 (1986).

13. S. L.Chow and D. Hollander, Linoleic acid absorption in the unanesthetized rat: Mechanism of transport and influence of luminal factors on absorption, Lipids 14:378 (1978).

14. P. J.H. Jones, P.B. Pencharz, and M.T. Clandinin, Absorbtion of ^{13}C-labeled stearic, oleic, and linoleic acids in humans: Application to breath tests, J. Lab. Clin. Med. 105:647 (1985).

15. W. Becker and A. Bruce, Retention of linoleic acid in carcass lipids of rats fed different levels of essential fatty acids, Lipids 21:121 (1986).

16. N. Zöllner, Dietary linolenic acid in man - an overview. Prog. Lipid Res. 21:177 (1986).

OMEGA-3 FATTY ACIDS FROM VEGETABLE OILS

J. Edward Hunter

The Procter and Gamble Company
6071 Center Hill Road, Cincinnati, Ohio   45224-1703

This article reviews the significance of vegetable oils as dietary sources of omega-3 fatty acids. In addition to fish and fish oils, certain vegetable oils, namely soybean and canola (or low erucic acid rapeseed) oils, may provide a significant source of dietary omega-3 fatty acids (1-6). The omega-3 fatty acid found in these vegetable oils is α-linolenic acid, an 18 carbon fatty acid with 3 double bonds (C18:3ω3). In contrast, the principal omega-3 fatty acids of fish oils, eicosapentaenoic (C20:5ω3) and docosahexaenoic (C22:6ω3) acids, have longer carbon chains and are more highly polyunsaturated than α-linolenic acid.

In the U.S. until recently, the only readily available edible vegetable oil source of α-linolenic acid was soybean oil, which in unhydrogenated form typically contains α-linolenic acid at about 7% of total fatty acids. Within the last 1 to 2 years in the U.S., a new vegetable oil source of α-linolenic acid, canola oil, has become commercially available as a salad and cooking oil. Canola oil typically contains α-linolenic acid at about 10% of total fatty acids. In Canada during the last decade, canola oil has become the most important vegetable oil source of α-linolenic acid, and soybean oil has been the second most important source. Compared to canola and soybean oils, fish oils usually have a higher total omega-3 fatty acid content (typically about 20-25% of total fatty acids) than canola or soybean oils. On the other hand, fish oils generally have poor palatability and are significant sources of cholesterol (2), whereas canola oil and soybean oil are highly palatable as salad and cooking oils and contain no cholesterol.

This article will focus on dietary aspects of α-linolenic acid. I will review first the relative levels of α-linolenic acid in various source oils as well as in finished food products. Then, using these data as well as data on production levels of various products, I will discuss estimates I have made of the per capita availability of α-linolenic acid in the U.S. and Canadian diets. These estimates then will be compared to the reported human dietary requirement for α-linolenic acid.

Table 1.   Vegetable Oil Sources
of α-Linolenic Acid

Oil	18:3 Level	
	g/100 g Oil	g/1 Tb (14 g) Serving
Linseed	53.3	7.5
Canola	11.1	1.6
Walnut	10.4	1.5
Wheat Germ	6.9	1.0
Soybean	6.8	1.0
Rice Bran	1.6	0.22
Corn	1.0	0.14
Cottonseed	0.5	0.07

Source: U.S. Department of Agriculture (USDA), Institute of Shortening and Edible Oils

Levels of α-Linolenic Acid in Vegetable Oils and in Finished Foods

There are, of course, other vegetable oil sources of α-linolenic acid besides canola and soybean oils (Table 1). These include linseed, walnut, wheat germ, and rice bran oils (2). None of these oils, however, are widely used in the U.S., and therefore oils other than canola and soybean are not regarded as important dietary sources of α-linolenic acid. Linseed oil is not approved for food use in the U.S., however, we are aware that linseed oil is used for edible purposes in the USSR, Poland, Czechoslovakia, and Hungary. Although reliable data on the usage of linseed oil for edible purposes are not available, we have estimated that the absolute quantity of linseed oil used for edible purposes is small, probably about 1 to 2% of total edible fats and oils used in Eastern Europe.

Table 2.   Principal Food Sources
of α-Linolenic Acid

Food	Source Oil	18:3 Level	
		g/100 g Fat	g/Serving
Salad and	Canola	10	1.4 (1 Tb = 14 g)
Cooking Oil	Soybean	7	1.0
Margarine (Tub)	Canola	4	0.45 (1 Tb = 14 g)[1]
	Soybean	3.5	0.39
Margarine (Stick)	Canola	2	0.22 (1 Tb = 14 g)[1]
	Soybean	2	0.22
Shortening	Canola	2	0.24 (1 Tb = 12 g)
(All Veg.)	Soybean	2	0.24

[1] 1 Tb Margarine = 11 g of Fat

Sources: Institute of Shortening and Edible Oils, POS Pilot Plant

In North American diets, the principal food sources of α-linolenic acid are salad and cooking oils, margarines, shortenings, salad dressing products, and foodservice fats and oils made from canola oil or soybean oil. Typical levels of α-linolenic acid in salad and cooking oils, margarines, and shortenings are shown in Table 2. Among these products, salad and cooking oils made from soybean or canola oils are largely unhydrogenated or lightly hydrogenated and thus are major food sources of α-linolenic acid. Margarines and shortenings are made from partially hydrogenated soybean or canola oils and thus contain smaller amounts of α-linolenic acid. Foodservice fats and oils are similar in their α-linolenic acid content to household shortenings. In Europe, fish oils frequently are used in manufacturing margarines and shortenings. However, because of their high oxidative instability, the fish oils used currently for such products are hydrogenated sufficiently that the products are not practical food sources of omega-3 fatty acids. In the U.S., fish oil and hydrogenated fish oil currently are not approved by the Food and Drug Administration as food ingredients, although the National Bureau of Marine Fisheries has submitted a petition requesting "generally recognized as safe" (GRAS) status for these materials. Despite uncertainty as to when the FDA will act on this petition, some fish oil manufacturers in the U.S. are actively investigating incorporation of fish oils into foods in ways that would contribute omega-3 fatty acids to the diet.

Table 3. α-Linolenic Acid in
Salad Dressing Products

Type	% Fat	18:3 Level	
		g/100 g	g/1 Tb (14 g) Serving
Mayonnaise	79	4.2	0.59
Salad Dressing (Spoon-Type)	33	2.0	0.28
Blue Cheese	52	3.7	0.52
Italian	48	3.3	0.46
Thousand Island	36	2.5	0.35
Vinegar and Oil	50	1.4	0.20

Source: USDA

The α-linolenic acid levels of several types of salad dressing products popular in the U.S. are shown in Table 3 (2). These products vary considerably in their total fat content, however, typically they contain about 0.2 to 0.6 g of α-linolenic acid per 1 tablespoon serving.

Other food sources of α-linolenic acid include walnuts, dairy products, beans, broccoli, and leafy vegetables like purslane, lettuce, and spinach (Table 4) (2). When considered on a serving size basis (7), however, these other foods generally would not contribute as much α-linolenic acid to the diets of Americans and Canadians as would vegetable oil products such as salad and cooking oils and salad dressings made from canola oil or soybean oil.

Table 4. Selected Food Sources
of α-Linolenic Acid

Food	% Fat	18:3 Level g/100 g Edible	g/Serving
Walnuts, English	62	6.8	1.02 (8-15 Halves, 15 g)
Butter	81	1.2	0.18 (1 Tb = 15 g)
Cheese, Cheddar	33	0.4	0.11 (1 Oz = 28 g)
Milk, Whole	3.3	0.1	0.24 (1 Cup = 240 g)
Purslane	0.85	0.4	0.4 (100 g)
Beans, Navy, Cooked	0.8	0.3	0.3 (100 g)
Broccoli or Spinach, Raw	0.4	0.1	0.1 (100 g)
Lettuce, Butterhead	0.2	0.1	0.1 (100 g)

Sources: USDA; Simopoulos & Salem, N. Engl. J. Med., 315:833, 1986

Estimation of α-Linolenic Acid Availability in U.S. and Canadian Diets

The first step in estimating the availability of α-linolenic acid
in U.S. and Canadian diets is recognition of the relative importance of
the key source oils in the U.S. and Canada. Table 5 compares the food
usage of vegetable oil sources of α-linolenic acid in the U.S. and
Canada in 1986, the last year for which data are available. In the
U.S., soybean oil was the most widely used vegetable oil and accounted
for 79% of total vegetable oil usage in 1986 (8). Corn oil was the
second most abundantly used vegetable oil in the U.S., however, its
usage was only about 10% of that of soybean oil. Canola oil usage in
the U.S. in 1986 was very small because canola oil was not introduced
into the U.S. market until the latter part of 1986. On the other hand,
in Canada, canola oil was the most widely used vegetable oil,
accounting for 58% of total vegetable oil usage in 1986 (9). Soybean
oil was the second most widely used vegetable oil in Canada.

Table 5. Vegetable Oil Usage in Specific
Products, U.S. and Canada, 1986

Source Oil	Salad/Cooking Oil U.S.	Canada	Shortening U.S.	Canada	Margarine U.S.	Canada	Totals U.S.	Canada
(% of Total Vegetable Oil Production)								
Canola	--	80	--	49	--	38	--	58
Soybean	78	5	81	30	87	51	79	26
Corn	9	6	NA[1]	--	10	5	7	4

[1] NA = Not Available

Sources: USDA, Statistics Canada

U.S. and Canadian usage of these source oils in salad and cooking oil, shortening, and margarine products in 1986 also is shown in Table 5. In the U.S., soybean oil accounted for about 78% of total vegetable oils used in salad and cooking oils in 1986. In Canada for comparison, canola oil was used for about 80% of salad and cooking oils. In the U.S., soybean oil continues to be the principal source oil for shortenings and margarines, whereas in Canada, both soybean and canola oils are used extensively for these products.

Availability of α-Linolenic Acid in the U.S. Diet

To estimate the availability of α-linolenic acid from vegetable oil sources in the U.S. diet, I obtained data on the market size, market share, and fatty acid composition of salad and cooking oil, salad dressing, margarine, shortening, and foodservice fat and oil products made with soybean oil and corn oil (10). I obtained such data for current products (i.e., for 1986) and, where available for historical perspective, for products manufactured during the 1960s and 1970s. The data were provided by member companies of U.S. trade associations, including the Institute of Shortening and Edible Oils, the National Association of Margarine Manufacturers, and the Association for Dressings and Sauces. The data have been expressed as per capita (11) availability of α-linolenic acid for a product category for a particular year.

Table 6. α-Linolenic Acid from Salad and Cooking Oils, U.S., 1960 - 1986

Year	Market Size (MM Pounds)	Typical [18:3][1] (%)	Total 18:3 (MM Pounds)	Per Capita Availability (g/Person/Day)
1960	455	0.5-1	2.2	0.02
1965	628	1-3.4	7.0	0.04
1970	763	3.4	16.1	0.10
1975	856	3.4	22.5	0.13
1980	1032	3.4	24.9	0.14
1985	1208	5-7	42.6	0.22
1986	1218	7	45.1	0.23

[1]Since 1965, products made from soybean oil.
Source: Institute of Shortening and Edible Oils

Salad and Cooking Oils. The production of salad and cooking oils and the availability of α-linolenic acid from these products since 1960 are shown in Table 6. Prior to the early 1960s, most salad and cooking oils in the U.S. were made from unhydrogenated cottonseed or corn oil. These products contributed low levels of α-linolenic acid to the diet because both oils typically contain α-linolenic acid at about 1% or less of total fatty acids. In the early 1960s, partially hydrogenated soybean oil products were introduced, in part due to increasing demand for economical liquid oils with relatively high levels of polyunsaturated fatty acids. These products typically had an α-linolenic acid content of about 3.4%.

The market size (or total sales) of salad and cooking oils has increased continuously since 1960. In a corresponding manner, the amounts of α-linolenic acid available for consumption also have increased with time. Since the mid 1980s, manufacturers have increased their production of household salad and cooking oils made from unhydrogenated soybean oil. One result has been a marked increase in the availability of α-linolenic acid from these products in 1985 compared to 1980.

Shortenings. Table 7 shows how α-linolenic acid contributed by household shortenings has changed over the years. Overall, the size of the shortening market has not changed much since 1960 although there has been a slight decline in sales since the early to mid 1970s. Similarly, the composition of household shortenings, including their content of α-linolenic acid, has not changed very much since 1960. In the 1960s, there was a slight increase in usage of vegetable fats at the expense of animal fats resulting in an increase in α-linolenic acid available from these products. Sales of household shortenings containing animal fats increased somewhat in the early 1980s, largely for price reasons, and this accounted for some of the decrease in α-linolenic acid availability from shortenings at that time. Overall, the per capita availability of α-linolenic acid from household shortenings has declined slightly from about 0.08 g/person/day in 1960 to about 0.05 g/person/day currently.

Table 7. α-Linolenic Acid from Household Shortenings, U.S., 1960 - 1986

Year	Market Size (MM Pounds)	Typical [18:3] Animal (%)	Typical [18:3] All Veg. (%)	Total 18:3 (MM Pounds)	Per Capita Availability (g/Person/Day)
1960	695	1	2	11.9	0.08
1970	749	1	2	13.2	0.08
1975	824	1	2	14.5	0.08
1980	709	1	2	12.4	0.07
1985	664	1	2	11.3	0.06
1986	657	1	2	10.0	0.05

Source: Institute of Shortening and Edible Oils

Margarines. α-Linolenic acid available from margarines is indicated in Table 8. In the 1960s, margarines were largely the stick variety with α-linolenic acid levels around 2% of total fatty acids (on a fat basis). Tub margarines were introduced in the late 1960s. These products contained unhydrogenated as well as partially hydrogenated vegetable oils. During the late 1960s to mid 1970s, these products gradually increased in their content of unhydrogenated oil and thus their α-linolenic acid content also increased. There is a broad range of linolenic acid levels in tub margarines, however, 3.5% of total fatty acids is considered typical of many products. The per capita availability of α-linolenic acid from margarines increased during the early 1960s and 1970s as margarine continued to replace butter as the principal table spread in the U.S. Since the mid 1970s, per capita availability of α-linolenic acid from margarines has remained fairly constant at about 0.2 g/person/day.

Table 8. α-Linolenic Acid from
Margarines, U.S., 1960 - 1986

Year	Total Margarine Fat Produced (MM Pounds)	Typical [18:3][1] Stick (%)	Typical [18:3][1] Tub (%)	Per Capita Availability (g/Person/Day)
1960	1356	2	-	0.14
1970	1776	2	2	0.19
1975	1875	2	3.5	0.21
1980	1979	2	3.5	0.22
1985	1914	2	3.5	0.20
1986	2096	2	3.5	0.22

[1]Margarines made from soybean oil.
Source: National Association of Margarine Manufacturers

Salad Dressing Products. Salad dressing products often are
classified as "spoon-type dressings" and "pourable-type dressings."
Spoon-type dressings include mayonnaise and a mayonnaise-like product
called "salad dressing." Federal regulations in the U.S. specify that
mayonnaise contain not less than 65% vegetable oil by weight and that
salad dressing contain not less than 30% vegetable oil by weight.
Pourable dressings may be two phase (such as vinegar and oil) or the
emulsified viscous type (such as French or Italian). There is a great
variety of products available of varying compositions with a wide range
in their oil content. Many pourable dressings typically contain about
50% oil by weight. In the U.S., mayonnaise and some pourable dressings
are made from a blend of unhydrogenated and partially hydrogenated
soybean oil. The majority of pourable dressings and also the
mayonnaise-like product "salad dressing" are made from unhydrogenated
soybean oil.

Table 9 shows estimates of availability of α-linolenic acid from
salad dressing products. Production of these products has increased
continuously since 1970 (12, 13). Prior to about 1970, cottonseed oil
and partially hydrogenated soybean oil were the principal source oils
used for salad dressing products, and these oils contributed relatively
small amounts of α-linolenic acid. By the mid 1970s there was
increasing use of unhydrogenated soybean oil in salad dressing
products, and since about 1975, the overall composition of these
products has changed relatively little. Considering contributions of
both spoonable and pourable-type products, the availability of
α-linolenic acid from these products since 1970 has increased from
about 0.1 to 0.6 g/person/day.

Table 9. α-Linolenic Acid from Salad
Dressing Products, U.S., 1970 - 1986

Year	Total Production (MM Gallons)	Total Oil Used (MM Pounds)	Total 18:3 (MM Pounds)	Per Capita Availability (g/Person/Day)
1970	218	872	16.2	0.10
1975	275	1113	70.1	0.40
1980	346	1438	90.1	0.49
1985	445	1755	109.9	0.57
1986	472	1867	116.6	0.60

Source: Association for Dressings and Sauces

Foodservice Fats and Oils. Foodservice fats and oils include deep frying fats, pan and grill fats, cake and pastry shortening, and bottled salad and cooking oils used by restaurants. Amounts of margarine and salad dressing products purchased by restaurants were included among the production data for these products which I discussed previously. In the foodservice industry, there is frequently a considerable amount of fat and oil discarded. This is particularly true in the case of fats used for deep frying since large amounts of such fats (we estimated as much as 50%) are discarded after use (10).

Total production of foodservice fats and oils increased between 1970 and the present (Table 10) reflecting continued growth of the fast food industry in the U.S. In conjunction with this increase in production, there was an increase in total ∝-linolenic acid availability due to increased usage of partially hydrogenated soybean oil in some foodservice fats and oils. ∝-Linolenic acid levels of individual foodservice fats and oils, however, generally have remained unchanged during this period. Overall, between 1970 and 1986 the

Table 10. ∝-Linolenic Acid from Foodservice
Fats and Oils, U.S., 1970 - 1986

Year	Total Fat Production	Typical [18:3] Solid	Typical [18:3] Liquid	Total 18:3[1]	Per Capita Availability
	(MM Pounds)	(%)		(MM Pounds)	(g/Person/Day)
1970	1492	1.5	2.5-3.4	13.7	0.08
1975	1655	1.5	2.5-3.4	19.7	0.11
1980	1793	1.5	2.5-3.4	19.4	0.11
1985	2079	1.5	2.5-3.4	24.1	0.13
1986	2122	1.5	2.5-3.4	23.7	0.12

[1] Adjusted for estimated 50% wastage of deep frying fat.
Source: Institute of Shortening and Edible Oils

estimated per capita availability of ∝-linolenic acid from foodservice fats and oils increased from about 0.08 to 0.12 g/day. This may be increasing further with the recent introduction of unhydrogenated soybean oil and canola oil products into the foodservice industry.

Total Availability of ∝-Linolenic Acid from Vegetable Oil Products, U.S.

The change in total ∝-linolenic acid availability from salad and cooking oils, household shortenings, margarines, salad dressing products, and foodservice fats and oils between 1970 and 1986 is presented in Table 11. Total per capita availability of ∝-linolenic acid from these vegetable oil products apparently has increased from about 0.6 to about 1.2 g/person/day between 1970 and the present.

Availability of α-Linolenic Acid in the Canadian Diet

In Canada, unlike in the U.S., data on market sizes and market shares of individual branded products are not readily available. Therefore, I made estimates of α-linolenic acid availability using data from the Agriculture Division of Statistics Canada on the annual domestic production of deodorized oils used for margarine, shortening, and salad and cooking oil (9). These data compiled by Statistics Canada are expressed as metric tons of various source oils, such as canola and soybean oil, used annually in the production of margarine oil, shortening oil, and salad and cooking oil. These production data include fats and oils used for making salad dressing products and for foodservice purposes. Statistics Canada also compiles data on the per capita disappearance of various food categories, such as salad and cooking oil, shortening, and margarine (14, 15).

Table 11. α-Linolenic Acid Availability from
Vegetable Oil Products, U.S., 1970 - 1986

Year	Salad & Cooking Oils	Household Shortenings	Margarines	Salad Dressing Products	Foodservice Fats & Oils	Totals
			(g/Person/Day)			
1970	0.10	0.08	0.19	0.10	0.08	0.55
1975	0.13	0.08	0.21	0.40	0.11	0.93
1980	0.14	0.07	0.22	0.49	0.11	1.03
1985	0.22	0.06	0.20	0.57	0.13	1.18
1986	0.23	0.05	0.22	0.60	0.12	1.22

The use of data from Statistics Canada for calculating the per capita availability of α-linolenic acid from salad and cooking oils is illustrated in Table 12. As noted previously (Table 5), canola oil now accounts for about 80% of total vegetable oils used for salad and cooking oils in Canada. Statistics Canada reported the per capita disappearance of salad and cooking oil for 1986 to be 15.1 g/person/day (14). Thus the contribution of canola oil to total salad and cooking oil production was about 80% of this, or about 12 g/person/day. Multiplying this figure by the typical level of α-linolenic acid in canola oil (i.e., 10%) indicates that canola oil contributed about 1.2 g of α-linolenic acid/person/day from salad and cooking oils in 1986. Considering also the contributions of soybean and corn oils to total salad and cooking oil production, the total availability of α-linolenic acid from salad and cooking oils in Canada in 1986 was about 1.3 g/person/day. I used a similar procedure to calculate the availability of α-linolenic acid from margarines and shortenings.

Table 12.   α-Linolenic Acid from Salad
and Cooking Oils, Canada, 1986

Source Oil	% of Total Veg. Oil Production	Per Capita Avail. of Oil (g/Person/Day)	Typical [18:3] (%)	Per Capita Availability (g/Person/Day)
Canola	79.8	12.03	10	1.20
Soybean	4.5	0.68	7	0.05
Corn	6.4	0.97	1	0.01

Total 18:3 from Salad and Cooking Oils = 1.26
(g/Person/Day)

Source: Statistics Canada

The results of these calculations using production data going back to 1973 are shown in Table 13.   These data indicate that total availability of α-linolenic acid from vegetable oil products has increased from about 1 g/person/day in 1973 to about 2 g/person/day currently.   Much of this increase can be accounted for by increased usage of canola oil for salad and cooking oil since 1973.   The increased usage of canola oil in these products was particularly dramatic during the early 1980s, as indicated in this table.

Table 13.   α-Linolenic Acid from Vegetable
Oil Products, Canada, 1973 - 1986

Year	Salad and Cooking Oil (g/Person/Day)	Shortening	Margarine	Totals
1973	0.53	0.27	0.24	1.04
1975	0.64	0.24	0.27	1.15
1980	0.68	0.20	0.27	1.15
1985	1.14	0.34	0.36	1.84
1986	1.26	0.34	0.35	1.95

Sources: Statistics Canada, Canola Council of Canada

Estimation of Total Dietary α-Linolenic Acid

In order to estimate the total amount of α-linolenic acid in U.S. and Canadian diets, the calculated contributions of vegetable oil products must be added to contributions from other foods. As indicated in Table 4, consumption of other foods containing α-linolenic acid, such as dairy products, beans, broccoli, and lettuce, could add about 0.5 to 1 g of α-linolenic acid per day. Thus the total availability of α-linolenic acid in the U.S. diet could be about 1.7 to 2.2 g/person/day and that in the Canadian diet, about 2.5 to 3 g/person/day.

Another approach to estimating the total intake of α-linolenic acid is to consider the fatty acid composition of human adipose tissue. Berry and Hirsch (16) have reported the level of α-linolenic acid in adipose tissue of healthy adult American males to be about 1.9% of total fatty acids. The range of values among 399 subjects was 1.2 to 3.1% of total fatty acids. Considering that the composition of unsaturated fatty acids of adipose tissue reflects that of the diet, an adult male (age 23 to 50) consuming his RDA of 2700 kcal/day (17) of a diet providing 38% of calories as fat (18) would ingest about 1.3 to 3.6 g of α-linolenic acid/day. The mean intake of this group of subjects would have been about 2.2 g/day. This mean and range of α-linolenic acid intakes are consistent with the estimates I have made from food production data.

Comparison of α-Linolenic Acid Availability with Reported Dietary Requirement

Holman and coworkers suggested a dietary requirement for α-linolenic acid for a 6 year old girl to be 0.54% of total calories (19). Bjerve and colleagues suggested a lower requirement of about 0.25% of calories for immobile adults (20). At an intake of 2700 kcal/day, the calculated requirement for α-linolenic acid would be 0.75 g/day according to Bjerve's estimate and 1.6 g/day using Holman's estimate. My calculated values for the availability of α-linolenic acid in both the U.S. and Canadian diets, namely about 2 to 3 g/person/day, clearly exceed this range of reported requirements. In view of this, it would appear that typical U.S. and Canadian diets contain sufficient amounts of α-linolenic acid to prevent symptoms associated with α-linolenic acid deficiency.

On the other hand, emerging research has suggested that modest increases in dietary α-linolenic acid, as could be achieved by substituting into the diet a salad and cooking oil made with canola oil, may be associated with certain cardiovascular benefits. Renaud and coworkers have reported that increasing dietary α-linolenic acid from about 1.2 g/day to about 3.3 g/day, in conjunction with reduction in dietary saturated fatty acids, was associated with reduced platelet aggregation, which in turn, may reduce risk of thrombosis (21). Also, Berry and Hirsch have found that increased adipose tissue levels of α-linolenic acid, which reflect increased intake, were associated with reduced blood pressure (16). These effects may be related to actions of prostaglandins derived from eicosapentaenoic acid, which humans can produce from α-linolenic acid (22).

Summary and Conclusions

In summary, the principal food sources of the omega-3 fatty acid α-linolenic acid are salad and cooking oil, salad dressing, shortening, margarine, and foodservice fat and oil products made from canola oil or soybean oil. Using food production data provided by various U.S. trade associations and by Statistics Canada, I have estimated the per capita availability of α-linolenic acid from vegetable oil products in the U.S. to be about 1.2 g/person/day and in Canada, about 2 g/person/day. The higher level of α-linolenic acid availability in Canada is largely accounted for by the widespread use of canola oil in Canada. Considering also contributions to dietary α-linolenic acid of other foods such as nuts, dairy products, and vegetables, it would appear that the total intake of α-linolenic acid in U.S. and Canadian diets adequately exceeds the reported nutritional requirement. Emerging research has suggested possible health benefits associated with modest increases in dietary α-linolenic acid, including reduced clotting tendency and reduced blood pressure.

Acknowledgments

I am very grateful to the following individuals who provided valuable information and suggestions for this presentation:

Mr. Reginald Bacchus, POS Pilot Plant, Saskatoon, Saskatchewan.
Mr. Bob Elbert, American Soybean Association, St. Louis, Missouri.
Ms. Karen Gray, Statistics Canada, Agriculture Division, Winnipeg, Manitoba.
Ms. Belva Jones, National Association of Margarine Manufacturers, Washington, D.C.
Ms. Jane Macdonald, Association for Dressings and Sauces, Atlanta, Georgia.
Mr. Theodore Mag, Canada Packers Inc., Toronto, Ontario.
Ms. Eileen McGregor, Canola Council of Canada, Winnipeg, Manitoba.
Mr. Robert Reeves and Members of the Technical Committee, Institute of Shortening and Edible Oils, Washington, D.C.

Bibliography

1. J. E. Hunter, n-3 fatty acids from vegetable oils. n-3 News, 2:1 (1987).
2. J. Exler and J. L. Weihrauch, Provisional Table on the Content of Omega-3 Fatty Acids and Other Fat Components in Selected Foods, U.S. Department of Agriculture, Washington, D.C. (1985) (publication no. HNIS/PT-103).
3. Consumer and Food Economics Institute, Composition of Foods: Fats and Oils: Raw, Processed, Prepared. Agriculture Handbook No. 8-4, Science and Education Administration, U.S. Department of Agriculture, Washington, D.C. (1979).
4. A. P. Simopoulos and N. Salem, Jr., Purslane: a terrestrial source of omega-3 fatty acids (letter to the editor). N. Engl. J. Med., 315:833 (1986).
5. J. E. Hunter, Vegetable oil sources of omega-3 fatty acids (letter to the editor). N. Engl. J. Med., 316:626 (1987).
6. R. Barnett, Hunting the wild purslane. American Health, 6:131 (1987).
7. J. A. T. Pennington and H. N. Church, Food Values of Portions Commonly Used, 14th ed., Harper & Row, New York (1985).
8. Institute of Shortening and Edible Oils, Inc., Food Fats and Oils, Washington, D.C. (1988).
9. Statistics Canada, Agriculture Division, Oils and Fats, Catalogue 32-006 (1974), (1976), (1981), (1986), and (1987).

10. J. E. Hunter and T. H. Applewhite, Isomeric fatty acids in the U.S. diet: levels and health perspectives. Am. J. Clin. Nutr., 44:707 (1986).

11. U.S. Department of Commerce, Statistical Abstract of the United States 1987, 107th ed., Bureau of the Census, Washington, D.C. (1986).

12. U.S. Department of Commerce, Mayonnaise, Salad Dressings, and Related Products, 1973, Bureau of Domestic Commerce, Washington, D.C. (1974).

13. Anonymous, State of the industry: dressings and sauces. Food Engineering, 59:86 (1987).

14. Statistics Canada, Agriculture Division, Apparent per capita food consumption in Canada. Catalogues 32-226 and 32-230 (1987).

15. M. Vaisey-Genser and N. A. M. Eskin, Canola Oil: Properties and Performance, Publication no. 60, Canola Council of Canada, Winnipeg, Manitoba (1987).

16. E. M. Berry and J. Hirsch, Does dietary linolenic acid influence blood pressure? Am. J. Clin. Nutr., 44:336 (1986).

17. Committee on Dietary Allowances, Food and Nutrition Board, Commission on Life Sciences, National Research Council, Recommended Dietary Allowances, 9th ed., National Academy Press, Washington, D.C. (1980).

18. W. W. Kim, W. Mertz, J. T. Judd, M. W. Marshall, J. L. Kelsay, and E. S. Prather, Effect of making duplicate food collections on nutrient intakes calculated from diet records. Am. J. Clin. Nutr., 40:1333 (1984).

19. R. T. Holman, S. B. Johnson, and T. F. Hatch, A case of human linolenic acid deficiency involving neurological abnormalities. Am. J. Clin. Nutr., 35:617 (1982).

20. K. S. Bjerve, I. L. Mostad, and L. Thoresen, Alpha-linolenic acid deficiency in patients on long-term gastric-tube feeding: estimation of linolenic acid and long-chain unsaturated n-3 fatty acid requirement in man. Am. J. Clin. Nutr., 45:66 (1987).

21. S. Renaud, F. Godsey, E. Dumont, C. Thevenon, E. Ortchanian, and J. L. Martin, Influence of long-term diet modification on platelet function and composition in Moselle farmers. Am. J. Clin. Nutr., 43:136 (1986).

22. E. A. Emken, R. O. Adlof, H. Rakoff, and W. K. Rohwedder, Metabolism of deuterium-labeled linolenic, linoleic, oleic, stearic and palmitic acid in human subjects. In: Proceedings of the Third International Conference on the Synthesis and Applications of Isotopically Labelled Compounds (in press).

PREPARATION OF FISH OIL FOR DIETARY APPLICATIONS

H.J. Wille and P. Gonus

Nestlé Research Centre, Nestec Ltd.
Vers-chez-les-Blanc
CH-1000 Lausanne 26 (Switzerland)

I.   INTRODUCTION

In recent years, evidence has been accumulated indicating that poly-unsaturated fatty acids (PUFAs) may be effective, to some extent, in prevention and therapy of cardiovascular diseases[1,4]. Growing interest has been accorded to PUFAs of the n-3 series such as eicosapentaenoic (EPA) and docosahexaenoic (DHA) acid. These acids are suggested to participate in the biosynthesis of prostaglandins, thromboxanes and leukotrienes; substances with high physiological activity influencing for instance blood platelet aggregation and contraction and dilatation of vascular glands.

Marine oils, especially fish oils, are important natural sources of these acids. Oils from species rich in these acids can contain about 30% of EPA and DHA together. Fish oils are composed of a great number of different fatty acids. At least fifty acids are present in detectable amounts[5,6], with eight principal fatty acids: C14:0, C16:0, C16:1, C18:1, C20:1, C20:5, C22:1 and C22:6. There is a wide variation in fatty acid composition between different species. Moreover, oil from individual species also varies, depending on sub-species, diet, season and fishing area. For instance, the iodine value of fish oil was found to increase with decreasing water temperatures and vice versa.

Table 1 shows the fatty acid compositions of oils from several species of fish presenting approximate values of the eight major fatty acids[5-7]. Herring and mackerel oils are rich in the monounsaturated fatty acids C20:1 and C22:1 and contain only few EPA and DHA. For dietary applications, the more appropriate oils seem to be menhaden, sardine or anchovy oils, with EPA and DHA contents of about 15% and 10%, respectively. When oils rich in DHA are needed, the oil of the white tuna may be used, which was reported to contain up to 28% of this acid[6].

Fish oils are obtained as by-products in fish meal production; they are extracted from offals remaining after filleting or from industrial fish. For good quality oils, the fish should be as far as possible undamaged, and should be chilled from the time of the catch. Sometimes the fishing trawlers are equipped on board with the machinery needed to produce fish oil. To begin the process, which is illustrated in Figure 1, minced fish is transferred to a cooker. By heating at about 100°C, the proteins are denatured and the oil is released. The fish slurry obtained is then passed over a strainer conveyor, removing some of the liquid

Table 1. Approximate contents of some fish oils in 8 principal fatty acids.

Fatty acid composition in %

Fatty acids	Herring	Mackerel	Menhaden	Sardine	Anchovy
C14:0	7	5	8	8	9
C16:0	14	18	20	17	19
C16:1	6	5	12	9	9
C18:1	10	12	15	12	12
C20:1	15	12	1	3	3
C20:5	5	7	14	17	17
C22:1	20	15	1	3	2
C22:6	6	8	8	9	9

phase, into a screw press which may be a single or a multiscrew design. The liquors produced in the strainer and the press are purified from fine solids by a decanter centrifuge. The oil and water phases are then separated in a further series of centrifuging operations, including washing with water to remove trace amounts of protein materials. The polished oil, maintained at 90-100°C during all the operations, is then cooled down for storage. Especially for dietetic products, extractions with supercritical CO_2 have been proposed[8,9], as alternative process.

The crude oil is not suitable in most cases for edible use because it contains a high level of impurities, such as proteins, phosphatides, carbohydrates, pigments, oxidation products, free fatty acids or trace metals. Normally, these impurities are removed by refining. This process consists of 4 principal stages as shown in Fig. 2. Refining begins with degumming, in which the oil is mixed with an acid like phosphoric acid to cause precipitation of phosphatides and other colloidal substances which can be removed by centrifugation. The next stage is the elimination of free fatty acids by neutralisation with caustic soda. This treatment with

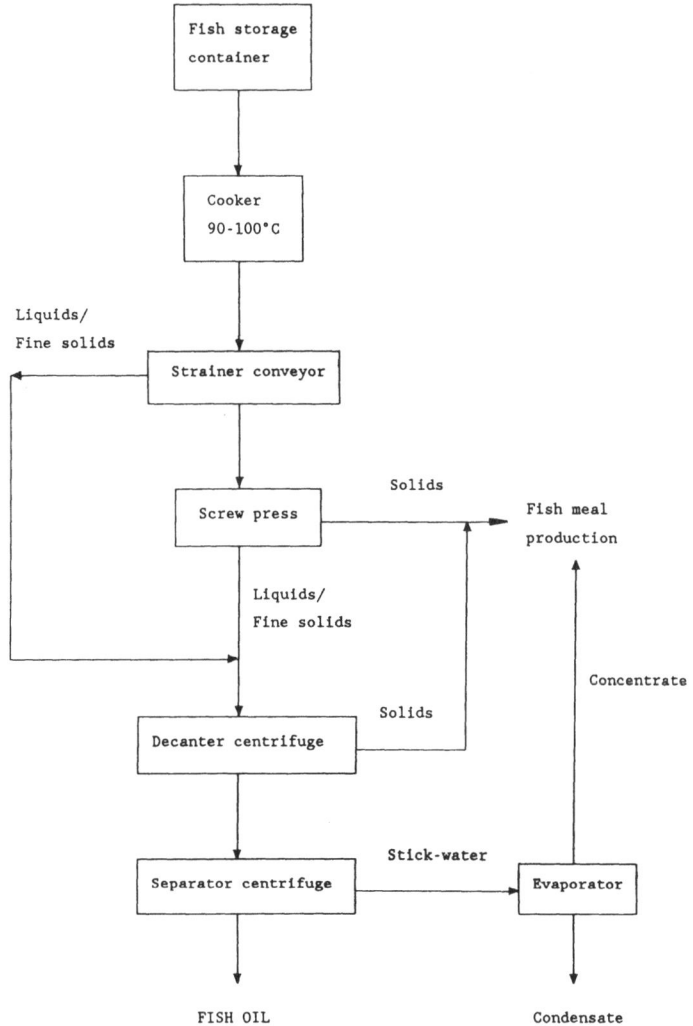

Figure 1. Flow-sheet of fish oil extraction

sodium hydroxide is usually carried out in a continuous line of separator centrifuges, allowing also to wash out other compounds like traces of gums, oxidised substances or further water soluble materials. Physical refining offers an alternative to alkali refining: here the free fatty acids are removed by deodorisation. In the next step, the oil is bleached by mixing under vacuum with adsorbents like natural or activated clays or active carbon. The spent bleaching earths are then filtered off, excluding air as far as possible because activated clays are potential oxidation

For dietetic applications, it has also been proposed to purify crude fish oil by extraction of the undesirable substances with supercritical carbon dioxide as alternative to usual refining practices[12].

As the polyunsaturated fatty acids EPA and DHA are very sensitive to oxidation and chemical alteration, they may be deteriorated under usual refining conditions. The objective of this work was to study the influence of certain refining parameters on fish oil quality.

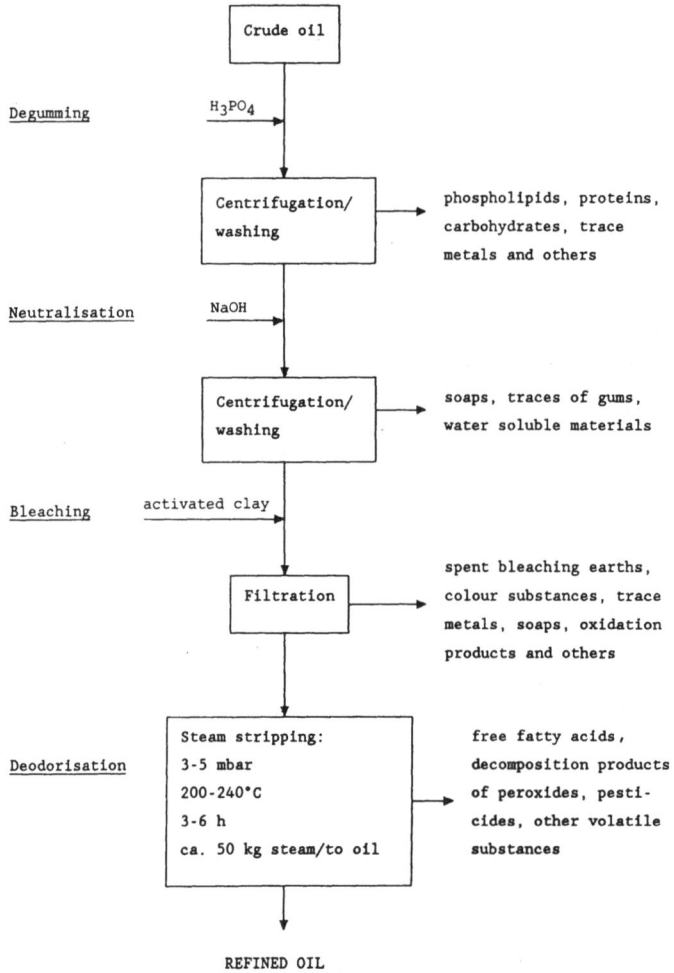

Figure 2. Flow-sheet with the principal operations of oil refining.

catalysts. By bleaching, not only coloured substances are removed but also traces of metals and soaps as well as oxidation products, especially hydroperoxides[10]. Furthermore, active carbon was reported to adsorb polycyclic aromatic hydrocarbons[11]. After bleaching, the oil is deodorised by steam distillation under vacuum at elevated temperatures of 200-240°C for several hours. The reason for steam stripping is to eliminate flavours and odours. Equally free fatty acids, aldehydes, ketones, pesticides and other volatile impurities are removed. During the cooling phase, an aqueous citric acid solution is sometimes added and the oil is pumped through a polishing filter before storage.

II.   RESULTS AND DISCUSSION

1.   INFLUENCE OF DEODORISATION CONDITIONS

a)   FATTY ACID COMPOSITION

For this investigation, commercially available crude or semi-refined fish oils were used, consisting of oils of different species but all with a high content of EPA and DHA, for instance ca. 15% and 10%, respectively.

Table 2 illustrates the effects of different refining operations on several oil characteristics. The oil was degummed with 0.3% of 75% H_3PO_4, neutralised with a 10% excess of 17% sodium hydroxide, bleached with 1% bleaching earth (Tonsil Optimum, Südchemie, München, FRG) at 90°C for 30 min. and deodorised for 5 hours at 230°C. Neutralisation diminished the free fatty acid (FFA) content and lightened the colour. The colour was then removed by bleaching and subsequent deodorisation decreased not only the FFA and colour values but had a disastrous effect on the EPA and DHA contents. The amount of EPA was reduced from 15% to 10.7% and that of DHA from 10.5% to 6%.

Table 2.   Effects of refining operations on free fatty acid (FFA) content, refractive index, colour and fatty acid composition.

	Fish oil			
	Crude	Neutralised	Bleached	Deodorised
FFA (%)	1.2	0.07	0.06	0.02
Refractive index n_D^{40}	1.4740	1.4743	1.4740	1.4758
Colour (Lovibond 1") red/yellow	3.0/30	2.0/19	0.2/2.3	0.2/0.5
n-3 PUFA content (%)				
EPA, C20:5	15.2	14.7	15.0	10.7
DHA, C22:6	10.9	10.4	10.5	6.0

The refractive index is known to be correlated to the iodine value in that more saturated oils have lower refractive indices. In this case, however, decreased amounts of EPA and DHA were accompanied with an increased refractive index. It is presumed that this might be due to a shifting of double bounds or to other isomerisations. Various deodorisations at different temperatures were carried out to identify deodorisation conditions which would avoid these unfavourable effects.

The influence of deodorisation time and temperature is demonstrated in Table 3. A short-time deodorisation at an elevated temperature of 240°C still led to a distinct deterioration of the n-3 PUFAs which was also evident in the refractive index. At 180°C and 170°C as deodorisation temperatures, the refractive indices were quite unchanged in comparison with the neutralised oil but the fatty acid determination indicated a very slight decrease in n-3 PUFAs at 180°C whereas at 170°C no changes were observed.

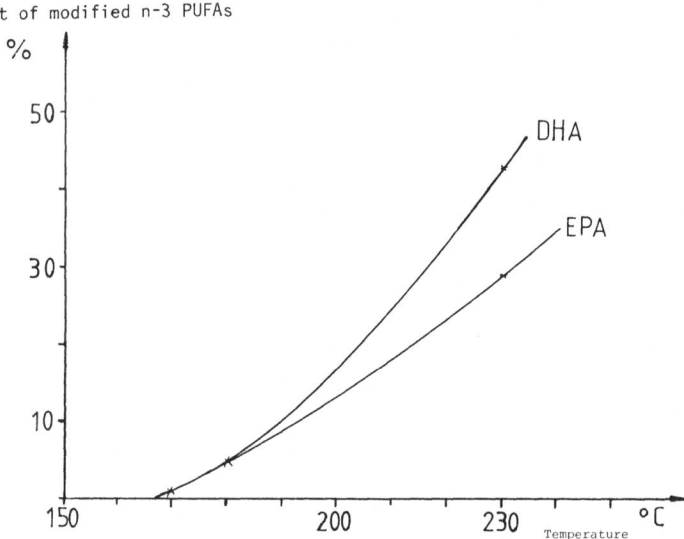

Amount of modified n-3 PUFAs

Diagram 1. Degradation of the n-3 PUFAs EPA and DHA in % as
a function of temperature.

Table 3. **Influence of deodorisation time and temperature on refractive index and fatty acid composition.**

	Neutralised	Deodorised		
		0.5h at 240°C	5h at 180°C	6h at 170°C
Refractive index n_D^{40}	1.4740	1.4750	1.4742	1.4741
n-3 PUFA content %				
EPA, C20:5	16.3	14.6	15.3	16.1
DHA, C22:6	10.6	8.9	10.1	10.5

The graph in Diagram 1 illustrates the dependence of n-3 PUFA degradation on temperature during batch deodorisation, signifying that, at elevated temperatures, DHA is more easily decomposed than EPA. These results suggest that, under batch deodorisation conditions, the temperature should not exceed 170°C. The detrimental effect of temperature could be reduced when shorter deodorisation times were used. But even a short time of 30 min. and at 240°C still caused a degradation of the n-3 PUFAs of ca. 15%.

Trials were otherwise carried out with a continuous thin-film deodorisation apparatus. In this case, the contact time of the oil with the heating surface was ≤ 1 min. with a heating medium of 230°C. Under these conditions, no changes in the refractive indices or fatty acid compositions were observed, thus emphasizing the applicability of this method when higher temperatures are required.

b)   PESTICIDES

As mentioned in Figure 2, refining also serves to eliminate pesticides and heavy metals. Organochlorine pesticides and polychlorinated biphenyls preferentially accumulate ·in the fatty fish species, which are precisely the fish generally used for oil production.

Table 4.

Residual contents of pesticides in a fish oil after neutralisation and deodorisation.

| | | Fish oil | |
		Neutralised	Deodorised
HCB	(ppb)	19	18
γ-HCH (Lindane)	(ppb)	13	13
total HCH	(ppb)	113	108
Heptachlor	(ppb)	8	4
Dieldrine	(ppb)	17	11
total DDT	(ppb)	45	43

For human consumption, the levels of these contaminants must be reduced by refining. It has been reported for vegetable oils that pesticides are principally eliminated during deodorisation[13,14]. The degree of removal depended on temperature, and to achieve a reduction below detection levels, deodorisation temperatures of 230-260°C were proposed. In fish oil, it was found that pesticide amounts were mainly reduced by hydrogenation and the residual traces were eliminated by deodorisation[15]. It was reported that, in commercially available encapsulated fish oils, which were not hydrogenated, small amounts of organochlorine pesticides could be detected[16]. This no doubt results from the milder refining conditions given to those oils. Table 4 resumes the results of pesticide measurements of a fish oil deodorised at 170°C for 6 hours. One can see that under these conditions the pesticide amounts were slightly reduced but still remained detectable, with a sum of residual contents of 0.2 ppm. If even lower concentrations are required, it will be favourable to use the continuous thin-film deodorisation technique mentioned above. Trials carried out using this apparatus showed that at 210°C and a vacuum of 2-3 mbar, the pesticide contents of a fish oil could be reduced to undetectable amounts.

c)   HEAVY METALS

The heavy metals cadmium, lead and mercury are reported to possess toxic qualities, with toxicities related to the chemical nature of the substances containing these metals[17]. The metals copper and iron, however, are known for their distinct pro-oxidant effects. For these reasons, it is essential to consider the concentrations of these elements. During oil processing, trace metals are eliminated, especially by neutralisation and bleaching. Table 5 presents trace metal concentrations determined for a fish oil after neutralisation and bleaching, respectively. The results show that after degumming with 0.3% H_3PO_4 (75%) and neutralisation with caustic soda the amounts of Cd, Hg and Pb were below the detection limits. Moreover, the concentrations of Cu and Fe were further reduced by bleaching with 1% bleaching earth at 90°C for 30 min. The metal levels found in this investigation were of the same order of magnitude as those reported for a refined menhaden oil[18] or for vegetable oils[13,17]. It seems that further reduction of trace metals should not be necessary.

Table 5.

Residual contents of heavy metals in a fish oil after neutralisation and bleaching.

| | | Fish oil | |
		Neutralised	Bleached
Cadmium	(ppb)	< 10	< 10
Copper	(ppb)	11	< 5
Iron	(ppb)	40	5
Mercury	(ppb)	< 50	< 50
Lead	(ppb)	< 20	< 20

2.   SILICA GEL TREATMENT

Different alternatives to the usual refining process have been described for the preparation of fish oil for health food. Among these, the purification by extraction with supercritical CO_2 may especially be mentioned[12,19], as also the treatment with activated silica gel in organic solvents as for instance hexane[20-22].

Bleaching is an essential stage in fish oil refining because not only coloured pigments and trace metals are removed, but also oxidation products which may decompose to give off-flavour components.

a)   DECOLOURIZING EFFECTS

In a series of trials, various bleaching methods for fish oil were tested. The effects of these different procedures on colour removal are illustrated in Table 6. The colours were measured with a Lovibond Tintometer in a 5 1/4" cell. After neutralisation, fish oils are generally still considerably coloured. By bleaching with 1% bleaching earth for 30 minutes at 90°C, the red colour component was preferably diminished. A second bleaching under the same process conditions but with 2% bleaching earth and 0.5% active carbon caused a further clearing up of the oil, both for the red and the yellow colour. Contacting this oil dissolved in hexane with activated silica gel and collecting the unadsorbed portion by washing with hexane gave a nearly colourless oil, which is demonstrated by the low Lovibond colour values. Determination of the fatty acid compositions gave similar values for all the samples, thus indicating that no fatty acid modifications had occurred as described for the deodorisation trials. In addition to the high bleaching activity, the silica gel treatment also efficiently eliminated the fishy odour of oil. It has been presumed[20] that mono- and diglycerides are sources of the odours, and are retained on the silica gel. Effects on the taste, however, were less evident.

Table 6. Effects of different bleaching procedures on colour removal.

Colour (Lovibond, 5 1/4" cell)

Fish oil

	Neutralisation	Bleaching A	Bleaching B	Silica gel treatment
Red	11	5.6	2.2	0.3
Yellow	> 40	> 40	20	2

Bleaching A: 1% Bleaching earth

Bleaching B: Second bleaching with 2% bleaching earth, 0.5% active carbon

Silica gel treatment: Contacting the oil dissolved in hexane with silica gel

b)   ULTRAVIOLET (UV) SPECTROPHOTOMETRY

During refining, especially bleaching, conjugated polyene acids can be formed. As fish oils are rich in PUFAs, such isomerisations may probably occur. Conjugated polyene acids can be measured by UV spectrophotometry as they have characteristic maxima in the region of 200-400 nm. Diagram 2 shows a UV-spectrum of a bleached fish oil with distinct maxima at 244, 274, 285 and 318 nm, which may be attributed to di- and triconjugated acids. Table 7 resumes the absorbance values at 4 wavelengths for differently treated fish oils. The treatment with bleaching earth caused a slightly increased absorption at the 3 higher wavelengths, indicating a higher concentration in triconjugated acids. Contacting fish oil with activated silica gel and eluting with hexane yielded a phase with reduced absorbance values. This suggests that triglycerides containing one or several conjugated polyene acids were more easily adsorbed on silica gel than triglycerides with non-modified fatty acids. This could be due to the higher polarity of the conjugated polyene acids. Finally the silical gel was eluted with a mixture of hexane and diethylether. The UV-absorption spectrum of the phase thus obtained showed intensified absorbances confirming the presumption that conjugated polyene acids are more easily adsorbed under these conditions. These results demonstrated a further beneficial effect of the silica gel treatment as, contrary to the bleaching with bleaching earth, the amount of conjugated polyene acids was not increased but reduced.

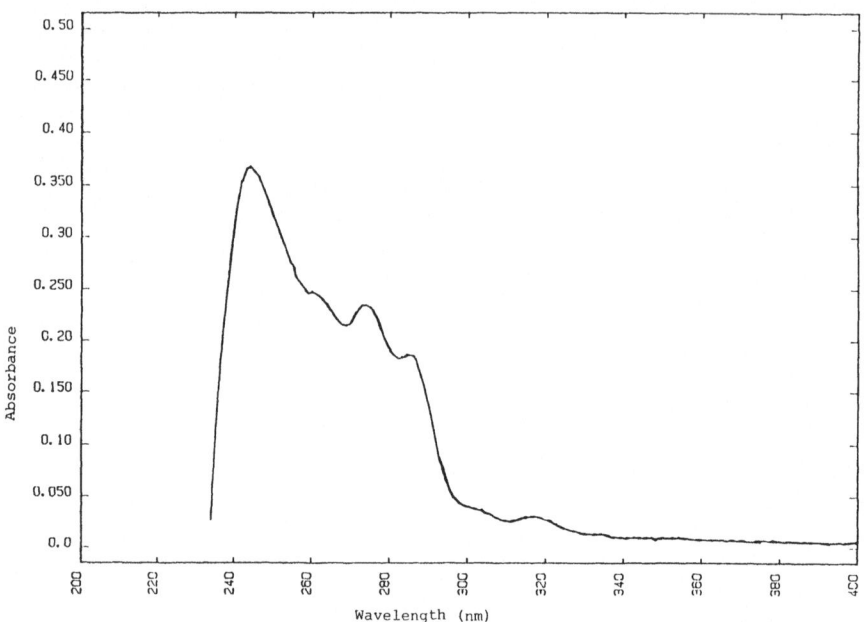

Diagram 2. UV spectrum of a bleeching-earth treated fish oil recorded with a US/VIS spectrophotometer, 8450 A Hewlet Packard (1 cm cell, solution of 0.05% oil in isooctane).

Table 7. **Effects of different bleaching procedures on UV-absorbance in the wavelength range 200-400 nm (1 cm cell, solutions of 0.05% oil in isooctane).**

UV-Absorbance

		Refining operations		
Wavelength (nm)	Raw material semi-refined fish oil	Bleaching 2 % bleaching earth 0.5% active carbon	Silica gel treatment A	Silica gel treatment B
244	0.38	0.37	0.34	0.56
274	0.23	0.24	0.17	0.32
285	0.18	0.19	0.14	0.25
318	0.02	0.03	0.01	0.03

Silica gel treatment A: phase eluded with hexane

Silica gel treatment B: phase eluded with a hexane/diethylether mixture

Furthermore, it could be shown by thin-layer chromatography that the silica gel treatment had removed mono- and diglycerides. The amounts of these substances in reference fish oil samples obtained by usual bleaching were estimated to be ca. 10%.

3.   CONCLUSIONS

For health-food applications selected fish oils, preferably with high concentrations of EPA and DHA, should be used. During refining of these oils, attention must be paid not only to organoleptic aspects, like the removal of fishy taste and odour, but also to the elimination of trace compounds such as pesticides, heavy metals, etc.

The results of the investigations described have demonstrated that the n-3 PUFAs were partially deteriorated at usual deodorisation temperatures in a batch process. To avoid these detrimental effects, deodorisation temperatures of < 180°C should be applied. Otherwise, to achieve under these conditions satisfactory organoleptic qualities, longer stripping times and sometimes also higher vapour amounts seemed to be necessary. As alternative to this batch process a thin-film deodorisation technique was presented. A further purification in respect to colour, odour and other polar impurities could be accomplished by silica gel treatment in organic solvents. Regarding the colour, heme pigments in particular must be eliminated as they are reported to be active catalysts of oil oxidation[23]. However, this technique is more sophisticated and expensive than a usual bleaching operation.

As fish oils are highly susceptible to oxidation, antioxidants may already be added during processing.

To protect oils thus refined and to retard off-flavour formation, stabilization with antioxidants is indispensable and storage under N_2 at low temperatures is required. For administration of these oils as dietary products, it might be appropriate for instance to encapsulate them in gelatine.

4.   BIBLIOGRAPHY

1.   J. Dyerberg, H.O. Bang, E. Stoffersen, S. Mancada and J.R. Vane, Eicosapentaenoic acid and prevention of thrombosis and atherosclerosis?, Lancet, II: 117 (1978).

2.   J. Dyerberg and H.O. Bang, Dietary fat and thrombosis, Lancet, I: 152 (1978).

3.   P.C. Weber, Which n-3 Fatty Acid Preparation for Clinical Intervention Studies?, n-3 News: Unsaturated Fatty Acids and Health II (3): 5 (1987).

4.   P.M. Herold and J.K. Kinsella, Fish oil consumption and decreased risk of cardiovascular disease: a comparison of findings from animal and human feeding trials, Am. J. Clin. Nutr., 43: 566 (1986).

5.   R.G. Ackman, Fatty Acid Composition of Fish Oils, in: "Nutritional Evaluation of Long-Chain Fatty Acids in Fish Oil", S.M. Barlow and M.E. Stansby eds., Academic Press, New York (1982).

6.   G. Piclet, Le Poisson, Aliment Composition - intérêt nutritionnel, Cah. Nutr. Diét., XXII (4): 317 (1987).

7. J.E. Kinsella, Edible Fish Oil Processing and Technology, in: "Food Science and Technology", Vol. 23: Seafoods and Fish Oils in Human Health and Disease", Marcel Dekker, New York (1987).

8. Japan Patent 83 JP-164865, Neutral lipid production from marine products - by contacting with carbon dioxide in supercritical state.

9. Japan Patent 86 JP-122176, Extraction of salmon oil using subcritical or supercritical carbon dioxide as extractant.

10. H. Pardun, E. Kroll and O. Werber, Determination of the Activity of a Bleaching Earth with Respect to its Application in the Quality Assessment of Oils, Fette, Seifen, Anstrichm., 70: 531 (1968).

11. A.N. Sagredos, D. Sinha-Roy and D. Thomas, On the Occurrence, Determination and Composition of Polycyclic Aromatic Hydrocarbons in Crude Oils and Fats, Fat Sci. Technol., 90: 76 (1988).

12. J. Spinelli, V.F. Stout and W.B. Nilsson, Purification of Fish Oils, U.S. Patent 4,692,280 (1987).

13. A. Thomas, The Removal of Trace Contaminants from Oils and Fats by Deodorization, Fette, Seifen, Anstrichm., 84: 133 (1982).

14. J.C. Florence, A. Monseigny and F. Zwoboda, Teneur en pesticides organochlorés des produits et sous-produits d'huilerie, Rev. franç. Corps Gras, 21: 359 (1974).

15. R.F. Addison and R.G. Ackman, Stepwise Removal of Chlorinated Hydrocarbons during Processing of Herring Oil for Edible Use, J. Am. Oil Chem. Soc., 54: 153 A (1977).

16. J.G. Ebel Jr., R.H. Eckerlin, G.A. Maylin, W.H. Gutenmann and D.J. Lisk, Polychlorinated biphenyls and p,p'-DDE in encapsulated fish oil supplements, Nutr. Rep. Int., 36: 413 (1987).

17. A. Seher, Influence of different methods of production and processing of edible oils on the nutritional quality, Nutrition, 11: 813 (1987).

18. C.M. Elson, E.M. Bem and R.G. Ackman, Determination of Heavy Metals in a Menhaden Oil after Refining and Hydrogenation Using Several Analytical Methods, J. Am. Oil Chem. Soc., 58: 1024 (1981).

19. R.W. Nelson, Liquid CO_2 extraction and fisheries research, Mar. Fish. Rev., 46: 28 (1982).

20. Japan Patent 85 Jp-234787: Purified fish oils preparations used as antithrombosis food - comprises reacting oils, containing glycerides with silical gel and collecting non-adsorbed portion.

21. Japan Patent 85 JP-196726: Purification of deep sea teleost fish oil -by addition of solvent to extracted oil then contacting with silica gel.

22. Japan Patent 86 JP-024667: Purification of fish oils and marine animal fats and oils - involves adding solvent and contacting with silica gel.

23. W.O. Lundberg, General deterioration reactions, in: "Fish Oils: their Chemistry, Technology, Stability, Nutritional Properties and Uses", M.E. Stansby ed., Avi Publishing Co., Westport (1967).

(N-3) AND (N-6) FATTY ACID METABOLISM

Howard Sprecher

Department of Physiological Chemistry
The Ohio State University
333 W. 10th Avenue,
Columbus, Ohio   43210

INTRODUCTION

Dietary (n-3) fatty acids mediate physiological processes in at least two different ways. The activities of the liver enzymes that metabolize fatty acids and synthesize triglycerides are modified by the type of fat that is fed. These changes include an elevated rate of fatty acid oxidation along with a reduced rate of fatty acid biosynthesis[1,2]. The mechanism of action of (n-3) acids at the enzyme level is still a matter of conjecture. Clarke and Armstrong[3] have recently reported that rats fed a fish oil diet had reduced levels of mRNA for fatty acid synthetase. These findings suggest that (n-3) fatty acids, or one of their metabolites, regulate enzyme synthesis rather than enzyme activity.

Secondly, (n-3) fatty acids mediate physiological processes by modifying the types and amounts of autocoids that are produced. These effects have been extensively studied in platelets, neutrophils and endothelial cells. The initial reaction in autocoid biosynthesis requires an agonist induced activation of the phospholipases that release (n-3) and (n-6) fatty acids from membrane phospholipids. The amounts of (n-3) and (n-6) fatty acids in phospholipids thus define the composition of the substrates that are potentially available for metabolism by lipoxygenases and cyclooxygenase. This brief review will focus on defining some of the differences and similarities that exist between (n-3) and (n-6) fatty acid metabolism.

Rate studies with liver microsomes[4] and isolated hepatocytes[5,6] are consistent with the concept that the activity of the 6-desaturase is rate limiting in converting linoleate and linolenate respectively to long chain (n-6) and (n-3) fatty acids. The amount of arachidonate in membrane lipids is however relatively independent of the amount of linoleate that is fed. This type of observation suggests that the energy dependent synthesis of arachidonate is tightly coupled with its requirement for membrane lipid biosynthesis. It has long been a matter of conjecture that the regulation of the three position specific desaturases, in part, regulate the types of fatty acids found in membrane phospholipids. None of the desaturases used in synthesizing long chain (n-6) or (n-3) fatty acids have been purified. Strittmatter and his colleagues[7,8] have shown that the level of mRNA for the 9-desaturase is modified by dietary manipulation. The level of mRNA coincides with the activity of the 9-desaturase which has a half-life of only four hours. Clearly a more comprehensive understanding of the factors regulating polyunsaturated fatty acid biosynthesis await the isolation of individual enzymes and subsequent studies defining what regulates the activity and synthesis of these important proteins.

The results in Table 1 compare individual reaction rates for the biosynthesis of (n-3) and (n-6) fatty acids. These results show that analogous acids from the (n-3) and (n-6) pathways are metabolized at similar rates. The results in Table 2 compare the unsaturated fatty acid composition of liver choline and ethanolamine phosphoglycerides of weanling male rats when they were fed either 3% by weight of ethyl linoleate, ethyl linolenate, or 1.5% by weight of each of the two ethyl esters for 5 weeks[9]. These compositional studies show that reaction rates for polyunsaturated fatty acid biosynthesis, by themselves, are poor predictors of the types of fatty acids found in these two liver phospholipids. Even though linoleate and linolenate are removed by desaturation at about the same rate the membrane lipids contain only trace amounts of linolenate even when linolenate is the sole source of dietary fat. When linolenate is incubated with hepatocytes it is incorporated into lipids[6]. Clearly, for reasons which remain to be established, linolenate is a poor substrate for liver phospholipid biosynthesis in vivo. Secondly, when rats were fed the three diets there were marked differences in the levels of 20- and 22-carbon polyenoic acids in these two phospholipids. When linoleate was fed both phospholipids contained large amounts of arachidonate. The low level of 22-carbon (n-3) acids in these lipids is due to the lack of an (n-3) precursor. When linolenate was fed there was the expected increase in 20:5(n-3) which was accompanied by a decline in the arachidonate content. However, the amount of 22-carbon (n-3) fatty acids was much greater than the 22-carbon (n-6) level when linoleate was fed. When equal amounts of linoleate and linolenate were fed there were only small changes in the levels of 22-carbon (n-3) acids. Interestingly, the sum of the 20- and 22-carbon (n-3) and (n-6) acids was a constant for all three dietary regimens.

The rate studies in Table 1 show that arachidonate and 20:5(n-3) are chain elongated at about the same rate. The compositional studies clearly show that arachidonate and 20:5(n-3) are metabolized quite differently. Twenty two carbon (n-6) and (n-3) fatty acids are substrates for mitochondrial and or peroxisomal degradation[10]. The low level of 22-carbon (n-6) acids in liver phospholipids could be explained if this process was selective for (n-6) versus (n-3) acids. This hypothesis seems unlikely since, in humans, the level of 20:5(n-3) in platelet phospholipids increased after feeding 22:6(n-3).[11] Moreover, prostaglandins of the 3-series were detected in the urine of humans after a single supplement of 22:6(n-3)[12]. Although the rates of chain elongation of arachidonate and 20:5(n-3) are similar, (Table 1) very little is known about the enzyme(s) which carry out these conversions. Our studies[13], as well as those of Cinti and his colleagues[14], are consistent with the hypothesis that two or more different malonyl-CoA dependent condensing enzymes channel their β-ketoacyl-CoA's into a common set of enzymes to complete the chain elongation process. However, little is known about the specificity of the enzymes which catalyze the chain elongation of 20-carbon polyenoic acids or whether these enzymes are the same as those which chain elongate 18:3(n-6) to 20:3(n-6) or 18:4(n-3) to 20:4(n-3). If a single condensing enzyme acts on 18- and 20-carbon polyenoic acids then obviously that enzyme is required two times for converting dietary 18:2(n-6) to 22:5(n-6) and 18:3(n-3) to 22:6(n-3). The last step in the synthesis of 22:5(n-6) and 22:6(n-3) requires the action of a 4-desaturase. The factors regulating this step are poorly understood. Indeed a recent review[15] questions whether the double bond at position-4 is introduced via the classical mechanism which apparently is operative for the 6-, 5-, and 9-desaturases. With the increased interest in (n-3) fatty acid dietary supplements it is clear that a better understanding is required of the factors regulating the conversion of 20:5(n-3) to 22:6(n-3) as well as the retroconversion of 22:6(n-3) back to 20:5(n-3) and how these processes differ from the analogous transformation of (n-6) acids.

Table 1

Reaction Rates for Desaturation and Chain Elongation of (n-6) and (n-3) Fatty Acids with Rat Liver Microsomes from Chow Fed Rats

Reaction	nmols/min/ mg protein	Reaction	nmols/min/ mg protein
9,12-18:2 ——> 6,9,12-18:3	0.21	9,12,15-18:3 ——> 6,9,12,15-18:4	0.27
6,9,12-18:3 ——> 8,11,14-20:3	2.54	6,9,12,15-18:4 ——> 8,11,14,17-20:4	2.46
8,11,14-20:3 ——> 5,8,11,14-20:4	0.40	8,11,14,17-20:4 ——> 5,8,11,14,17-20:5	0.34
5,8,11,14-20:4 ——> 7,10,13,16-22:4	1.14	5,8,11,14,17-20:5 ——> 7,10,13,16,19-22:5	1.31

Table 2

The (n-6) and (n-3) Fatty Acid Composition in Percent of Liver Choline and Ethanolamine Phosphoglycerides. Rats were Fed either Linoleate, Linolenate or a Mix of the Two Ethyl Esters

	PC			PE		
	Dietary Component					
	18:2	18:3	18:2/18:3	18:2	18:3	18:2/18:3
18:2(n-6)	11.3	3.6	9.7	5.1	1.3	3.6
18:3(n-3)	–	0.8	0.2	–	–	–
20:3(n-6)	0.8	0.4	1.1	0.5	–	0.3
20:4(n-6)	25.6	3.8	12.6	35.8	4.5	18.6
20:5(n-3)	–	18.7	5.5	–	19.3	5.4
22:4(n-6)	0.2	–	–	0.9	–	0.1
22:5(n-6)	2.7	–	0.1	5.8	0.1	0.3
22:5(n-3)	–	1.6	1.1	–	3.5	2.4
22:6(n-3)	0.9	7.6	9.0	2.4	17.6	19.4
20 plus 22-carbon (n-6) acids	28.5	3.8	13.8	42.5	4.6	19.0
20 plus 22-carbon (n-3) acids	0.9	27.9	15.6	2.4	40.4	27.2
Sum of 20 and 22-carbon (n-3) and (n-6) acids	29.4	31.7	29.4	44.9	45.0	46.2

FATTY ACID METABOLISM IN PLATELETS

When rats or humans are fed (n-3) fatty acids their platelet lipids contain elevated levels of 20:5(n-3) along with reduced levels of arachidonate. Platelet lipids, unlike liver lipids, do not contain high levels of 22-carbon (n-6) or (n-3) fatty acids[9]. Clearly different mechanisms exist between cell types to regulate the uptake, activation and acylation of fatty acids into membrane lipids. In part these differences may be due to the types of fatty acids released from liver for subsequent metabolism by extrahepatic cells and tissues[16]. However, unique specificities exist for fatty acids in any given cell. For example, the inositol phosphoglycerides in platelets from chow fed animals contain high levels of arachidonate. When (n-3) acids are added to the diet only small amounts of 20:5(n-3) are incorporated into this lipid. Surprisingly, when (n-3) acids are included in the diet the inositol phosphoglycerides from liver contain high levels of both 22:6(n-3) and 22:5(n-3) but only low levels of 20:5(n-3)[9].

Exogenous (n-3) and (n-6) fatty acids are metabolized quite differently by platelet cyclooxygenase. All four 20- and 22-carbon (n-6) fatty acids are metabolized into thromboxanes[17-20]. Adrenic acid-i.e. 22:4(n-6) is also metabolized into prostaglandins by both kidney microsomes[21] and endothelial cells[22]. Conversely, in the (n-3) family only 20:5(n-3) is metabolized into a thromboxane[23] although 20:4(n-3) is a substrate for vesicular gland cyclooxygenase[24]. Table 3 tabulates the monohydroxy fatty acids that are produced when (n-3) and (n-6) fatty acids are added to platelets. Studies by Rappoport and his colleagues[30] have shown that purified rabbit reticulocyte lipoxygenase metabolized arachidonic acid into both 12- and 15-HETE. This finding suggests that abstraction of a proton from the ω8 and ω11 methylene carbons, by a single platelet lipoxygenase, could account for the synthesis of all the hydroxy fatty acids from 18- and 20-carbon (n-3) and (n-6) acids as well as those produced from 22:4(n-6) and 22:5(n-6). Conversely, the synthesis of the 14- and 11-hydroxy fatty acids from 22:5(n-3) and 22:6(n-3) requires proton abstraction from the ω11 and ω14 carbon atoms respectively. The product analysis thus suggests that platelets may possibly contain a second lipoxygenase which acts on 22:5(n-3) and 22:6(n-3). If this is true then its function is totally undefined since to date the agonist induced release of 22-carbon (n-3) fatty acids from platelet phospholipids has not been demonstrated.

FATTY ACID METABOLISM IN NEUTROPHILS

It is well established that 20:5(n-3) is incorporated into neutrophil lipids and upon agonist induced release it

is metabolized into LTB_5 which is only about 10-20% as
active in recruiting neutrophils as is LTB_4. Studies in the
laboratories of Wykle[31] and Murphy[32] suggest that the ether
containing cholinephosphoglycerides are a common precursor
for both leukotriene and platelet activating factor (PAF)
biosynthesis. When phospholipase A_2 is activated, the
arachidonate is metabolized into LTB_4 while the resulting
lyso-PAF is acetylated to give PAF. We maintained rats on
either a 5% by weight corn oil diet or a diet in which 50%
of the corn oil was replaced by fish oil. The purpose of
this study was to define whether there was selective
acylation of (n-3) fatty acids into specific subclasses of

Table 3

Hydroxy Acids Produced from (n-6) and (n-3) Polyunsaturated
Fatty Acids by Human Platelets

Acid	Position of Proton Removal OOH-end	ω-end	Product	Reference
6,9,12-18:3	8	11	10-OH-6,8,12-18:3	25
	11	8	13-OH-6,9,11-18:3	25
8,11,14-20:3	10	11	12-OH-8,10,14-20:3	17
5,8,11,14-20:4	10	11	12-OH-5,8,10,14-20:4	18
	13	8	15-OH-5,8,11,13-20:4	29
7,10,13,16-22:4	12	11	14-OH-7,10,12,16-22:4	19
4,7,10,13,16-22:5	12	11	14-OH-4,7,10,12,16-22:5	20
8,11,14,17-20:4	10	11	12-OH-8,10,14,17-20:4	26
5,8,11,14,17-20:5	10	11	12-OH-5,8,10,14,17-20:5	23
7,10,13,16,19-22:5	9	14	11-OH-7,9,13,16,19-22:5	28
	12	11	14-OH-7,10,12,16,19-22:5	28
4,7,10,13,16,19-22:6	9	14	11-OH-4,7,9,13,16-19-22:6	27
	12	11	14-OH-4,7,10,12,16,19-22:6	27

phospholipids. The results in Tables 4 and 5 define the (n-
6) and (n-3) fatty acid composition at the sn-2 position of
the subclasses of the choline- and ethanolamine- containing
phosphoglycerides. The amount of arachidonate in each lipid
subclass varied when corn oil was the dietary fat. When
fish oil was added to the diet some of the arachidonate in
all lipid classes was replaced by (n-3) fatty acids. Small
amounts of 22-carbon (n-3) fatty acids were also
incorporated into all four lipid subclasses. Neutrophil
lipids contain relatively large amounts of palmitic and
linoleic acids at their sn-2 position. The levels of these

two acids were not altered by including fish oil in the diet. These compositional studies are thus consistent with enzymatic studies[33] showing that neutrophils contain two different pools of phospholipids. This conclusion is further supported by defining the molecular species composition of individual subclasses. Table 6 shows the content of selected molecular species from diacyl-GPC. When rats were fed corn oil the arachidonate paired preferentially with palmitate. The ratio of 16:0-20:4/18:0-20:4 was 1.2. When fish oil was added to the diet there was a reduction in the level of arachidonate but the 16:0-

Table 4

The (n-6) and (n-3) Fatty Acid Composition of Neutrophil Choline Phosphoglycerides. Rats were Fed Either Corn Oil or a Mix of Fish Oil and Corn Oil*

| | 1,2-Diacyl-GPC | | 1-0-Alkyl-2-Acyl-GPC | |
	Corn Oil	Corn Oil/ Fish Oil	Corn Oil	Corn Oil/ Fish Oil
18:2(n-6)	28.0	28.0	14.6	15.0
18:3(n-3)	-	-	1.0	1.7
20:3(n-6)	3.0	2.0	0.9	1.5
20:4(n-6)	25.2	11.8	33.6	20.9
20:5(n-3)	-	4.8	-	5.9
22:4(n-6)	3.8	4.0	1.6	3.1
22:5(n-6)	0.4	-	2.2	2.4
22:5(n-3)	-	3.0	-	1.7
22:6(n-3)	-	1.0	-	0.7

* Results expressed as mol % at the sn-2 position

20:4/18:0-20:4 ratio was not altered. The ratio of 16:0-20:5/18:0-20:5 was 1.4 in the fish oil fed animals. In general these studies show that arachidonate and 20:5(n-3) are metabolized in identical ways. Moreover, the addition of fish oil did not alter the amounts of 16:0-16:0 or 16:0-18:2. Under these dietary conditions the (n-3) fatty acids did not replace either 16:0 or 18:2(n-6) at the sn-2 position.

Table 5

The (n-6) and (n-3) Fatty Acid Composition of Neutrophil Ethanolamine Phosphoglycerides. Rats were Fed Either Corn Oil or a Mix of Fish Oil and Corn Oil[*]

| | 1,2-Diacyl-GPE | | 1-0-Alk-1'-Enyl-2-Acyl-GPE | |
	Corn Oil	Corn Oil/ Fish Oil	Corn Oil	Corn Oil/ Fish Oil
18:2(n-6)	9.2	10.4	8.3	11.5
20:3(n-6)	3.4	4.4	-	-
20:4(n-6)	41.0	30.0	55.8	41.8
20:5(n-3)	-	4.2	-	8.2
22:4(n-6)	8.2	4.2	3.9	3.4
22:5(n-6)	1.0	-	1.0	1.7
22:5(n-3)	-	9.4	-	2.9
22:6(n-3)	-	2.8	-	1.7

* Results expressed as mol % at the sn-2 position

Table 6

Mole Percent of Various Molecular Species of 1,2-Diacyl-sn-Glycerophosphocholines from Rat Neutrophils. Rats were Fed Either Corn Oil or a Mix of Fish Oil-Corn Oil

| Molecular Species | Dietary Fat | |
	Corn Oil	Corn Oil/Fish Oil
16:0 - 20:5	-	2.2
18:0 - 20:5	0.1	1.5
16:0 - 20:4	9.1	5.4
18:0 - 20:4	7.3	4.1
16:0 - 18:2	13.3	12.1
16:0 - 16:0	11.9	12.9

Acknowledgements

These studies were supported by NIH grants DK 20387 and DK 18844.

REFERENCES

1. P.J. Nestel, S. Wong and D.L. Topping, Dietary long chain polyenoic acids: 1. Suppression of triglyceride formation in rat liver: 2. Attenuation in man of the effects of dietary cholesterol on lipoprotein cholesterol, in: "Health Effects of Polyunsaturated Fatty Acids in Seafoods", A.P. Simopoulos, R.R. Kifer and R.E. Martin, eds., Academic Press, Orlando, Florida, pp. 211-246 (1986).

2. P.J. Nestel, D. Topping, J. Marsh, S. Wong, H. Barrett, P. Roach and B. Kambouris, Effects of polyenoic fatty acids (n-3) on lipid and lipoprotein metabolism, in: "Polyunsaturated Fatty Acids and Eicosanoids", W.E.M. Lands, ed., American Oil Chemists' Society, Champaign, Illinois, pp. 94-102 (1987).

3. S.D. Clarke and M.K. Armstrong, Suppression of rat liver fatty acid synthetase mRNA level by dietary fish oil, Federation Proceedings Abstracts, Abstract 3235: 1988.

4. J.T. Bernert and H. Sprecher, Studies to determine the role rates of chain elongation and desaturation play in regulating the unsaturated fatty acid composition of rat liver lipids, Biochim. Biophys. Acta 398: 354 (1975).

5. B.O. Christophersen, T-A. Hagve and J. Norseth, Studies on the regulation of arachidonic acid synthesis in isolated liver cells, Biochim. Biophys. Acta 712: 305 (1982).

6. T-A. Hagve and B.O. Christophersen, Linolenic acid desaturation and chain elongation and rapid turnover of phospholipid (n-3) fatty acids in isolated rat liver cells, Biochim. Biophys. Acta 753: 339 (1983).

7. M.A. Thiede, J. Ozols and P. Strittmatter, Construction and sequence of cDNA for rat liver stearoyl-coenzyme A desaturase, J. Biol. Chem. 261: 13230 (1986).

8. M.A. Thiede and P. Strittmatter, The induction and characterization of rat liver stearyl-CoA desaturase mRNA, J. Biol. Chem. 260: 14459 (1985).

9. T.W. Weiner and H. Sprecher, Arachidonic acid, 5,8,11-eicosatreinoic acid and 5,8,11,14,17-eicosapentaenoic acid, dietary manipulation of the levels of these acids in rat liver and platelet phospholipids and their incorporation into human platelet phospholipid, Biochim. Biophys. Acta 792: 293 (1984).

10. W-H. Kunau and F. Bartnik, Studies on the partial degradation of polyunsaturated fatty acids in rat-liver mitochondria, Eur. J. Biochem. 48: 311 (1974).

11. C. von Schacky and P.C. Weber, Metabolism and effects on platelet function of the purified eicosapentaenoic and docosahexaenoic acid in humans, J. Clin. Invest. 76: 2446 (1985).

12. S. Fischer, A. Vischer, V. Preac-Mursil and P.C. Weber, Dietary docosahexaenoic acid is ketroconverted in man to eicosapentaenoic acid, which can be quickly transformed to prostaglandin I_3, Prostaglandins 34: 367 (1987).

13. J.T. Bernert and H. Sprecher, An analysis of partial reactions in the overall chain elongation of saturated and unsaturated fatty acids in rat liver microsomes, J. Biol. Chem. 252: 6736 (1977).

14. M.N. Nagai, L. Cook, R. Prasad and D. Cinti, Do rat hepatic microsomes contain multiple NADPH-supported fatty acid chain elongation pathways or a single pathway? Biochem. Biophys. Res. Commun. 140: 74 (1986).

15. R. Jeffcoat and A.J. James, The regulation of desaturation and elongation of fatty acids in mammals, in: Fatty Acid Metabolism and its Regulation", Vol. 7, New Comprehensive Biochemistry, S. Numa, ed., Elsevier, New York, pp. 85-112 (1984).

16. J.D. Lefkowith, V. Flippo, H. Sprecher and P. Needleman, Paradoxical conservation of cardiac and renal arachidonate content in essential fatty acid deficiency, J. Biol. Chem. 260: 15736 (1985).

17. P. Falardeau, M. Hamberg, and B. Samuelsson, Metabolism of 8,11,14-eicosatrienoic acid in platelets, Biochim. Biophys. Acta 491: 193 (1976).

18. M. Hamberg and B. Samuelsson, Prostaglandin endoperoxides: Novel transformation of arachidonic acid in human platelets, Proc. Natl. Acad. Sci. USA 71: 3400 (1974).

19. M. VanRollins, L. Horrocks and H. Sprecher, Metabolism of 7,10,13,16-docosatetraenoic acid to dihomo-thromboxane, 14-hydroxy-7,10,12-nonadecatrienoic acid and hydroxy acids by human platelets, Biochim. Biophys. Acta 833: 272 (1985).

20. M. Milks and H. Sprecher, Metabolism of 4,7,10,13,16-docosapentaenoic acid by platelet cyclooxygenase and lipoxygenase, Biochim. Biophys. Acta 835: 29 (1985).

21. Sprecher, H., M. VanRollins, F. Sun, A. Wyche, and P. Needleman, Dihomo-prostaglandin and thromboxanes: A novel prostaglandin family from adrenic acid that may specifically be synthesized in the kidney, J. Biol. Chem. 257: 3912 (1982).

22. W.B. Campbell, J.R. Falck, J.R. Okita, A.R. Johnson and K.S. Callahan, Synthesis of dihomoprostaglandin from adrenic acid (7,10,13,16-docosatetraenoic acid) by human endothelial cells, Biochim. Biophys. Acta 837: 67 (1985).

23. M. Hamberg, Transformation of 5,8,11,14,17-eicosapentaenoic acid in human platelets, Biochim. Biophys. Acta 618: 389 (1980).

24. E.H. Oliw, H. Sprecher and M. Hamberg, Isolation of two novel prostaglandins in human seminal fluid, J. Biol. Chem. 261: 2675 (1986).

25. M. Hamberg, ω-Oxygenation of 6,9,12-octadecatrienoic acid in human platelets, Biochem. Biophys. Res. Commun. 117: 593 (1983).

26. M.M. Careaga and H. Sprecher, Metabolism of 8,11,14,17-eicosatetraenoic acid by human platelet lipoxygenase and cyclooxygenase, Biochim. Biophys. Acta 920: 94 (1987).

27. M.I. Aveldano and H. Sprecher, Synthesis of hydroxy fatty acids from 4,7,10,13,16,19-[1-^{14}C]docosahexaenoic acid, J. Biol. Chem. 258: 9339 (1983).

28. M.M. Careaga and H. Sprecher, Synthesis of two hydroxy fatty acids from 7,10,13,16,19-docosapentaenoic acid by human platlets, J. Biol. Chem. 259: 14413 (1984).

29. Y-K. Wong, P. Westlund, M. Hamberg, E. Granstrom, PH-W. Chao and B. Samuelsson, 15-Lipoxygenase in human platelets, J. Biol. Chem. 260: 9162 (1985).

30. R.W. Bryant, J.M. Bailey, T. Schewe and S.M. Rappoport, Positional specificity of a reticulocyte lipoxygenase. Conversion of arachidonic acid to 15S-hydroperoxy-eicosatetraenoic acid, J. Biol. Chem. 257: 6050 (1982).

31. F.H. Chilton, J.M. Ellis, S.C. Olson and R.L. Wykle, 1-O-Alkyl-2-arachidonoyl-sn-glycero-3-phosphocholine. A common source of platelet-activating factor and arachidonate in human polymorphonulcear leukocytes, J. Biol. Chem. 259: 12014 (1984).

32. F.H. Chilton and R.C. Murphy. Remodeling of arachidonate-containing phosphoglycerides within the human neutrophil, J. Biol. Chem. 261: 7771 (1986).

33. C.L. Swendsen, F.H. Chilton, J.T. O'Flaherty, J.R. Surles, C. Piantadosi, M. Waite and R.L. Wykle. Human neutrophils incorporate arachidonic acid and saturated fatty acids into separate molecular species of phospholipids, Biochim. Biophys. Acta 919: 79 (1987).

THE EFFECT OF DIETARY FISH OIL SUPPLEMENTATION AND <u>IN VITRO</u> COLLAGEN STIMULATION ON HUMAN PLATELET PHOSPHOLIPID FATTY ACID COMPOSITION

Harold M. Aukema and Bruce J. Holub

Department of Nutritional Sciences
University of Guelph
Guelph, Ont, Canada

INTRODUCTION

A large body of evidence has recently accumulated linking the ingestion of n-3 fatty acids of marine origin to decreased cardiovascular disease risk (reviews by Herold & Kinsella, 1986, Leaf & Weber, 1988, and Weaver & Holub, 1988). Consumption of eicosapentaenoic acid (EPA) plus docosahexaenoic acid (DHA) reduces blood platelet-vessel wall interactions, thereby decreasing the thrombotic potential by reducing the level of esterified arachidonic acid (AA) in platelet phospholipid and replacing it with n-3 polyunsaturated fatty acids, including EPA, DHA, and docosapentaenoic acid (DPA). In addition, EPA and DHA may also competitively inhibit AA metabolism at the cyclooxygenase level and thus decrease the amount of AA metabolites (Needleman et al., 1979). Upon platelet activation, AA is released from the platelet phospholipid by a combination of phospholipase A_2 and phospholipase C-mediated activities (Rittenhouse-Simmons & Deykin, 1981). While phospholipase A_2 liberates AA directly, phospholipase C produces diacylglycerol from inositol glycerophospholipids which can either be further metabolized to release AA or phosphorylated to form phosphatidic acid (PA). The released AA is metabolized by cyclooxygenase to 2-series eicosanoids, including thromboxane A_2 (Hamberg et al., 1975), a potentiator of platelet aggregation and a vasoconstrictor. The predominant n-3 fatty acid in the platelet phospholipid of seafood consumers, EPA, is also released from phospholipid and can be metabolized by cyclooxygenase to 3-series eicosanoids including the inactive thromboxane A_3 (Needleman et al., 1979; Hamberg, 1980; Fischer & Weber, 1983). While DHA does inhibit platelet aggregation (Rao et al., 1983), it has been shown not to be released in significant quantities from platelet phospholipid upon thrombin stimulation (Fischer et al., 1984).

Although much work on thrombin-induced phospholipid changes in platelets has been done using radioactive tracers or mass analysis (for review, see Mauco, 1987), there has been little work on collagen-stimulated platelets in which the mass changes in fatty acids have been documented (Broekman et al., 1980; Takamura et al., 1987). Furthermore, any collagen-induced fatty acid changes in the platelet phospholipids from fish oil consumers have not been investigated to date; nor has there been any mass work on EPA-enriched platelets in which agonist-induced fatty acid changes in the diacyl class of ethanolamine glycerophospholipid (PE) have been separated from the changes in the alkenylacyl PE class. The alkenylacyl PE class is particularly enriched in the n-3 fatty acids, especially EPA, upon fish oil supplementation (Holub et al, 1988); therefore, the potential fatty acid changes in this class upon platelet activation are of interest. As well, in light of the recent reports in other tissues of C phospholipases which are not phosphoinositide specific (Besterman et al., 1986; Daniel et al., 1986; Glatz et al., 1987) and the very minor amount of EPA in the platelet phosphatidylinositol (PI) in contrast to the PE or choline glycerophospholipid (PC) of fish oil consumers, the mass AA/EPA ratio in the PA that accumulates in collagen-stimulated platelets may be used to assess its phospholipid origin.

In the present study, we used platelets from fish oil consumers to document the quantitative fatty acid and phospholipid changes in the EPA-containing phospholipids, including the changes in alkenylacyl PE and PA, that occur in collagen-stimulated platelets. The presence of EPA relative to AA in the newly formed PA was used to indicate the phosphoinositides as the predominant phospholipid origin of PA in collagen-stimulated human platelets.

MATERIALS AND METHODS

MaxEPA capsules were provided by Seven Seas Health Care Ltd., Marfleet, U. K. Heparin and 2',7'-dichlorofluorescein were obtained from Sigma Chemical Co. (St. Louis, MO.). Bovine serum albumin (fraction V, fatty acid free) was from Boehringer Mannheim Canada (Dorval, QUE.). Apyrase was isolated from potatoes (Molnor & Lorand, 1961) and dissolved in 0.9 % NaCl. Merck silica gel 60 HR plates were from British Drug House (Toronto, ONT.). Collagen was obtained from Hormone-Chemie (Munchen, West Germany). Siliconized glassware or polypropylene centrifuge tubes were used throughout the platelet isolation and incubation. All reagents used were of analytical grade.

The experimental protocol was approved by the Human Subjects Committee of the University of Guelph and informed consent was obtained from 5 male volunteers who had a mean age of 29.4 years and mean weight of 73.1 kg. Subjects consumed 20 MaxEPA (a fish oil concentrate) capsules per day with their meals for 7 weeks, providing an intake of 3.6 g of EPA and 2.4 g of DHA per day. Otherwise, they maintained a normal diet except that they were requested to refrain from consuming seafood during the entire 7 week period. During the last 2 weeks of the study, subjects were asked not to take any alcohol or anti-inflammatory drugs. On day 49, blood samples from each subject were drawn into evacuated collection bottles containing 1/6 volume of acid-citrate-dextrose anticoagulent (Aster & Jandl, 1964). Washed platelet suspensions were prepared according to the method of Mustard et al. (1972), the final platelet suspension being a Tyrode's buffer containing albumin (0.35%) and apyrase (3 μg/ml). Platelets were counted using a Coulter Counter model ZM, Coulter Electronics of Canada, Ltd. (Burlington, ONT), and the final platelet concentration was adjusted to 2×10^9 platelets/ml.

One ml aliquots of platelets were incubated with 20 μg of collagen for 3 minutes in a shaking water bath at 37°C; control platelets were incubated for the corresponding time period without collagen addition. Reactions were stopped with chloroform/methanol (1:2, v/v) and lipids were extracted twice by the method of Bligh and Dyer (1959). The individual phospholipids were separated by a modification of the two-dimensional thin-layer chromatography system developed by Mitchell et al. (1986), including an acid hydrolysis step between dimensions. The solvent system in the first dimension consisted of chloroform/methanol/concentrated NH_4OH (65/35/5.5, v/v/v). In this system, PE ran higher than serine glycerophospholipid (PS), PI, PA, and PC. After drying in a nitrogen chamber for 50 minutes, the latter phospholipids were covered with glass while the part of the plate containing PE was exposed to the fumes from concentrated HCl at a distance of 5-6 cm for 10 minutes (Horrocks, 1968; Tessner & Wykle, 1987). After being ventilated in the nitrogen chamber for another 30 minutes, the plates were developed in the second solvent system which contained chloroform/methanol/88% formic acid (55/25/5, v/v/v). Lipid classes were visualized under ultraviolet light after plates had been sprayed with 2',7'-dichlorofluorescein in methanol/water (1/1, v/v) and exposed to ammonia vapour. The area representing PE contained mainly diacyl PE (ca 94%) with little alkylacyl PE (ca 6%) (Mueller et al., 1983; Natarajan et al., 1983); this area will be referred to as diacyl PE. Since the acid exposure occurred after development in the first solvent system, the 1-lyso (2-acyl) PE derived from the acid hydrolysis of alkenylacyl PE was separate from the endogenous lyso PE. The separated classes of n-3 containing phospholipid, namely total PC, alkenylacyl PE (i.e. 1-lyso (2-acyl) PE), diacyl PE, total PI, and total PS, were analyzed by gas-liquid chromatography, as described (Holub & Skeaff, 1987), except that methylation of fatty acids was done in 3-4 hours. Duplicate samples of control and collagen stimulated platelets were analyzed for each subject.

RESULTS

The composition of the EPA-containing phospholipids of resting platelets isolated from MaxEPA consumers is shown in Table 1. Sphingomyelin has been excluded because it is essentially devoid of EPA, DHA, and AA (Ahmed et al., 1984). The relative composition of the individual phospholipids are in good general agreement with other data from normal human subjects (Broekman et al., 1980; Mahadevappa & Holub, 1982; Mueller et al., 1983) and from fish oil consumers (Ahmed & Holub, 1984), confirming that fish oil supplementation does not significantly affect the

Table 1. Phospholipid composition of platelets isolated from fish oil consumers

Phospholipid	Mole percent of total PC/PE/PI/PS
PC	53.2 ± 1.0
alkenylacyl PE	16.2 ± 0.9
diacyl PE	11.6 ± 0.6
PI	5.9 ± 0.3
PS	13.1 ± 0.1

Values given as mean ± SEM for 5 subjects.

relative individual phospholipid composition of platelets (Ahmed & Holub, 1984). As well, Table 1 shows that the alkenylacyl PE class, a major EPA reservoir, makes up 58% of the total PE, which is in agreement with data from others (Mueller et al., 1983; Broekman et al., 1976) and our own observations on platelets from non-EPA consumers (unpublished data); thus, the alkenylacyl PE to diacyl PE ratio also appears not to be significantly affected by fish oil supplementation.

Table 2 gives the fatty acid composition of the individual platelet phospholipids as well as the 2 major PE classes, the alkenylacyl and diacyl PE. There is considerable variation not only between individual phospholipids, but also between these 2 PE classes, with the alkenylacyl PE being most enriched in EPA. There are also marked differences between the individual phospholipids with respect to the collagen-induced loss of fatty acids (Table 3). Besides the significant loss of AA and EPA from PC upon collagen stimulation, DHA also was lost (p=.06). There were no significant differences between the amounts of fatty acid in PS in control versus collagen-stimulated platelets. As well, while there was a decrease in the amount of EPA and AA in both PE classes upon collagen stimulation, this did not reach statistical significance.

The relative contribution of the individual AA- and EPA-containing phospholipids to the collagen-dependent mean decreases of AA and EPA are shown in Table 4. While 50% of the AA loss is from PI, only 3% of the EPA loss is due to PI degradation. Conversely, the PC and PE account for more than 95% of the EPA loss as compared to only 50% of the AA loss. The PA mass data for stearic acid, AA, EPA, and total PA in resting and collagen-stimulated platelets are shown in Table 5. The PA formed in

Table 2. Fatty acid composition of individual phospholipids of platelets isolated from MaxEPA consumers

Fatty acid	PC	alkenyl-acyl PE	diacyl PE	PI	PS
16:0	31.9±0.5	0.6±0.1	8.2±0.4	1.8±0.2	0.4±0.0
18:0	12.0±0.2	0.5±0.0	30.6±0.4	43.3±0.5	40.2±0.4
18:1	26.0±0.7	3.7±0.2	16.4±0.6	5.5±0.3	27.8±1.2
18:2(n-6)	7.1±0.6	1.6±0.2	4.4±0.4	0.4±0.1	0.6±0.1
20:0	1.0±0.1	tr	1.2±0.1	0.5±0.1	1.8±0.1
20:1	1.3±0.1	tr	1.1±0.1	0.4±0.1	tr
20:3(n-6)	0.9±0.1	0.7±0.1	0.8±0.1	0.5±0.1	1.5±0.2
20:4(n-6)	8.6±0.4	52.6±0.9	23.8±1.0	44.4±0.2	16.8±0.9
20:5(n-3)	3.5±0.4	17.0±0.9	4.2±0.8	0.9±0.1	0.7±0.1
22:4(n-6)	0.3±0.0	3.6±0.4	0.8±0.1	tr	0.5±0.0
22:5(n-3)	1.0±0.1	10.5±0.4	2.4±0.1	0.4±0.1	1.6±0.1
22:6(n-3)	1.6±0.1	8.3±0.5	3.8±0.3	0.2±0.0	2.6±0.2

Data given as mol% ± SE for 5 subjects. Fatty acids not contributing at least one mol% to any phospholipid class have been omitted from the table.

Table 3. Relative collagen-induced losses of AA and EPA from the individual human platelet phospholipids of MaxEPA consumers

Phospholipid	Percentage loss[a]	
	AA	EPA
PC	8.3[b] (3.1±1.2)	7.7[b] (1.1±0.5)
alkenylacyl PE	5.3 (1.2±1.8)	4.7 (0.7±0.8)
diacyl PE	2.9 (0.9±1.3)	0.7 (0.2±0.3)
PI	25.7[b](5.7±0.7)	17.1[b](0.1±0.0)
PS	-	-

Data from 5 subjects. Numbers in parenthesis are losses in nmols/2 x 10^9 platelets.

[a]Calculated relative to controls.

[b]Significantly different from controls(p<.05). Differences were tested by the two-tailed, single-sample Student's t-test (Steele and Torrie, 1980).

response to collagen was represented mainly by 18:0 and AA, although traces of 18:1, and other fatty acids, including DPA and DHA were detected in this fraction as well as EPA.

Table 6 gives the AA/EPA mass ratios of the PA formed upon collagen exposure and the ratios found in the resting platelet PC, alkenylacyl PE, diacyl PE, PI, and PS. The latter five phospholipids exhibited AA/EPA mass ratios in resting and stimulated conditions which were very similar. These findings are in overall agreement with previous radiolabelled and mass work from our laboratory which suggested that there is no marked selectivity for either AA or EPA loss from thrombin-stimulated platelet phospholipids (Weaver & Holub, 1986; Mahadevappa & Holub, 1987).

DISCUSSION

In the platelets of fish oil consumers, the phospholipid origins of the AA losses are markedly different than the origins of the EPA losses upon collagen stimulation (Table 4); this

Table 4. Relative contribution of individual AA- and EPA-containing phospholipids to the collagen dependent loss of AA and EPA in the platelets of EPA consumers

Phospholipid	Percentage contribution	
	AA	EPA
PC	27	54
alkenylacyl PE	15	34
diacyl PE	7	9
PI	50	3
PS	-	-
Total	100	100

Values based on mean nmol decreases as given in Table 3.

Table 5. PA mass and mol% data in resting and collagen stimulated
platelets isolated from MaxEPA consumers

| Fatty acid | nmol/2 x 10^9 platelets | | |
	Control	Stimulated	Net Change
18:0	0.56±0.03	2.41±0.26	1.85±0.27[a]
20:4(n-6)	0.25±0.20	1.99±0.22	1.74±0.21[a]
20:5(n-3)	0.01±0.01	0.03±0.01	0.03±0.01[a]
Total PA	0.76±0.03	2.88±0.32	2.12±0.30[a]

[a]$p < .05$
Data given as mean ± SE from 5 subjects. Differences were
tested by the two-tailed, paired Student's t-test (Steele and
Torrie, 1980).

heterogeneity of phospholipid sources was also found in platelets
rich in n-3 fatty acids activated by thrombin (Mahadevappa &
Holub, 1987). Under the conditions employed in the present study,
PI contributed to half of the AA loss, but very little to the EPA
loss. On the other hand, PC contributed to approximately half of
the EPA loss, but also contributed to 27% of the AA loss, while PS
lost little, if any, AA or EPA upon collagen stimulation.
Although the loss of AA and EPA from either PE class was not
statistically significant, the role of alkenylacyl and/or diacyl
PE as a source of releasable fatty acids cannot be ruled out.
Using radiolabelled AA, Rittenhouse-Simmons et al. (1977), and
Purdon et al. (1987), have shown that diacyl PE releases
radioactivity from thrombin-stimulated platelets, while
alkenylacyl PE increases in radioactivity. In contrast to these
radiolabelled studies, the degradation of alkenylacyl PE has
recently been documented in two mass studies of agonist-stimulated
platelets (Kambayashi et al., 1987; Takamura et al., 1987),
including collagen-stimulated platelets (Takamura et al., 1987).
Earlier work by Broekman (1980) had implicated alkenylacyl PE
degradation by the presence of fatty aldehydes in the accumulated
lyso PE in activated platelets. However, the fatty acid losses
from alkenylacyl PE may be masked by the transacylation of fatty
acids from PC to alkenylacyl PE as has been seen with
radiolabelled AA in activated human (Kramer & Deykin, 1983) and
rat platelets (Colard et al., 1984). Recent work has shown that

Table 6. AA/EPA mass ratios of individual phospholipids in
platelets of MaxEPA consumers

Phospholipid	AA/EPA
PC (resting)	2.3
alkenylacyl PE (resting)	3.0
diacyl PE (resting)	5.6
PI (resting)	54.5
PS (resting)	23.1
PA (collagen-induced accumulation)	58.0

EPA can also actively participate in the thrombin dependent transacylation pathway (Weaver & Holub, 1987). In the case of collagen-stimulated platelets, this turnover of fatty acids through alkenylacyl PE may be important for fatty acid release for the production of eicosanoids. Since this class is a major reservoir of both AA and EPA (Table 2), alkenylacyl PE may have an important role in collagen-induced activation, which is partly dependent on eicosanoid production. In addition to the fatty acid loss from alkenylacyl PE, the heterogeneous losses of AA and EPA from platelet PC and PI may play an important role in the modulation of platelet reactivity seen in fish oil consumers, since AA and EPA are precursors of eicosanoids with opposing effects on platelet aggregation.

With respect to the recent findings of C phospholipases (Besterman et al., 1986; Daniel et al., 1986; Glatz et al., 1987) which are not phosphoinositide specific, the origin of the PA in collagen stimulated platelets is of interest. The similarity of the stearic acid/AA ratio in the resting PI and newly formed PA has been used to indicate that PA is derived from PI in thrombin-stimulated platelets (Broekman et al., 1981). The presence of n-3 fatty acids in PA (Table 5) afforded the possibility of examining the question of the phospholipid origin of the collagen-induced rise in PA. The AA/EPA ratio of PI was 54.5, while the ratios for PC, alkenylacyl PE, and diacyl PE were 2.3, 3.0, and 5.6, respectively (Table 6). The ratio for the stimulated PA was 58.0, and thus it appears that PI and/or the polyphosphoinositides (PI-4-phosphate and PI-4,5-bisphosphate) are the predominant, but not necessarily exclusive sources of the newly-formed PA. Thirty percent of the mass AA and EPA loss from PI upon collagen stimulation (Table 3) can be attributed to the AA and EPA accumulation in PA.

The presence of detectable levels of DPA plus DHA in collagen-stimulated platelet PA was observed herein (data not shown). The moderate loss of DHA from PC found with collagen exposure is of interest since it has been shown that significant amounts of [^{14}C]DHA are not released from thrombin-stimulated platelet phospholipid (Fischer et al., 1984). These results support the possibility that there may be differences in the phospholipase-mediated hydrolysis of platelet phospholipids for collagen verses thrombin. For example, although Takamura et al. (1987) did not report any data on DPA or DHA, they reported a significant loss of linoleic acid from platelet PC upon thrombin stimulation, but not when collagen was the agonist. As well, it has been shown that platelets contain an enzyme or enzyme system that catalyzes the epoxidation of PI-AA and then hydrolyzes this product when thrombin is the agonist, but when collagen is used, this enzyme appears not to be activated (Ballou et al., 1987). In light of the fact that DHA is a potent _in vitro_ inhibitor of platelet aggregation (Rao et al., 1983), this finding may be of some relevance in understanding how platelet aggregation induced by collagen is markedly reduced in fish oil consumers (Siess et al., 1980; Skeaff & Holub, 1988), in contrast to thrombin (Skeaff & Holub, 1988).

In conclusion, the present study on platelets from fish oil consumers has indicated the heterogeneity of the distribution of AA and EPA lost from various platelet phospholipids upon collagen stimulation. The enrichment of alkenylacyl PE in EPA upon fish oil consumption may also bear relevance to the altered platelet reactivity. The relative importance of these compositional changes in terms of altered eicosanoid synthesis, changes in membrane fluidity, as well as altered receptor-mediated signal transduction, and their dampening of the thrombotic tendency remains to be studied.

ACKNOWLEDGEMENT

This work was supported in part by a grant from the Heart and Stroke Foundation of Ontario.

REFERENCES

Ahmed, A.A., and Holub, B.J., 1984, Alteration and recovery of bleeding times, platelet aggregation and fatty acid composition of individual phospholipids in platelets of human subjects receiving a supplement of cod-liver oil, Lipids, 19:617-624.

Aster, R.H., and Jandl, J.H., 1964, Platelet sequestration in man, I. Methods, J. Clin. Invest., 43:843-855.

Ballou, L.R., Lam, B.K., Wong, P.Y., and Cheung, W.Y., 1987, Formation of cis-14,15-oxido-5,8,11-eicosatrienoic acid from phosphatidylinositol in human platelets, Proc. Nat. Acad. Sci. U.S.A., 84:6990-6994.

Besterman, J.M., Duronio, V., and Cuatrecasas, P., 1986, Rapid formation of diacylglycerol from phosphatidylcholine: a pathway for generation of a second messenger, Proc. Nat. Acad. Sci. U.S.A., 83:6785-6789.

Bligh, E.G., and Dyer, W.J., 1959, A rapid method of total lipid extraction and purification, Can. J. Biochem. Physiol., 37:911-917.

Broekman, M.J., Handin, R.I., Derksen, A., and Cohen, P., 1976, Distribution of phospholipids, fatty acids and platelet factor 3 activity among subcellular fractions of human platelets. Blood, 47:963-971.

Broekman, M.J., Ward, J.W., and Marcus, A.J., 1980, Phospholipid metabolism in stimulated human platelets: changes in phosphatidylinositol, phosphatidic acid, and lysophospholipids, J. Clin. Invest., 66:275-283.

Broekman, M.J., Ward, J.W., and Marcus, A.J., 1981, Fatty acid composition of phosphatidylinositol and phosphatidic acid in stimulated platelets: persistence of arachidonyl-stearyl structure, J. Biol. Chem., 256:8271-8274.

Colard, O., Breton, M., and Bureziat, G., 1984, Arachidonoyl transfer from diacyl phosphatidylcholine to ether phospholipids in rat platelets, Biochem. J., 222:657-662.

Daniel, L.W., Waite, M., and Wykle, R.L., 1986, A novel mechanism of diglyceride formation: 12-O-tetradecanoylphorbol-13-acetate stimulates the cyclic breakdown and resynthesis of phosphatidylcholine, J. Biol. Chem., 26:9128-9132.

Fischer, S., and Weber, P.C., 1983, Thromboxane A3 (TxA3) is formed in human platelets after dietary eicosapentaenoic acid, Biochem. Biophys. Res. Comm., 116:1091-1099.

Fischer, S., Schacky, C.v., Siess, W., Strasser, Th., and Weber, P.C., 1984, Uptake, release and metabolism of docosahexaenoic acid (DHA, c22:6ω3) in human platelets and neutrophils, Biochem. Biophys. Res. Comm., 120:907-918.

Glatz, J.A., Muir, J.G., and Murray, A.W., 1987, Direct evidence for phorbol ester-stimulated accumulation of diacylglycerol derived from phosphatidylcholine, Carcinogenesis, 8:1943-1945.

Hamberg, M., 1980, Transformations of 5,8,11,14,17-eicosapentaenoic acid in human platelets, Biochim. Biophys. Acta, 618:389-398.

Hamberg, M., Svensson, J., and Samuelsson, B., 1975, Thromboxanes: a new group of biologically active compounds derived from prostaglandin endoperoxides, Proc. Natl. Acad. Sci. U.S.A., 72:2994-2998.

Herold, P.M., and Kinsella, J.E., 1986, Fish oil consumption and decreased risk of cardiovascular disease: a comparison of findings from animal and human feeding trials, Amer. J. Clin. Nutr., 43:566-598.

Holub, B.J., and Skeaff, C.M., 1987, Nutritional regulation of cellular phosphatidylinositol, in: "Methods in Enzymology-Cellular Regulators: Calcium Lipids", P. Conn and A.R. Means, eds., Academic Press, New York.

Holub, B.J., Celi, B., and Skeaff, C.M., 1988, The alkenylacyl class of ethanolamine phospholipid represents a major form of eicosapentaenoic acid (EPA)-containing phospholipid in the platelets of human subjects consuming a fish oil concentrate, Thromb. Res., 50:135-143.

Horrocks, L.A., 1968, The alk-1-enyl group content of mammalian myelin phosphoglycerides by quantitative two-dimensional thin-layer chromatography, J. Lipid Res., 9:469-472.

Kambayashi, J., Kawasaki, T., Tsujinaka, T., Sakon, M., Oshiro T., and Mori, T., 1987, Active metabolism of phosphatidyl-ethanolamine plasmalogen of stimulated platelets, analyzed by high performance liquid chromatography, Biochem. Int., 14:241-247.

Kramer, R.M., and Deykin, D., 1983, Arachidonoyl transacylase in human platelets, J. Biol. Chem., 258:13806-13811.

Leaf, A., and Weber, P., 1988, Cardiovascular effects of n-3 fatty acids, N. Eng. J. Med., 318:549-557.

Mahadevappa, V.G., and Holub, B.J., 1982, The molecular species composition of individual diacylphospholipids in human platelets, Biochim. Biophys. Acta, 713:73-79.

Mahadevappa, V.G., and Holub, B.J., 1987, Quantitative loss of individual eicosapentaenoyl-relative to arachidonoyl-containing phospholipids in thrombin-stimulated human platelets, J. Lipid Res., 28:1275-1280.

Mauco, G., 1987, Phospholipids: release of arachidonate for prostaglandin and thromboxane synthesis, in: "Platelet Responses and Metabolism, Volume III: Response-Metabolism Relationships", H. Holmsen, ed., CRC Press, Boca Raton, FL.

Mitchell, K.T., Ferrell, J.E.,jr, and Huestis, W.H., 1986, Separation of phosphoinositides and other phospholipids by two-dimensional thin-layer chromatography, Anal. Biochem., 158:447-453.

Molnor, J., and Lorand, L., 1961, Studies on apyrase, Arch. Biochem. Biophys., 93:353-363.

Mueller, H.W., Purdon, A.D., Smith, J.B., and Wykle, R.L., 1983, 1-0-alkyl-linked phosphoglycerides of human platelets: distribution of arachidonate and other acyl residues in the ether-linked and diacyl species, Lipids, 18:814-819.

Mustard, J.G., Perry, D.W., Ardlie, N.G., and Packham, M.A., 1972, Preparation of suspensions of washed platelets from humans, Br. J. Haematol., 22:193-204.

Natarajan, V., Zuzart-Augustin, M., Schmid, H.H.O., and Graf, G., 1983, The alkylacyl and alkenylacyl glycerophospholipids of human platelets, Thromb. Res., 30:119-125.

Needleman, P., Raz, A., Minkes, M.S., Ferrendelli, J.A., and Sprecher, H., 1979, Triene prostaglandins: prostacyclin and thromboxane biosynthesis and unique biological properties, Proc. Natl. Acad. Sci. U.S.A., 76:944-948.

Purdon, A.D., Patelunas, D., and Smith, J.B., 1987, Resolution of radiolabeled molecular species of phospholipid in human platelets: effect of thrombin, Lipids, 22:116-120.

Rao, G.H.R., Radha, E., and White, J.G., 1983, The effect of docosahexaenoic acid (DHA) on arachidonic acid metabolism and platelet function, Biochem. Biophys. Res. Commun., 117:549-555.

Rittenhouse-Simmons, S., and Deykin, D., 1981, Release and metabolism of arachidonate in human platelets, in: "Platelets in Biology and Pathology", vol. 2, J.L. Gordon, ed., Elsevier North-Holland Inc, New York.

Rittenhouse-Simmons, S., Russell, F.A., and Deykin, D., 1977, Mobilization of arachidonic acid in human platelets: kinetics and Ca^{2+} dependency, Biochim. Biophys. Acta, 488:370-380.

Schacky, C.v., Siess W., Fischer S., and Weber, P.C., 1985, A comparative study of eicosapentaenoic acid metabolism by human platelets in vivo and in vitro, J. Lipid Res., 26:457-464.

Siess, W., Scherer, B., Bohlig, B., Roth, P., Kurzmann, I., and Weber, P.C., 1980, Platelet-membrane fatty acids, platelet aggregation, and thromboxane formation during a mackerel diet, Lancet, 1:441-444.

Skeaff, C.M., and Holub, B.J., 1988, The effect of fish oil consumption on platelet aggregation responses in washed human platelet suspensions, Thromb. Res., (in press).

Steele, R.G.D., and Torrie, J.H., 1980, "Principles and Procedures of Statistics: A Biometrical Approach", 2nd ed, McGraw-Hill Book Co. Toronto, ONT.

Takamura, H., Narita, H., Park, H.J., Tanaka, D., Matsuura, T., and Kito, M., 1987, Differential hydrolysis of phospholipid molecular species during activation of human platelets with thrombin and collagen, J. Biol. Chem., 262:2262-2269.

Tessner, T.G., and Wykle, R.L., 1987, Stimulated neutrophils produce an ethanolamine plasmalogen analog of platelet-activating factor, J. Biol. Chem., 262:12660-12664.

Weaver, B.J., and Holub, B.J., 1986, The relative degradation of [14C]eicosapentaenoyl and [3H]arachidonoyl species of phosphatidylinositol and phosphatidylcholine in thrombin-stimulated human platelets, Biochem. Cell. Biol., 64:1256-1261.

Weaver, B.J., and Holub, B.J., 1987, The thrombin-dependent enrichment of alkenylacyl ethanolamine phosphoglyceride with [14C]-eicosapentaenoic acid and [3H]-arachidonic acid in prelabelled human platelets, Biochem. Cell. Biol., 65:405-408.

Weaver, B.J., and Holub, B.J., 1988, Health effects and metabolism of dietary eicosapentaenoic acid, Prog. Food Nutr. Sci., 12:in press.

IN VITRO STUDIES ON DOCOSAHEXAENOIC ACID IN HUMAN PLATELETS

M. Lagarde, M. Croset and M. Hajarine

INSERM U 205, INSA B 406

69621 Villeurbanne Cedex, France

INTRODUCTION

Arachidonic acid (5,8,11,14-20:4 or 20:4w6) is the main polyunsaturated fatty acid of blood and vascular cells and it has been recognized as the precursor of numerous oxygenated derivatives exhibiting a variety of biological activities (1-3). In blood platelets, 20:4w6 is almost exclusively located in glycerophospholipids, from where it is fastly mobilized upon cell activation by aggregating agents like thrombin, collagen or the calcium ionophore A 23187 (4-6). Such a mobilization is the limiting step of prostaglandin endoperoxides and thromboxane A_2 formation, the potent pro-aggregatory molecules, as well as of that of 12-lipoxygenase products.

Several other polyunsaturated fatty acids (PUFAs) may arise under certain nutritional conditions and then modulate platelet functions, either in altering the 20:4w6 cascade or by the action of their own oxygenated metabolites (7,8). Among these PUFAs, eicosapentaenoic acid (5,8,11,14,17-20:5 or 20:5w3) and docosahexaenoic acid (4,7,10,13,16,19-22:6 or 22:6w3) are prominent components of fish fat which have been assumed to be associated with the decreased platelet functions observed after high fish fat intake (9,10). Although these two PUFAs are found in equivalent amounts in such fat, only few platelet studies concerned 22:6w3 as compared to those relating to 20:5w3. The present paper will report some biological effects and metabolism of 22:6w3 versus 20:5w3 as observed in the in vitro experiments.

RESULTS AND DISCUSSION

In order to simulate the physiological situation, each PUFA was pre-coated onto defatted human albumin at the molecular ratio PUFA/albumin of 0.1 or 0.5 and the complex was incubated with platelets for 2 hours at 37°C, and then removed to provide PUFA-rich platelets (11,12). With the molecular ratio of 0.5, around 8 nmoles of 20:5w3 or 22:6w3 were incorporated into 10^9 platelets of which more than

90% were located in glycerophospholipids. This represents a 2% enrichment of the total fatty acids, which is in the range of enrichments observed in the usual fish fat diets. In the presence of equal amounts of $20:4\omega6$, the incorporation of both $20:5\omega3$ and $22:6\omega3$ into total phospholipids was partially inhibited, and $22:6\omega3$ more actively acylated phosphatidylethanolamine at the expense of phosphatidylcholine, at the opposite of what observed with $20:5\omega3$ for which the percentage repartition in glycerophospholipid subclasses was not changed (11). When submitted to aggregation by threshold doses of thrombin, collagen or U-46619 (the stable analogue of PGH_2 which exhibits a TXA_2 mimetic activity), $20:5\omega3$- and $22:6\omega3$-rich platelets were significantly less responsive than control platelets, and $22:6\omega3$ enrichment appeared the most potent one to induce such an inhibition. The inhibition of U-46619-induced aggregation was quite superimposable to that induced by thrombin or collagen (12), suggesting that, at least under these conditions, the enrichment with $20:5\omega3$ or $22:6\omega3$ was likely inhibiting the TXA_2 response rather than its generation from the endogenous pool of $20:4\omega6$. This might indicate that modifying the lipid composition of platelets in this way would alter the final steps of the aggregation, possibly at the level of TXA_2 receptor sites.

In addition, the measurement of TXB_2 and HHT, the two main cyclooxygenase products of endogenous $20:4\omega6$ (under 0.1 U/ml thrombin stimulation), revealed a significant inhibition of both $20:5\omega3$- and $22:6\omega3$-rich platelets, the later being again most inhibited. The formation of 12-HETE, the lipoxygenase metabolite, was however not affected, suggesting that the inhibition observed occurred at the cyclooxygenase level rather than at the endogenous $20:4\omega6$ liberation step (13). The cyclooxygenase / thromboxane synthase system being exclusively particulate and located in the dense tubular system of platelets (14) whereas the lipoxygenase is mainly cytosolic (15), these results provide an additional example of the modulation of membrane function by fatty acid modifications. It may be however argued that the liberation of $20:5\omega3$ and $22:6\omega3$ by thrombin could be responsible for a competitive inhibition of endogenous $20:4\omega6$ cyclooxygenation. This may be excluded at the light of the following results, at least for $22:6\omega3$ which was virtually not freed from the phospholipid pool.

In using radiolabelled fatty acids instead of unlabelled ones, we studied their liberation and their subsequent oxygenation upon stimulation by thrombin and the calcium ionophore A 23187. $20:5\omega3$ was significantly liberated from total phospholipids and its lipoxygenase product, 12-HEPE, as well as TXB_3 could be measured, 12-HEPE being the prominent product. At the opposite, $22:6\omega3$ was virtually not liberated from phospholipids as previously reported (16), although a very small amount of its lipoxygenase products could be detected (17). These products have been characterized as the 11-and 14-hydroxy derivatives of $22:6\omega3$ by liquid chromatography / mass spectrometry (18) as they were first described from exogenous $22:6\omega3$ (19). As mentioned above, the quasi absence of $22:6\omega3$ liberation from membrane phospholipids allows to exclude any competitive inhibition of the $20:4\omega6$ cyclooxygenation, strongly suggesting that 22:6w3 present in membrane phospholipids may inhibit the cyclooxygenase / thromboxane synthase complex of the dense tubular system, at the level of protein-lipid interactions. In contrast, since 20:5w3 was liberated from phospholipids, it might then decrease the thromboxane A2 formation in competing with endogenous 20:4w6 liberated simultaneously. Such a competition is however unlikely to occur very efficiently since the amount of 20:5w3 liberated is low compared to 20:4w6.

Finally, the phospholipid metabolism of the pre-incorporated 20:5w3 and 22:6w3 was investigated in the four glycerophospholipid subclasses, namely phosphatidyl-choline (PC), -ethanolamine (PE), -inositol (PI) and -serine (PS), during platelet stimulation by thrombin (0.1 and 1 U/ml) and the calcium ionophore A 23187 (0.5 and 2 μM). Both acids were liberated from PI as observed for 20:4w6 but the extent of such a liberation is very weak as compared to other modifications. On the other hand, 20:5w3 liberation from PI is questionable since the percentage of in vivo incorporation of this acid into PI is very low (20,21). Whereas 20:5w3 was freed from PC with a resulting appearance of its oxygenated products, 22:6w3 was also liberated from PC but reciprocally re-incorporated into PE, and was not made available for the oxygenation process. In addition, the amount of 22:6w3 transfered from PC to PE was greater than that of 20:5w3 liberated from PC (17). The absence of 22:6w3 freeing from total phospholipids then does not exclude an active phospholipid subclass exchange but can explain the very weak formation of lipoxygenase products of 22:6w3, although this acid is fairly well lipoxygenated when exogenously added to platelets (13). The physiological relevance of the platelet lipoxygenase derivatives of 22:6w3 is then questionable. Like most other lipoxygenase products of polyunsatured fatty acids, they exhibit an anti-thromboxane A2 activity. As compared to 12-HETE, various lipoxygenase derivatives of 22:6w3, the platelet lipoxygenase products 11- and 14-OH-22:6 and the soybean lipoxygenase product 17-OH-22:6, were significantly more potent for inhibiting the platelet aggregation induced by the TXA2 mimetic U-46619, whereas 12-OH-20:5 was equipotent. Amongst the lipoxygenase derivatives of 22:6w3, 14-OH-22:6 was the most potent. In addition, we found that 12-HETE as well as 14-OH-22:6 were similarly inhibitors of TXA2-induced contraction of rabbit aorta (22). This indicates that whether they are produced, during platelet incorporation for instance, platelet lipoxygenase products of 22:6w3 might antagonize TXA2 activity.

CONCLUSION

These results show that 22:6w3 should be accounted for in the mechanism of action of platelet inhibition after fish fat intake. In vitro experiments revealed that 22:6w3 might be even more active in such an inhibition than 20:5w3 and that the metabolic behaviour of the two fatty acids is quite different. Whereas 20:5w3 resembles 20:4w6 in several aspects, 22:6w3 is not freed from phospholipids during platelet activation by thrombin or the calcium ionophore A 23187 but it is very efficiently transfered from PC to PE. Although 22:6w3 is not freed, the appearance of oxygenated metabolites of endogenous 20:4w6 is markedly reduced, suggesting that 22:6w3 might inhibit the enzymes involved in the 20:4w6 cascade at the protein-lipid interaction level instead of competing with 20:4w6 at the enzyme active sites. The potent antagonizing effect of lipoxygenase products of 22:6w3 on the TXA2 activity also points out interesting structure-activity relationships between these products and TXA2 receptor.

ACKNOWLEDGMENTS

This work was supported by INSERM and INRA (AIP).

REFERENCES

1. Samuelsson, B., Goldyne, M., Granström, E., Hamberg, M., Hammarström, S. and Malmsten, C. Prostaglandins and thromboxanes. Ann. Rev. Biochem. 47:997 (1978).

2. Hammarström, S. Leukotrienes. Ann. Rev. Biochem. 52:355 (1983).

3. Needleman, P., Turk, J., Jakschik, B.A., Morrison, A.R. and Lefkowith, J.B. Arachidonic acid metabolism. Ann. Rev. Biochem. 55:69 (1986).

4. Smith, J.B. The prostanoids in hemostasis and thrombosis. Am. J. Pathol. 99:742 (1980).

5. Irvine, R.F. How is the level of free arachidonic acid controlled in mammalian cells? Biochem. J. 204:3 (1982).

6. Lagarde, M. Roles of cyclooxygenase and lipoxygenase metabolites in platelets. In. "Platelets in Biology and Pathology III", D.E. Mc Intyre, & J.L Gordon, eds. Elsevier, Amsterdam p.269 (1987).

7. Willis, A.L. Nutritional and pharmacological factors in eicosanoid biology. Nutr. Rev. 39:289 (1981)

8. Lagarde, M. Metabolism of fatty acids by platelets and the functions of various metabolites in mediating platelet function. Progr. Lipid Res. 27:135 (1988).

9. Goodnight, S.H., Harris, W.S., Connor, W.E. and Illingworth, D.R. Polyunsaturated fatty acids, hyperlipidemia, and thrombosis. Arteriosclerosis 2:87 (1982).

10. Dyerberg, J. Linolenate-derived polyunsaturated fatty acids and prevention of atherosclerosis. Nutr. Rev. 44:125 (1986).

11. Hajarine, M. and Lagarde, M. Studies on polyenoic acid incorporation into human platelet lipid stores: interactions with linoleic and arachidonic acids. Biochim. Biophys. Acta 877:299 (1986).

12. Croset, M. and Lagarde, M. In vitro incorporation and metabolism of icosapentaenoic and docosahexaenoic acids in human platelets. Effect on aggregation. Thromb. Haemostas. 56:57 (1986).

13. Croset, M., Guichardant, M. and Lagarde, M. Different metabolic behavior of long chain n-3 polyunsaturated fatty acids in human platelets. Biochim. Biophys. Acta 1988, In press.

14. Carey, F., Menashi, S. and Crawford, N. Localization of cyclooxygenase and thromboxane synthetase in human platelet intracellular membranes. Biochem. J. 204:847 (1982).

15. Lagarde, M., Croset, M., Authi, K.S. and Crawford, N. Subcellular localization and some properties of lipoxygenase activity in human blood platelets. Biochem. J. 222:495 (1984).

16. Fischer, S., Von Schacky, C., Siess, W., Strasser, T. and Weber, P.C. Uptake, release and metabolism of docosahexaenoic acid (DHA, C22:6w3) in human platelets and neutrophils. Biochem. Biophys. Res. Commun. 120:907 (1984).

17. Hajarine, M. and Lagarde, M. Liberation and oxygenation of polyenoic acids in stimulated platelets. Biochimie 1988. In press.

18. Guichardant, M., Lagarde, M., Lesieur M. and De Maack, F. Thermospray-mass spectrometric analysis of underivatized monohydroxy fatty acids. Application to stimulated platelets. J. Chromatogr. Biomed. Appl. 425:25 (1988).

19. Aveldano, M. and Sprecher, H. Synthesis of hydroxy fatty acids from 4,7,10,13,16,19-[1-^{14}C]-docosahexaenoic acid by human platelets. J. Biol. Chem. 258:9339 (1983).

20. Ahmed, A.A. and Holub, B.J. Alterations and recovery of bleeding times, platelet aggregation and fatty acid composition of individual phospholipids in platelets of human subjects receiving a supplement of cod liver oil. Lipids 19:617 (1984).

21. Galloway, J.H., Cartwright, I.J., Woodcoek, B.E., Greaves, M., Russel, G.G. and Preston, F.E. Effects of dietary fish oil supplementation on the fatty acid composition of the human platelet membrane: demonstration of selectivity in the incorporation of eicosapentaenoic acid into membrane phospholipid pools. Clin. Sci. 68:449 (1985).

22. Croset, M., Sala, A., Folco, G.C., and Lagarde, M. Inhibition by lipoxygenase products of TXA$_2$ responses of platelets and vascular smooth muscle. 14-hydroxy from 22:6n-3 is more potent than 12-HETE. Biochem. Pharmacol. 37:1275 (1988).

ALPHA-LINOLENIC ACID AND THE

METABOLISM OF ARACHIDONIC ACID

P. Budowski

Faculty of Agriculture
The Hebrew University of Jerusalem
Rehovot, Israel

INTRODUCTION

Recognition of α-linolenic acid (LnA)* as an EFA in higher animals has been late in forthcoming, in sharp contrast to the abundance of reports demonstrating the essentiality of LA in many animal species, including humans.[1] Evidence for an essential role in LnA in vision of Rhesus monkeys has been described,[2] and cases of human LnA deficiency have been reported[3].

Apart from an absolute requirement for LnA, this FA appears to exert a restraining action on the metabolism of AA.[4,5] Thus, the dietary balance between the two EFA may be of nutritional significance. This question is further highlighted by the recent surge of interest in possible health benefits accruing from the consumption of ω3-FA from seafoods,[6-9] which has led several authors to question the adequacy of the dietary ratio of ω6 to ω3-FA in Western industrialized countries.

---

* Abbreviations used: FA - fatty acids; FFA - free FA; PUFA - polyunsaturated FA; EFA - essential FA; LA - linoleic acid; LnA - α-linolenic acid; AA - arachidonic acid; EPA - eicosapentaenoic acid; DHA - docosahexaenoic acid; PG - prostaglandins; NE - nutritional encephalomalacia. Unsaturated FA are designated by the shorthand <u>omega</u> notation, e.g. 20:4ω6 for AA, which shows the carbon chain length, number of double-bonds and position of the double-bond proximal to the terminal methyl group.

The main purpose of the present report is to develop the concept of
unrestrained AA metabolism and to discuss the modulating effect of LnA on
the metabolism of AA, as illustrated especially in a chick model. The
origin and implications of the present-day imbalance between LA and LnA will
also be discussed.

METABOLISM OF AA

AA is the main LA-derived FA in animal tissues. Both LA and LnA are
produced _de novo_ only in plants, but animals, especially liver, are able to
convert these parent FA to derived FA by alternating desaturation and chain

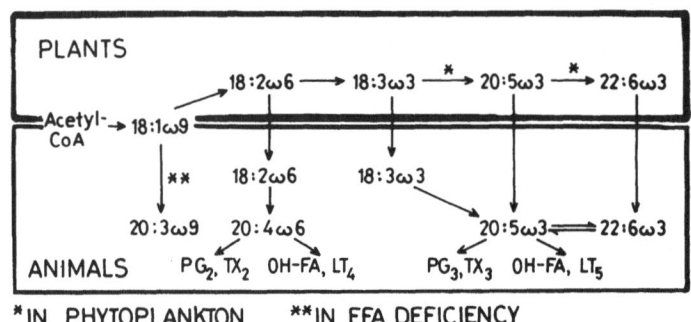

Fig. 1. Major pathways of linoleic and α-linolenic acid
in plants and animals

elongation, thus forming two distinct series of FA referred to as ω6- and
ω3-FA families. No interconversion is possible between these two families
in the animal body. The conversion of the parent FA to their long-chain
derivatives assumes special importance in the case of herbivores or vegans
who receive LA or LnA as the only PUFA in their diet. The derived FA play
important roles in the structure and function of membranes, and some of
them, 20:3ω6, 20:4ω6 and 20:5ω3, act as precursors for the complex series

of eicosanoids formed via the cyclo-oxygenase and lipoxygenase pathways
(Fig. 1).

Some primitive plants are also able to convert LA and LnA to
long-chain PUFA, a process which plays a role in the food web of marine
organisms and in the accumulation of EPA and DHA in the lipids of fish in
the colder regions of the oceans.

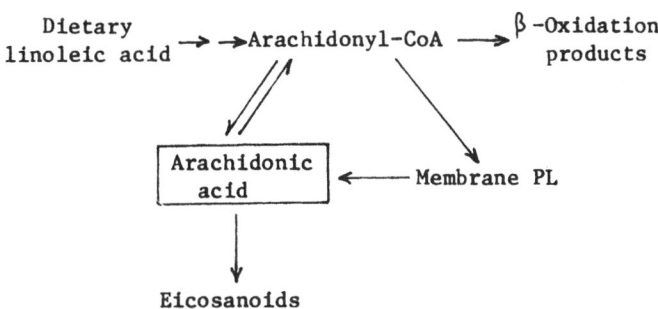

Fig. 2. Simplified scheme of arachidonic acid metabolism

Fig. 2 represents a simplified scheme of AA metabolism. The FFA pool
from which AA is drawn for eicosanoid synthesis is fed by membrane
phosphoglycerides, after hydrolysis at the sn-2 position. But there is
also a direct link between the dietary PUFA input and the FFA pool which
bypasses membrane phospholipids.[10]

α-Linolenic acid interferes with several steps in this metabolic
scheme. It especially inhibits the synthesis of AA,[11,12] at the initial
desaturation step of LA.[13] It also competes, via its long-chain derived
FA, with AA for incorporation into phosphoglycerides[14] and decreases the
formation of AA-derived eicosanoids such as thromboxane A_2[15] and leukotriene
B_4.[16,17]

THE CONCEPT OF UNRESTRAINED AA METABOLISM

According to the classical concept developed by Burr and Burr,[18,19] EFA
produces a typical threshold effect. For instance, with an increasing input

of dietary LA, EFA deficiency signs are no longer observed and a normal state is achieved at a critical supply level of LA, as judged by various criteria, such as rate of growth, skin symptoms or the $20:3\omega9/20:4\omega6$ ratio, and nothing much happens as the LA input is increased beyond that level. On the other hand, eicosanoid production is <u>dose-dependent.</u> The dose-response relationship is complex and there may be ups and downs, but as a rule, the higher the EFA input, the greater the eicosanoid production. This is well illustrated by the results of a trial in which female volunteers received a formula diet with varying LA content but a constant level of LnA.[20,21] As the daily ingestion of LA was increased from 0 to 50 g, the excretion of PGE plus PGF metabolites in 24-h urine at the end of the two-week experimental period increased in a nearly linear manner.

If eicosanoid production from AA increases with an increasing supply of LA, the possibility that excessive eicosanoid production might have detrimental consequences should be examined. Some examples of unrestrained AA metabolism are provided by animal experiments in which the synthesis of AA from LA is by-passed by direct injection of AA into the bloodstream. Intravenous AA led to rapid death in rabbits due to platelet aggregation in the lungs,[22] whereas intrajugular injection of AA produced strokes in rats.[23] Ingestion of a daily dose of 6 g ethyl arachidonate by volunteers caused a pronounced increase in platelet reactivity to ADP in four male volunteers, two of whom were taken off the experiment before the end of the 3-week period, because their platelet aggregability was judged to have reached dangerously high levels.[24]

CHICK NUTRITIONAL ENCEPHALOMALACIA

Chick NE is a particularly useful model for demonstrating the catastrophic consequences of excessive AA metabolism and the restraining action of LnA. NE is one of the classical syndromes of vitamin E deficiency, long known in the field as crazy-chick disease. The popular name derives from the fact that the target organ is the cerebellum, which controls motor coordination. The chick becomes ataxic and may die within a few hours of the appearance of the first symptoms. The cerebellum, but not the cerebral hemispheres, exhibits oedema and haemorrhages. Microscopic and ultra-microscopic observations reveal extensive cellular and fibrillar damage, but one finding of particular interest, also reported by many other investigators, is the presence of thrombi in the capillaries and post-capillary blood vessels of the cerebellum.[25] The involvement of the blood coagulation system in the pathogenesis of NE is seen from the protective effect of the anticoagulant, dicoumarol[26] and the shortening of bleeding time with the onset of NE.[27]

Vitamin E deficiency is a necessary condition for the experimental induction of NE, but it is not sufficient. A dietary supply of LA is also required[28], and AA is even more effective[29]. In our own work, we have used safflower oil (S) as a source of LA, linseed oil (L), rich in LnA, and cod-liver oil (C), which provides EPA and DHA. In order to remove the tocopherols, the oils were either aerated at high temperatures,[25,26,30] or converted to the methyl esters which were then vacuum-distilled.[26,27] The S diet produced a high incidence of NE[27], whereas the L diet did not cause a single case of NE during the experimental period. Moreover, the addition of L to the S diet revealed a pronounced protective effect of LnA against NE, and C was less effective in this respect. On the other hand, C produced a greater increase in the ratio of $\omega3$ to $\omega6$-FA in the cerebellar ethanolamine phosphoglycerides, the major lipid fraction in brain lipids of chicks. When the diet was switched from S to L during the third week, when cases of NE multiplied rapidly, there was an immediate cessation in the appearance of further cases of NE. These results indicate that the PUFA composition of brain membrane phospholipids does not reflect the effect of the dietary PUFA on NE.

A closer examination of the effect of S and L on the changes in FA profiles of cerebellum and cerebrum of chicks over a four-week period[31,32] revealed a progressive mutual replacement of $\omega6$ and $\omega3$-FA, according to the dietary treatment, with the total PUFA content remaining constant, in spite of the extreme difference in the supply of dietary PUFA. Strict regulation of the total PUFA content of brain lipids in the rat was reported as early as 1971 by Galli et al.[33] and was observed by others since then. In the present chick experiment, however, the rate of PUFA accumulation in the cerebellum was considerably increased during the third week, an effect that was not seen in the cerebrum. The pathogenesis of NE may therefore be pictured in the following manner.[31] Under conditions of vitamin E deficiency, the S diet, which is virtually devoid of LnA, results in an excessive oxygenation of AA. During the vulnerable period of cerebellar development, a stroke-like syndrome is produced by the generation of thrombi in the small blood vessels of the cerebellum, involving ischaemia and anoxia, aneurisms and oedema. Extensive tissue damage causes activation of phospholipase A_2 and increased formation of free AA, which results in a self-accelerating process. We recently found that the phospholipase A_2 activity in the cerebellum of the young chick is over twice that of the cerebrum (S. Greenberg-Levy et al., unpublished), which tends to increase the vulnerability of the cerebellum. LnA exerts its protective effect principally by interfering with the 6-desaturation step of AA synthesis. The increased levels of EPA and DHA in the L treatment compete with the incorporation of AA in membrane phosphoglycerides and inhibit the oxygenation of AA to eicosanoids. Additional effects due to IPA-derived eicosanoids may also play a role in the protective action of LnA.

Other animal models support the concept that linseed oil is antithrombotic, compared to oils rich in LA. In rats previously made hypercholesterolaemic by a diet rich in saturated fat and cholesterol, venous thrombosis and platelet stickiness were corrected by a supplement of linseed oil, while maize oil exacerbated the condition.[34,35] When thrombus formation was induced in rats fed on saturated fat by intravenous injection of a low dose of ADP, the inclusion of linseed oil in the diet caused a greater reduction in thrombosis than did cottonseed oil.[36] Linseed oil was also more effective than maize and safflower oil in reducing platelet aggregability in rabbits[37] and in rats.[38] Others[39,40] reported that the administration of purified esters of LnA reduced prostanoid production in rats, and the same effect was observed when linseed oil was given to rats and rabbits.[41,42] Trials with formula diets showed that urinary excretion of PG metabolites in female volunteers was depressed in a linear fashion with increasing intakes of linseed oil, when the LA intake was kept constant.[43] Two male volunteers who ingested a daily dose of 60 ml linseed oil over a period of 48-53 days showed a greatly increased platelet aggregability, as measured by the threshold dose of collagen required for the induction of aggregation.[44] Therefore, LnA exerts modulating effects on AA metabolism even in humans in which EPA formation from LnA is very slight.

ORIGIN AND IMPLICATIONS OF THE PRESENT-DAY IMBALANCE BETWEEN LA AND LnA

Today the balance of LA and LnA in the usual diets of Western industrialized countries is heavily weighted in favour of LA. In the U.S., where statistics have been available for the past 80 years, there has been a 15-fold increase in the consumption of salad and cooking oils, i.e vegetable oils rich in LA.[45] This change is reflected in a threefold increase in the per capita intake of LA. A similar, though less conspicuous, trend exists in other countries. Several factors have contributed to this change in consumption pattern.

Two technological innovations that took place at the turn of the century marked the beginning of the modern vegetable oil industry.[46] One was the invention of the continuous screw press, tradenamed "Expeller" by V.D. Anderson; the other was the steam-vacuum deodorization process developed by D. Wesson, a process which today still constitutes the final and crucial stage in the refining of vegetable oils. Both inventions together made possible the industrial production of cottonseed oil, and later other vegetable oils, for edible purposes. Before that time, cottonseed was a valueless by-product of the fiber industry and was sometimes burned as fuel. After World War I, solvent extraction of oilseeds came into increasing use and rendered the large-scale production of vegetable oils more efficient and more economic. Catalytic hydrogenation, which had been discovered at the end of the last century by P. Sabatier, was applied to the hardening of vegetable oils, after the use of Ni as a catalyst was patented during the first few years of this century. A special application of this process consists in the partial but highly selective hydrogenation of soya-bean oil, which reduces the LnA content of the oil while still leaving a high level of LA. This is done because LnA causes a great many organoleptic problems in soybean oil. For the same reason, the search is now on for new strains of soya-beans low in LnA.[47]

All these developments, together with the awareness of the public of the hypocholesterolaemic effects of LA-rich vegetable oils and margarines, have led to the present-day dietary EFA imbalance, in which LA exceeds LnA by roughly one order of magnitude.

Nowhere is the mean per capita intake of LA as high as in Israel, where high-LA oils constitute by far the major source of separated fats, while the consumption of animal fats is very low, as seen from dietary surveys and food balance sheets.[48,49] The mean LA content of subcutaneous adipose tissue fat of Israelis is around 25%,[50-52] and the LnA content is about 1.5%. In the U.S., the mean LA content in adipose tissue is now about 16%,[51,53] and much lower values have been reported for European countries.[54]

The food available to early man (and to the few surviving hunter-gatherer populations) was generally low in fat. Indeed, fat was at a premium: a good example is provided by fragmented bones of bisons (buffalo) at a 15th-century site in New Mexico, where hunting societies slaughtered the animals and tried to get at the bone marrow.[55,56] Foods from non-cultivated edible plants and wild animals not only have less fat than cultivated seeds and domesticated animals, but their lipids, being largely of the structural type, also contain a higher proportion of PUFA, with a good overall balance between ω6 and ω3-FA.[4,5] During the long evolutionary history of the genus Homo spanning several millions of years, genetic changes must have occurred, partly in response to dietary influences,[57] but the human genetic constitution has remained relatively unchanged since Homo sapiens sapiens made his appearance, about 40,000 years ago.

With the beginnings of agriculture, some 10,000 years ago, a gradual change in the food consumption pattern set in, a change that gained momentum with the industrial revolution. The improvement in the production, processing, transport, distribution and storage of foods first led to a rapid rise in the consumption of animal fats and, more recently, to a change in the type of fat consumed, resulting in the present EFA imbalance. There is a strong suspicion that these recent dietary changes make an important contribution to some of the chronic illnesses afflicting Western societies, illnesses for which prior genetic adaptation has poorly prepared us and which are virtually unknown among the few surviving hunter-gatherer populations.[57]

CONCLUSIONS

The conclusions to be drawn from this discussion are that (a) LnA is an important modulator of the metabolism of LA and its main derived FA, AA; (b) the dietary LA-LnA imbalance does not allow the modulatory effect of LnA to receive its full expression and facilitates unrestrained AA metabolism; (c) an excessive supply of LA over LnA interferes with the conversion of LnA to such long-chain ω3-FA as DHA, for which an absolute requirement has now been established; and (d) the dietary ratio of LA to LnA should be considered by nutritionists in addition to the better known P/S ratio.

REFERENCES

1. R. T. Holman, Essential fatty acid deficiency, in "Progress in the Chemistry of Fats and Other Lipids," R. T. Holman, ed., vol. 9, pp. 274-348, Pergamon, Oxford (1968).
2. M. Neuringer and W. E. Connor, n-3 Fatty acids in the brain and retina: evidence for their essentiality, Nutr. Rev. 44:285 (1986).
3. K. S. Bjerve, I. Løvold-Mostad, and L. Thoreson, Alpha-linolenic acid deficiency in patients on long-term gastric tube feeding: estimation of linolenic acid and long-chain unsaturated n-3 fatty acid requirement in man, Am. J. Clin. Nutr. 45:66 (1987).
4. P. Budowski, Dietary linoleic acid should be balanced by alpha-linolenic acid; a discussion of the nutritional implications of the dietary ratio of polyunsaturated fatty acids, in "Advances in Diet and Nutrition," C. Horwitz, ed., pp. 199-206, Libbey, London (1985).

5. P. Budowski and M.A. Crawford, α-Linolenic acid as a regulator of the metabolism of arachidonic acid: dietary implications of the ratio, n-6:n-3 fatty acids, Proc. Nutr. Soc. 44;221 (1985).

6. W. E. M. Lands, Biochemical observations on long-chain fatty acids from fish oil and their effect on prostaglandin synthesis in animals and humans, in "Nutritional Evaluation of Long-Chain Fatty Acids in Fish Oil," S. M. Barlow and M. E. Stansby, eds., pp. 267-282, Academic Press, London (1982).

7. W. E. M. Lands, "Fish and Human Health," Academic Press, Orlando, (1986).

8. A. P. Simopoulos, R. R. Kifer, and R. E. Martin, eds., "Health Effects of Polyunsaturated Fatty Acids in Seafoods," Academic Press, Orlando (1986).

9. J. E. Kinsella, "Seafood and Fish Oils in Human Health and Disease," Marcel Dekker, New York (1987).

10. A. G. Hassam, A. L. Willis, J. P. Denton, P. Stevens, and M. A. Crawford, The effect of essential fatty acid deficiency on the levels of prostaglandins and their fatty acid precursors in the rabbit, Lipids 14:78 (1979).

11. L. J. Machlin, Effect of dietary linolenate on the proportion of linoleate and arachidonate in liver fat, Nature 194:868 (1962).

12. H. Mohrhauer and R. T. Holman, Effect of linolenic acid upon the metabolism of linoleic acid, J. Nutr. 81:67 (1963).

13. R. Brenner and R. O. Peluffo, Effect of saturated and unsaturated fatty acids on the desaturation in vitro of palmitic, stearic, oleic, linoleic and linolenic acids, J. Biol. Chem. 241:5213 (1966).

14. B. J. Weaver and R. J. Holub, The inhibition of arachidonic acid incorporation into human platelet phospholipids by eicosapentaenoic acid, Nutr. Res. 5:31 (1985).

15. W. Siess, B. Scherer, B. Böhling, O. Roth, I. Kurzman, I., and P. C. Weber, Platelet membrane fatty acids, platelet aggregation and thromboxane formation during a mackerel diet, Lancet i:441 (1980).

16. T. Terano, J. A. Salmon, and S. Moncada, Effect of orally administered eicosapentaenoic acid (EPA) on the formation of leukotriene B_4 and leukotriene B_5 by rat leukocytes, Biochem. Pharmacol. 33:3071 (1984).

17. S. M. Prescott, The effect of eicosapentaenoic acid on leukotriene B production by human neutrophils, J. Biol. Chem. 259:7615 (1984).

18. G. O. Burr and M. M. Burr, A new deficiency disease produced by the rigid exclusion of fat from the diet, J. Biol. Chem. 82:345 (1929).

19. G. O. Burr and M. M. Burr, On the nature and role of fatty acids in nutrition, J. Biol. Chem. 86:587 (1930).

20. N. Zöllner, O. Adam, and G. Wolfram, The influence of linoleic acid intake on the excretion of urinary prostaglandin metabolites, Res. Exp. Med. (Berlin) 175:149 (1979).

21. O. Adam, G. Wolfram, and N. Zöllner, Prostaglandin formation in man during intake of different amounts of linoleic acid in formula diets, Ann. Nutr. Metab. 26:315 (1982).

22. M. J. Silver, W. Hoch, J. J. Kocsis, C. M. Ingerman, and J. B. Smith, Arachidonic acid causes sudden death in rabbits, Science 183:1085 (1974).

23. T. Furlow and N. Bass, Stroke in rats produced by carotid injection of sodium arachidonate, Science 187:658 (1975).

24. M. D. Seyberth, O. Oelz, T. Kennedy, B. J. Sweetman, J. C. Frölich, M. Heimberg, and J. A. Oates, Increased arachidonate in lipids after administration to man: effects on prostaglandin synthesis, Clin. Pharmacol. Ther. 18:521 (1975).

25. Y. Dror, P. Budowski, J. J. Bubis, U. Sandbank, and M. Wolman, Chick nutritional encephalopathy induced by diet rich in oxidized oil and deficient in tocopherol, in "Progress in Neuropathology," vol. 3, H. M. Zimmerman, ed., pp. 343-357, Grune & Stratton, New York (1976).

26. P. Budowski, I. Bartov, Y. Dror, and E. N. Frankel, Lipid oxidation products and chick nutritional encephalopathy, Lipids 14:768 (1979).

27. P. Budowski, C. M. Hawkey, and M. A. Crawford, L'effet protecteur de l'acide α-linolénique sur l'encephalomalacie chez le poulet, Ann. Nutr. Aliment. 34:389 (1980).

28. H. Dam, G. K. Nielsen, I. Prange, and E. Søndergaard, Influence of linoleic and linolenic acids on symptoms of vitamin E. deficiency in chicks, Nature 182:802 (1958).

29. H. Dam, and E. Søndergaard, The encephalomalacia-producing effect of arachidonic and linoleic acids, _Z. Ernährungswiss._ 2:217 (1962).

30. P. Budowski and S. Mokadi, Detection of free-radical damage in the vitamin E-deficient chick, _Biochem. Biophys. Acta_ 52:609 (1961).

31. P. Budowski and M. A. Crawford, Effect of dietary linoleic and α-linolenic acid on the fatty acid composition of brain lipids in the young chick, _Prog. Lipid Res._ 25:615 (1986).

32. P. Budowski, M. Leighfield, and M. A. Crawford, Nutritional encephalomalacia in the chick: an exposure of the vulnerable period for cerebellar development and the possible need for both ω6- and ω3-fatty acids, _Br. J. Nutr._ 58:511 (1987).

33. C. Galli, H. I. Trzeciak, and R. Paoletti, Effects of dietary fatty acids on the fatty acid composition of brain ethanolamine phosphoglyceride: reciprocal replacement of n-6 and n-3 polyunsaturated fatty acids, _Biochem. Biophys. Acta_ 248:449 (1971).

34. A. Nordøy, The influence of saturated fat, cholesterol, corn oil and linseed oil on experimental venous thrombosis in rats, _Diath. Haemorrh._ 13:244 (1965).

35. A. Nordøy, The influence of saturated fat, cholesterol, corn oil and linseed oil on the ADP-induced platelet adhesiveness in the rat, _Diath. Haemorrh._ 13:543 (1965).

36. A. Nordøy, J. T. Hamlin, A. B. Chandler, and H. Newland, The influence of dietary fats on plasma and platelet lipids and ADP-induced platelet thrombosis in the rat, _Scand. J. Haemat._ 5:458 (1968).

37. F. W. Vas Dias, M. G. Gibney, and T. G. Taylor, The effect of polyunsaturated fatty acids of the n-3 and n-6 series on platelet aggregation and platelet and aorta fatty acid composition in rabbits, _Atherosclerosis_ 43:245 (1982).

38. M. Ishinaga, M. Kakuta, H. Narita, and M. Kito, Inhibition of platelet aggregation by dietary linseed oil, _Agr. Biol. Chem._ 47:903 (1983).

39. D. H. Hwang and A. E. Carroll, Decreased formation of prostaglandins derived from arachidonic acid by dietary linolenate in rats, _Am. J. Clin. Nutr._ 33:590 (1980).

40. H. S. Hansen, B. Fjalland, and B. Jensen, Extremely decreased release of prostaglandin E_2-like activity from chopped lung of ethyl linolenate-supplemented rats, _Lipids_ 18:691 (1983).

41. F. ten Hoor, E. A. M. de Deckere, E. Haddeman, G. Hornstra, and J. F. A. Quadt, Dietary manipulation of prostaglandin and thromboxane synthesis in heart, aorta and blood platelets of the rat, in "Advances in Prostaglandin, and Thromboxane Research," vol. 8, B. Samuelsson, P. W. Ramwell and R. Paoletti, eds., pp. 1771-1781, Raven Press, New York (1980).

42. G. Hornstra, E. Haddeman, J. Kloeze, and P. M. Verschuren, Dietary fat-induced changes in the formation of prostanoids of the 2 and 3 series in relation to arterial thrombosis (rat) and atherosclerosis (rabbit), Adv. Prost. Thromb. Leukotr. Res. 12:193 (1983).

43. O. Adam, G. Wolfram, and N. Zöllner, Wirkung der Linol- und Linolensäure auf die Prostaglandinbildung und Nierenfunktion beim Menschen, Fette Seifen Anstrichmittel 86:180 (1984).

44. P. Budowski, N. Trostler, M. Lupo, N. Vaisman, and A. Eldor, Effect of linseed oil ingestion on plasma fatty acid composition and platelet aggregation in healthy volunteers, Nutr. Res. 4:343 (1984).

45. R. L. Rizek, S. O. Welsh, R. M. Marston, and E. M. Jackson, Levels and sources of fat in the U.S. food supply and in diets of individuals, in "Dietary Fats and Health," E. C. Perkins and W. J. Visek, eds., pp. 13-43, American Oil Chemists'Society, Champaign (1983).

46. H. G. Kirschenbauer, "Fats and Oils," (2nd ed.), Reinhold Publ. Corp., New York (1960).

47. Anon., Researchers report gains in hunt for low-linolenic soybeans, J. Am. Oil Chem. Soc. 59:882A (1982).

48. S. Bavly, R. Poznanski, and N. Kaufmann, Levels of Nutrition in Israel 1975/76. Ministry of Education and Culture, and Hebrew University-Hadassah Medical School, Jerusalem, 1980.

49. Central Bureau of Statistics, Jerusalem, Food Balance Sheets, 1985/1986.

50. S. H. Blondheim, T. Horne, R. Davidovich, J. Kapitulnik, S. Segal, and N. Kaufman, Unsaturated fatty acids in adipose tissue of Israeli Jews, Isr. J. Med. Sci. 12:658 (1976).

51. R. G. Schwartz, T. Horne, and S. H. Blondheim, Fatty acid saturation in subcutaneous fat of young Americans and Israelis, Isr. J. Med. Sci. 15:778 (1979).

52. M. G. Enig, P. Budowski, and S. H. Blondheim, Trans-unsaturated fatty acids and human subcutaneous fat in Israel, Human Nutr.: Clin. Nutr. 38C:223 (1984).

53. E. M. Berry and J. Hirsch, Does dietary linolenic acid influence blood pressure? Am. J. Clin. Nutr. 44:336 (1986).

54. R. A. Riemersma, D. A. Wood, and S. Butler, Linoleic acid in adipose tissue and coronary heart disease, Br. Med. J. 292:1423 (1986).

55. J. D. Speth, "Bison Kills and Bone Counts," Univ. Chicago Press, Chicago (1983).

56. J. D. Speth, Early hominid subsistence strategies in seasonal habitats, J. Archeological Sci. 14:13 (1987).

57. S. B. Eaton and M. Konner, Paleolithic nutrition. A consideration of its nature and current implications, N. Engl. J. Med. 312:283 (1985).

COMPARATIVE UPTAKE IN RATS AND MAN OF ω3 AND ω6 FATTY ACIDS

Wulf Becker

Nutrition Laboratory
National Food Administration
Uppsala, Sweden

INTRODUCTION

The tissue uptake and metabolism of the various ω3 and ω6 fatty acids are influenced by a number of factors. Of great importance is the fat content and the fatty acid composition of the diet. Both linoleic and α-linolenic acid are metabolised by the same enzyme system and the affinity for the enzyme system depends upon the degree of unsaturation and chain-length. This means that the ratio of linoleic to α-linolenic acid in the diet, rather than the absolute amount, strongly influences their conversion and incorporation into lipids (Alling et al., 1972, 1974; Becker, 1985). Taking these factors into account, the problems in making relevant comparisons of essential fatty acid (EFA) metabolism within and between species are obvious.

Using whole-body autoradiography we have studied the uptake and tissue distribution of several ^{14}C-labelled EFAs in rats (Becker et al., 1983, 1985; Becker and Bruce, 1985). The animal is injected with a radiolabelled fatty acid, killed after various time periods, rapidly frozen whereafter sagittal sections are cut through the whole animal. The sections are freeze-dried at -20^0C and then pressed against a photographic film for exposure. Quantitative data concerning the activity can be obtained by different techniques. A typical whole-body autoradiogram is shown in fig. 2.

In other studies the distribution of some EFAs were measured by direct measurements of radioactivity in the tissue lipids (Becker, 1985; Becker and Månsson, 1986; Becker and Nilsson, unpublished). Table 1 shows which EFAs have been studied with these techniques.

In humans there are rather few studies with labelled acids and these and other feeding studies give only limited information on the tissue uptake of EFAs (Nichaman et al., 1967; Ormsby et al., 1963).

GENERAL OBSERVATIONS OF TISSUE UPTAKE OF EFAs IN RATS

In rats fed sufficient amounts of EFAs, a major part of a single dose of both linoleic and α-linolenic acid is oxidized to carbon dioxide within 24 hours. (Becker, 1984, Leyton et al., 1987). γ-Linolenic acid and especially 20:3ω3, 20:4ω6 and 22:6ω3 are oxidized at a much slower rate and a larger part of each of these acids is retained in the body (Leyton et al., 1987; Sinclair, 1975).

The results of our and other studies showed that after an intravenous injection of a tracer dose of fatty acids bound to albumin or to an emulsion, fatty acids were rapidly taken up from the blood by various tissues of which the liver, brown fat, adrenals and kidneys were qualitatively the most important. After 5 to 20 min, little radioactivity was seen in the blood, indicating that most of the given dose has been taken up by the tissues.

Our data (Becker and Bruce, 1985) show that, for most fatty acids, the general pattern of distribution in rats fed a diet with a sufficient EFA-content showed many similarities. Table 2 summarizes the uptake of radioactivity after 1 hr, which in most cases did not differ much from that seen after 5 min. The pattern was characterized by a high initial

Table 1. Essential fatty acids studied

EFA	Technique	
^{14}C-18:2ω6	i.v.	autoradiography
"	p.o.	LSC[1] of tissue lipids
^{14}C-20:4ω6	i.v.	autoradiography
^{14}C-18:3ω6	i.v.	autoradiography
"	i.v.	LSC of tissue lipids
^{14}C-20:5ω3	i.v.	autoradiography
^{3}H-20:5ω3	i.v.	LSC of tissue lipids
^{14}C-22:6ω3	i.v.	autoradiography

[1]LSC = liquid scintillation counting

uptake of radioactivity in the brown fat and liver, moderate to high levels in the gastric mucosa, adrenal and kidney cortex. Low levels were generally found in the brain, blood, white fat, thymus, lungs, and testes. After 18 h the level of radioactivity from linoleic, α-linolenic and docosahexaenoic acid generally had decreased in most tissues whereas little or no change was found with arachidonic acid. For some EFAs the activity in the white fat had increased.

Linoleic and α-linolenic acid

Our studies showed that the tissue distribution of linoleic and α-linolenic acid was rather similar (fig. 1, table 2). Other studies have shown major differences in the uptake into tissue lipids. Linoleic acid is normally a major component of nearly all tissue lipids, whereas only small amounts of α-linolenic acid generally are found in the tissue lipids, especially the phospholipids. These differences can largely be explained by a low level of α-linolenic acid in most experimental diets and by its stronger affinity to the enzyme system for desaturation and elongaton. Thus after application of labelled α-linolenic acid, more of the label incorporated into liver phospholipids is associated with higher acids than is the case after application of linoleic acid (Sinclair, 1975). In rats given linseed oil a marked increase of the α-linolenic level was seen in the body triglycerides and adipose tissue lipids (Roshanai and Sanders, 1985, Becker and Bruce, 1986). In the phospholipids the level of α-linolenic acid did not change, whereas the level of eicosapentaenoic (EPA) and docosahexaenoic (DHA) acid increased in some tissues.

Table 2. Tissue distribution of radioactivity in rats studied by whole-body autoradiography 1 hr after the intravenous injection of ^{14}C-labelled fatty acids*.

Tissue	Linoleic acid	Arachidonic acid	Linolenic acid	EPA	DHA
			1 hr		
Brain	+	+	+	+	+
White fat	+	+	+	+	+
Skeletal muscles	+	++	++	+(+)	++
Myocardium	+	+++	+	+	+++(+)
Gastric mucosa	++	++(+)	+(+)	++	++
Intestinal mucosa	++	++	++	+	++
Bone marrow	++	++	++	+	+(+)
Kidney cortex	++	+++	+++	++	+++
Adrenal cortex	++	++++	++	++	++++
Liver	++	++++	++	++	++++
Brown fat	++++	++++	++++	++(+)	++++

* For each fatty acid, the radioactivity in a tissue was related to that in the brain which was set to 1. "+" corresponds to a relative activity of 1-4, "++" 4-12; "+++" 12-30; "++++" >30. (+) Indicate that the relative activity is close to the lower limit of the interval. From Becker and Bruce, 1985.

In a recent study with rats, Leyton and co-workers (1987) found that α-linolenic acid was oxidized faster and to a larger extent than linoleic acid. This finding could also explain why α-linolenic acid levels in tissues are low. In a similar study, Sinclair (1975) found no differences in carcass retention of linoleic or α-linolenic acid in suckling rats.

In EFA-deficient rats, the distribution pattern of linoleic acid resembled much that of arachidonic acid (fig. 2). This was associated with a proportionally larger body retention of ω6 acids in the phospholipids, especially in the liver and skeletal muscles. (Becker and Månsson, 1985, Becker and Bruce, 1986). Data by Poovaiah and co-workers (1976) indicated a similar effect of EFA-deficiency on the conversion of α-linolenic acid to DHA.

Arachidonic acid

Following oral dosing a much larger proportion of arachidonic acid than linoleic acid is retained in the tissue lipids, e.g. the liver (Sinclair, 1975, Leyton et al., 1987). Our autoradiographic studies also showed a high uptake in the myocardium, kidney cortex, adrenal cortex and skeletal muscles (fig. 3, Becker et al., 1985). The larger retention of arachidonic than of linoleic acid is largely due to a higher incorporation of arachidonic acid into the phospholipids (Sinclair, 1975).

An interesting finding was the high uptake of arachidonic acid in the myocardium. Earlier studies have shown that arachidonic acid has a marked influence on coronary resistance and flow. (Blass et al., 1978; Needleman, 1976) and that this is associated with prostaglandin (PG) production. The role of arachidonic acid as a potent PG precursor might partly explain the high uptake in our study.

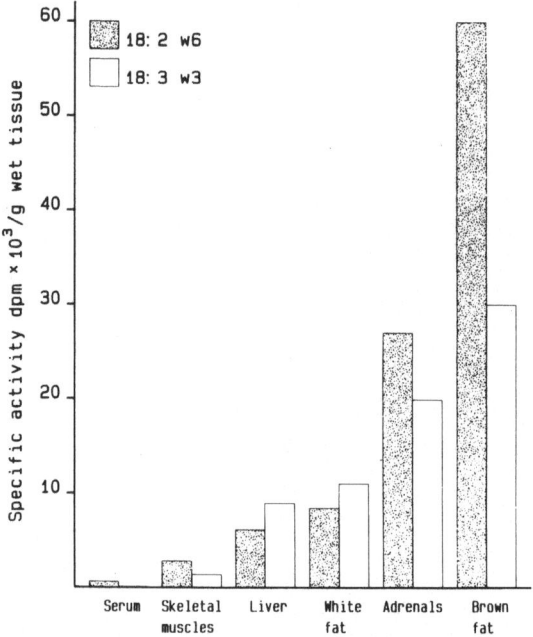

Fig. 1. Specific activities in tissues of rats 20 hr after oral administration of $1-^{14}$C-linoleic acid or 18 hr after i.v. administration of $1-^{14}$C-linolenic acid. Data from Becker (1985) and Becker and Nilsson (1988).

Eicosapentaenoic (EPA) and docosahexaenoic acid (DHA)

The autoradiographic studies showed that the uptake of EPA was rather similar to that of α-linolenic acid after 1 h. After 18 h a marked concentration of radioactivity from EPA in the adrenal cortex was the most important finding (Becker and Bruce, 1985). The latter finding was confirmed in a subsequent study, in which α-linolenic and EPA were injected simultaneously to rats (fig. 4, Becker and Nilsson, 1988). In most tissues more radioactivity from EPA than from α-linolenic acid was recovered in the tissues, except in the white and brown fat. The larger retention was generally associated with higher incorporation into the phospholipids, mostly in phosphatidyl choline and after 18 h also in phosphatidyl ethanolamine. The high initial uptake of ^{14}C from α-linolenic acid in the liver followed by the rapid decrease in activity indicate a faster oxidation and turn-over of α-linolenic acid.

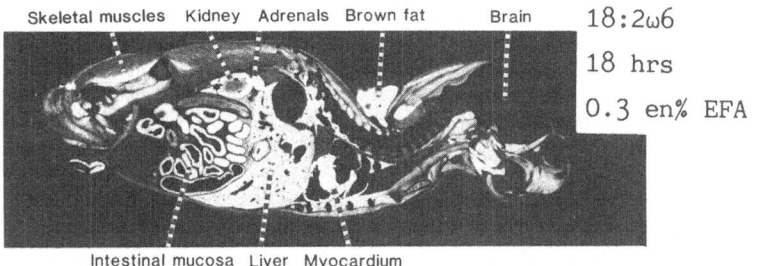

Skeletal muscles Kidney Adrenals Brown fat   Brain

18:2ω6

18 hrs

0.3 en% EFA

Intestinal mucosa Liver Myocardium

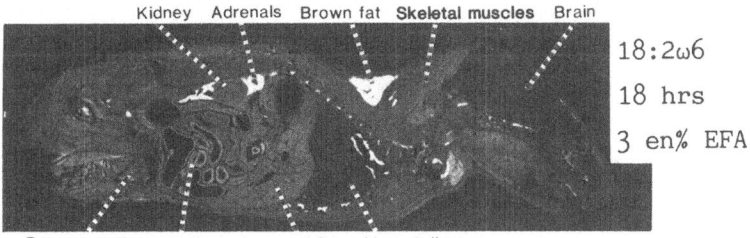

Kidney Adrenals Brown fat **Skeletal muscles** Brain

18:2ω6

18 hrs

3 en% EFA

Bone marrow Intestinal mucosa Liver Myocardium

Fig. 2. Autoradiograms of rats fed a diets with a low (A) or normal (B)
EFA content 18 hr after an i.v. injection of albumin-bound
1-^{14}C-linoleic acid. From Becker and Bruce (1985) with
permission from Prog. Lip. Res.

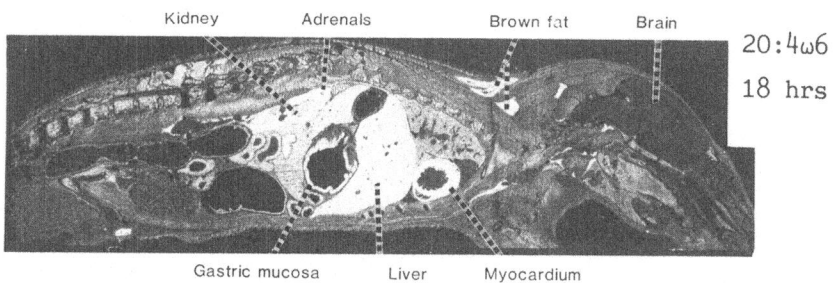

Kidney   Adrenals   Brown fat   Brain

20:4ω6

18 hrs

Gastric mucosa   Liver   Myocardium

Fig. 3. Autoradiogram of a rat fed a diet with a normal EFA content 18 hr
after an i.v. injection of albumin-bound 1-^{14}C-arachidonic acid.
From Becker et al. (1986) with permission of Ann. Nutr. Metab.

Earlier studies have shown that EPA is present in low levels in most tissues in rats fed diets with vegetable oils as the fat source (Becker and Bruce, 1985; Roshanai and Sanders, 1985; Bruckner et al., 1984; Horrobin et al., 1984). Small daily supplements of fish oils rich in EPA and DHA to rats lead to increased levels of EPA in serum, platelet and erythrocyte phosphoglycerides (Roshanai and Sanders, 1985; Bruckner et al., 1984), kidney and liver phospholipids (Croft et al., 1987), aorta and lung microsomes (Bruckner et al., 1984), heart and skeletal muscle lipids (Jackson et al., 1988; Hølmer and Beare-Rodgers, 1985), compared to rats fed diets without supplements. In the adipose tissue, brain and retinal lipids, EPA was not retained to any significant extent (Roshanai and Sanders, 1985).

No data have, however, been found in the literature on the levels of EPA in the adrenals, and the finding in the present findings might indicate some specific function in this tissue. Notably, an marked secondary uptake of radioactivity in the adrenal cortex was also found when α-linolenic acid was given to EFA-deficient rats (Becker et al., 1985). This could indicate a significant biosynthesis of EPA from α-linolenic acid in the EFA-deficient rat.

The pattern of radioactivity after administration of DHA differered in some respects from that of EPA and α-linolenic acid. It more resembled that found earlier for arachidonic acid (Becker et al., 1985), with a high initial labelling in the brown fat, liver, adrenal cortex, myocardium and kidney (fig. 5).

The remarkable uptake of radioactivity in the myocardium is in agreement with other studies showing a preferential incorporation of DHA into rat myocardial phospholipids from dietary fish oil (Roshanai and Sanders, 1985; Hølmer and Beare-Rodgers, 1985; Gudbjarnason et al., 1978). Accumulation of DHA in the heart lipids also occurs when the ratio of linoleic acid to α-linolenic acid in the diet is low (Charnock et al., 1985a).

Studies by Gudbjarnason and coworkers (1978) and Charnock and co-workers (1985, a, b) showed that the level of DHA in the myocardial lipids influenced heart rate and coronary function in rats. According to Gudbjarnason and coworkers (1978) rats fed diets containing cod liver oil were more sensitive to noradrenaline stimulation than rats fed "standard" diets with vegetable oil. Sanders (1985) suggested that this effect could have been due to a high content of peroxides in the fish oil rather than DHA per se.

DHA is also a major component of the brain and retinal lipids (Sinclair, 1975; Alling et al., 1974; Alling et al., 1984). Administration of linseed oil or fish oil to weanling rats for several weeks lead to a significant increase of DHA in the brain phosphoglycerides, compared to in rats given safflower oil (Roshanai and Sanders, 1985). Earlier studies with labelled DHA showed a larger uptake from a single dose into the brain lipids of 16-17 day-old suckling rats than in adult rats (Sinclair, 1975). The low uptake of radioactivity in the our study indicate a slow incorporation of DHA into the brain lipids after weaning.

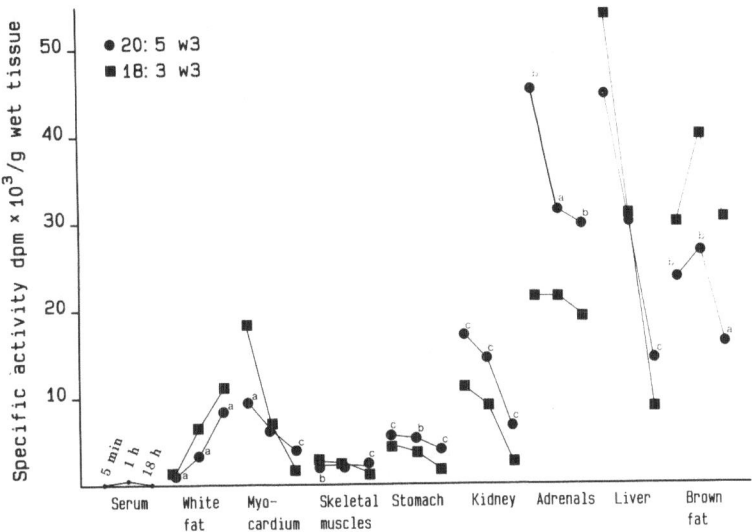

Fig. 4. Tissue distribution of radioactivity in rats 5 min, 1 hr and 18 hr after i.v. application of albumin-bound ^{14}C-linolenic acid and ^{3}H-EPA (Becker and Nilsson, 1988).

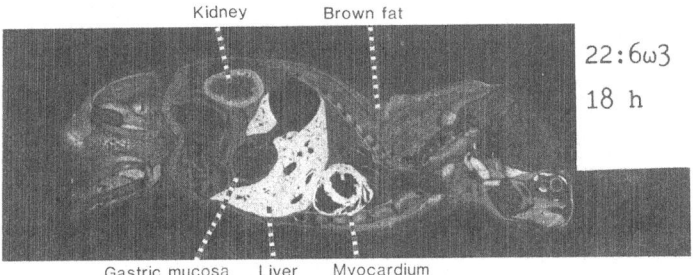

Fig. 5. Autoradiogram of a rat fed a EFA-normal diet 18 hr after an i.v. injection of albumin-bound 1-^{14}C-docosahexaenoic acid. From Becker and Bruce (1985), with permission from Prog. Lip. Res.

In man, we have rather limited knowledge of the uptake and tissue distribution of the various EFAs. As for rats the level and composition of fat in the diet influences the uptake and metabolism of EFAs in the tissues. I most human diets, linoleic acid is the major EFA, and this is often the case in diets for experimental animals. Of the ω3-acids, α-linolenic acid is most important.

Both in humans and rats fed diets with a sufficient EFA content the conversion of linoleic acid to arachidonic acid is reported to be slow (Nichaman et al., 1967; Sinclair, 1975). In vitro experiments, however, indicate that the activities of the Δ6 and Δ5 desaturases are lower in humans compared to in rats and that this conversion is slower in man (Cunnane et al., 1984; Stone et al., 1979). Other data indicate that the same is true for the conversion of α-linolenic acid to long-chained ω3 acids (Sinclair, 1975; Adam et al., 1984; Singer et al., 1986).

Linoleic and α-linolenic acid

Linoleic acid is a major EFA in many tissues of Western man whereas α-linolenic acid generally is found only in small amounts. This of course largely reflects the fatty acid composition of the common diet.

There is generally a strong connection between the dietary level of linoleic acid and the level in the adipose tissue both in humans (Field and Clandinik, 1984) and rats (Becker and Bruce, 1986). Ormsby and collaborators (1963) calculated that 8-16 % of the activity from a single oral dose of labelled linoleic acid was present in the adipose tissue of humans 24 hr after the administration. In rats there is also a direct relation between the α-linolenic level of adipose tissue or in body triglycerides and the dietary supply (Sanders and Roshanai, 1985; Becker and Bruce, 1986). The same probably applies to humans.

In table 3 the level of some EFAs found in structural lipids of human and rat tissues are given. Linoleic and α-linolenic acid are generally found only in traces in the phosphoglycerides of brain but in other tissues linoleic acid is present in significant amounts. The data suggest that a significant part of deposited α-linolenic acid is converted to 20 and 22 carbon ω3 PUFAs.

Arachidonic acid

Arachidonic acid is a major component of tissue cholesterol esters and phospholipids of humans, but small amounts are normally found in triglycerides (table 3). It is difficult to determine the origin of the tissue arachidonate, but data from vegans suggest that a large part is derived from dietary linoleic acid (Sanders and Younger, 1981). The small amount of arachidonic acid present in the diet of omnivores is probably largely deposited into the phospholipids.

EPA and DHA

Low levels of EPA is generally found in tissue lipids of Western man whereas DHA is present in significant levels in membrane lipids (phospholipids) especially of liver, heart and brain (table 3). However, Greenland Eskimos have high levels of long chained ω3 fatty acids in serum and platelet lipids reflecting a high intake of marine foods (Dyerberg et al., 1975; Dyerberg and Bang, 1979).

Table 3. Fatty acids (wt %) in tissues of man and rats.

Fatty acid	Brain EPG		Liver PL		Heart TL		Muscle CPG	
	Man[1] n=6	Rat[2] n=5	Man[3] n=10	Rat[3,4] n=5-7	Man[5] n=23	Rat[6] n=6	Man[7] n=12	Rat[8] n=4
18:2 ω6	0.7	-	17	13	15	21	38	20
18:3 ω3	-	2.9	0.4	0.3	-	-	-	-
20:4 ω6	20	16	13	20-30	6-10	17	5	13
20:5 ω3	-	-	0.7	0.5	0.2	0.1	-	-
22:6 ω3	19	18	7	8	1-2	13	1	10

[1] Martinez and Ballabriga (1987)
[2] Alling et al. (1974)
[3] Horrobin et al. (1984)
[4] Alling et al. (1976), values in mol %
[5] Heckers et al. (1977)
[6] Charnock et al. (1986), values for total phospholipids
[7] Bruce (1974b)
[8] Bruce (1974a), values in mol %

Administration of fish oils rich in EPA and DHA leads to substantial increases in the EPA levels in serum, erythrocyte and platelet lipids of man (Kinsella, 1987; Boberg et al., 1986). The increases in DHA levels were generally less pronounced. This is partly in agreement with results from studies on rats (Croft et al., 1987; Roshanai and Sanders, 1985).

In both humans and rats short term administration of EPA does not lead to significant changes of the DHA level in the serum lipids (von Schacky and Weber, 1985; Hamazaki et al., 1987). Retroconversion of DHA to EPA occurs both in humans and rats (von Schacky and Weber, 1985; Schlenk et al., 1969).

Conclusions

Many factors affect the tissue uptake and metabolism of EFAs. In rats studies with several EFAs showed some common characteristics with respect to tissue uptake and distribution. Specific features were observed for some acids, e.g. a strong uptake of arachidonic and DHA in the liver, kidney cortex and myocardium and of EPA in the adrenal cortex. In EFA-deficiency the uptake of linoleic acid resembled that of arachidonic acid with a strong radioactive concentration in the myocardium.

Other studies have shown that the metabolism of the two parent fatty acids linoleic and α-linolenic acids differs in several aspects, mainly with respect to conversion and incorporation into structural lipids. Thus, whereas linoleic acid is a major fatty acid of most tissue lipids α-linolenic acid seldom occurs in significant levels in structural lipids. The higher members of the ω3 and ω6 family are mainly incorporated into phospholipids.

Data from studies on humans indicate that the desaturase system is less efficient than in rats, but data on EFA incorporation into blood lipids after feeding different fats indicate similarities between the two species.

REFERENCES

Adam, O., Wolfram, G., and Zöllner, N., Wirkung der Linol - und
Linolensäure auf die Prostaglandinbildung und die Nierfunktion beim
Menschen. Fette Seifen-Anstrichmittel., 1984, 86, 180-183.

Alling, C., Bruce, Å., Karlsson, I., Sapia, O., and Svennerholm, L., 1972,
Effect of maternal essential fatty acid supply on fatty acid
composition of brain, liver, muscle and serum in 21-day-old rats. J .
Nutr. 102, 773-782.

Alling, C., Bruce, Å., Karlsson, I., and Svennerholm, L., 1974, The effect
of different dietary levels of essential fatty acids on lipids of
rat cerebrum during matuation. J. Neurochem. 23, 1263-1270.

Alling, C., Bruce, Å., Karlsson, I., and Svennerholm, L., 1976, The effect of
different dietary levels of essential fatty acids on liver and
serum lipids in rat. Nutr. Metabol. 20, 440-451.

Alling, C., Becker, W., Jones A. W., and Änggård, E., 1984, Effects of
chronic ethanol treatment on lipid composition and prostaglandins
in rats fed essential fatty acid deficient diets. Alcholism: Clin.
Exp. Res. 8, 238-242.

Becker, W., 1984, Distribution of ^{14}C after oral administration of 1-^{14}C-
linoleic acid in rats fed different levels of essential fatty
acids. J. Nutr., 114, 1690-1696.

Becker, W., 1985, Thesis, Acta Univ. Ups. Abstracts of Uppsala
Dissertations from the Faculty of Medicine. 513, Uppsala.

Becker, W., and Bruce, Å., Retention of linoleic acid in carcass lipids of
rats fed different levels of essential fatty acids. Manuscript.

Becker, W., Bruce, Å., and Larsson, B., 1983, Autoradiographic studies
with albumin-bound 1-^{14}C-linoleic acid in normal and essential
fatty acid-deficient rats. A pilot study. Ann. Nutr. Metab., 27,
415-424.

Becker, W., and Månsson J. E., 1985, Incorporation of ^{14}C into tissue
lipids after oral administration of 1-^{14}C-linoleic acid in rats fed
different levels of essential fatty acids. J. Nutr., 115, 1248-1258.

Becker, W., and Bruce, Å, 1985, Autoradiographic studies with fatty
acids and some other lipids: a review. Prog. Lipid Res., 24,
325-346.

Becker, W., Mohammed, A., And Slanina, P., 1985, Uptake of radiolabelled
α-linolenic, arachidonic and oleic acid in tissues of normal
and essential fatty acid deficient rats - an autoradidographic
study. Ann. Nutr. Metab., 29, 65-75.

Becker, W., and Nilsson Å., 1988, Unpublished.

Blass, K-E., P., Mentz, P., and Förster, W., 1978, Effects of unsaturated
fatty acids on canine coronary flow. Acta Biol. Med. Germ., 37,
765-767.

Boberg, M., Vessby, B., and Selenius, I., 1986, Effects of dietary
supplementation with n-6 and n-3 long-chain polyunsaturated fatty
acids on serum lipoproteins and platelet function in hypertriglyceridaemic patients. Acta Med. Scand., 220, 153-160.

Bruce, Å., 1974a, Changes in the concentration and fatty acid composition
of phospholipids in rat skeletal muscle during postnatal development. Acta Physiol. Scand., 90, 743-749.

Bruce, Å., 1974b, Skeletal muscle lipids. III. Changes in fatty acid composition of individual phosphoglycerides in man from fetal to middle
age. J. Lip. Res., 15, 109-113.

Bruckner, G. G., German, B., Lokesh, B., and Kinsella, J. E., 1984, Biosynthesis of prostanoids, tissue fatty acid composition and thrombotic parameters in rats fed diets enriched with docosahexaenoic (22:6n3) or eicosapentaenoic (20:5n3) acids. Thrombosis Res., 34, 479-497.

Charnock, J. S., McLennan, P. L., Abeywardena, M. Y., and Russel, G. R., 1985a, Altered levels of n-6/n-3 fatty acids in rat heart and storage fat following variable dietary intake of linoleic acid. Ann. Nutr. Metab. 29:279-288.

Charnock, J. S., McLennan, P. L., Abeywardena, M. Y., and Dryden, W. F., 1985b, Diet and cardiac arrhythmia: effects of lipids on age-related changes in myocardial function in the rat. Ann. Nutr. Metab., 29, 306-318.

Charnock, J. S., Abeywardena, M. Y., and McLennan, P. L., 1986, Comparative changes in the fatty-acid composition of rat cardiac phospholipids after long-term feeding of sunflower seed oil- or tuna fish oil supplemented diets. Ann. Nutr. Metab. 30, 393-406.

Croft, K. D., Beilin, L. J., Legge, F. M., and Vandongen, R., 1987, Effects of diets enriched in eicosapentaenoic or docosahexaenoic acids on prostanoid metabolism in the rat. Lipids, Vol 22, 647-650.

Cunnane, S. C., Keeling, P. W. N., Thompson, R. P. H., and Crawford, M. A., 1984, Linoleic acid and arachidonic acid metabolism in human periferal blood lymphocytes: comparison with the rat. Br. J. Nutr., 51, 209-217.

Dyerberg, J., Bang, H. O., and Hjørne, N., 1975, Fatty acid composition of the plasma lipids in Greenland Eskimos. Am. J. Clin. Nutr., 28, 958-966.

Dyerberg, J., and Bang, H. O., 1979, Haemostatic function and platelet polyunsaturated fatty acids in Eskimos. Lancet., ii, 433-435.

Field, J. C., and Clandinin, M. T., 1984, Modulation of adipose tissue fat composition by diet: a review. Nutr. Res., 4, 743-755.

Gudbjarnason, S., Doell, B., and Oskarsdottir, G., 1978, Docosahexaenoic acid in cardiac metabolism and function. Acta. Biol. Med. Germ., 37, 777-784.

Hamazaki, T., Urakaze, M., Makuta, M., Ozawa, A., Soda, Y., Tatsumi, H., Yano, S., and Kumagai, A., 1987, Intake of different eicosapentaenoic acid-containing lipids and fatty acid pattern of plasma lipids in the rats. Lipids., 22, 994-998.

Heckers, H., Körner, M., Tüschen, T. W. L., and Melcher, F. W., 1977, Occurence of individual trans-isomeric fatty acids in human myocardium, jejunum, and aorta in relation to different degrees of atherosclerosis. Atherosclerosis, 28, 389-398.

Holman, R. T., Smythe, L., and Johnson, S., 1979, Effect of sex and age on fatty acid composition on human serum lipids. Am. J. Clin. Nutr., 32, 2390-2399.

Hølmer, G., and Beare-Rogers, J. L., 1985, Linseed oil and marine oil as sources of (N-3) fatty acids in rat heart. Nutr. Res., 5, 1011-1014.

Horrobin, D. F., Huang, Y-S., Cunnane, S. C., and Manku, M. S., 1984, Essential fatty acids in plasma, red blood cells and liver phospholipids in common laboratory animals as compared to humans. Lipids 19, 806-811.

Jackson, M. J., Roberts, J., and Edwards, R. H. T., 1988, Dietary fish oil supplementation modifies rat skeletal muscle fatty acid composition but does not influence the response of muscles to experimental damage. Proc. Nutr. Soc., 47, 32A.

Kinsella, J. E., 1987, "Seafoods and fish oils in human health and disease", Marcel Dekker, inc., New York and Basel.

Leyton, J., Drury, P. J., and Crawford, M. A., 1987, Differential oxidation of saturated and unsaturated fatty acids in vivo in the rat. Br. J. Nutr., 57:383-393.

Martinez, M., and Ballabriga, A., 1987, Effects of parenteral nutrition with high doses of linoleate on the developing liver and brain. Lipids, 22, 133-138.

Needleman, P., 1976, The synthesis and functions of prostaglandins in the heart. Fed. Proc., 35, 2376-2381.

Nichaman, M. Z., Olson, R. E. and Sweeley, C. C., 1967, Metabolism of linoleic acid-1-^{14}C in normolipemic and hyperlipemic humans fed linoleate diets. Am. J. Clin. Nutr., 20, 1070-1083.

Ormsby, J. W., Schnatz, J. D., and Williams, R. H., 1963, The incorporation of linoleic-1-^{14}C acid in human plasma and adipose tissue. Metabolism, 12, 812-820.

Poovaiah, B. P., Tinoco, J., and Lyman, R. L., 1976, Influence of diet on conversion of 1-^{14}C-linolenic acid to docosahexaenoic acid in the rat. Lipids, 11, 194-202.

Roshanai, F., and Sanders, T. A. B., 1985, Influence of different supplements of n-3 polyunsaturated fatty acids on blood and tissue lipids in rats receiving high intakes of linoleic acid. Ann. Nutr. Metab., 29, 189-196.

Sanders, T. A. B., and Younger, K. M., 1981, The effect of dietary supplements of ω3 polyunsaturated fatty acids on the fatty acid composition of platelets and plasma choline phosphoglycerides. Br. J. Nutr., 45, 613-616.

Sanders, T. A. B., 1985, The importance of eicosapentaenoic and docosahexaenoic acids. In: "The Role of Fats in Human Nutrition" (eds P. B. Padley, J. Podmore, J.P. Brun, R. Burt, B.W. Nicols) pp. 101-116, Ellis Horwood, Chichester.

von Schacky, C. and Weber, P.C., 1985, Metabolism and effects on platelet function of the purified eicosapentaenoic and docosahexaenoic acids in humans. J. Clin. Invest., 76, 2446-2450.

Schlenk, H., Sand, D. M., and Gellerman, J. L., 1969, Retroconversion of docosahexaenoic acid in the rat. Biochim. Biophys. Acta, 187, 201-207.

Sinclair, A.J., 1975, Incorporation of radioactive polyunsaturated fatty acids into liver and brain of developing rat. Lipids 10, 175-184.

Singer, P., Wirth, M., and Berger, I., 1986, Zur mangelnden Synthese von Arachidon- und Eicosapentaensäure aus Linol- und α-Linolensäure in klinischen Experimenten. Akt. Ernähr., 11, 203-207.

Stone, K.J., Willis, A.L., Kirtland, S.J., Kernoff, P. B. A., and McNicol, G. P., 1979, The metabolism of dihomo-γ-linolenic acid in man. Lipids, 14, 174-180.

Vanier, M-T., 1974, Contribution a l'étude des lipides cérébraux an cours du développement chez le foetus et le jenne enfant. Thesis. Lyon.

DIETARY POLYUNSATURATED FATTY ACIDS IN RELATION TO NEURAL DEVELOPMENT

IN HUMANS

Manuela Martinez

Autonomous University of Barcelona, Hospital Infantil

Vall d'Hebron, 08035 Barcelona, Spain

INTRODUCTION

Among polyunsaturated fatty acids (PUFA), docosahexaenoic acid (22:6ω3) is conspicuous in neural membranes and the retina. The most abundant fatty acid of the ω6 series is arachidonate (20:4ω6), followed by adrenic (22:4ω6) and docosapentaenoic (22:5ω6) acids. The proportion of these PUFA in neural tissues seems to be genetically determined and is quite constant in the mammalian brain, despite wide species differences in other organs[1]. Only with great difficulty does nutritional aggression modify the PUFA composition of the brain. However, during the period of maximum growth speed – the so-called brain "growth spurt"[2] – the brain can become vulnerable to nutritional insults, and this has been repeatedly proved in the experimental animal fed on diets very poor or unbalanced in PUFA[3-8]. In contrast, there is little information concerning the influence of dietary PUFA on the developing human brain.

For many years, the content of linoleic acid (18:2ω6) has been the sole concern when considering the supply of essential fatty acids (EFA) to the developing human, and even the essentiality of α-linolenic acid (18:3ω3) has been questioned[9]. This has led to the commercial production of formulas very unbalanced in ω3/ω6 fatty acids, for oral as well as for parenteral administration to the human infant. However, several experimental studies have lately demonstrated that such unbalanced diets, some of them with ω3/ω6 ratios as low as 1:200 or even lower, can deplete the tissues of ω3 fatty acids[10-12], and there are some indications that a ω3 deficiency can also be produced in the human infant[13,14] and adult[15].

Before speaking of nutritional influences on the developing human, it is necessary to know how normal development proceeds, since the biochemical composition of all organs changes with maturation. In the case of fatty acids, it is worth tracing the most important developmental profiles underlying the period of brain "growth spurt", in an attempt to predict the moment of maximum vulnerability for the most important PUFA of the brain, as has been done for other lipids[16,17]. In quantitative terms, the accretion of PUFA is known to be very rapid during the third trimester of human gestation[18], and the proportion of 22:6ω3 increases steadily with maturation in the ethanolamine phosphoglycerides (EPG) of the brain[19]. By studying a large enough number of cases with different gestational ages, it is possible to trace the exact profiles of the two main PUFA, 22:6ω3 and 20:4ω6, in brain EPG during perinatal life. These two fatty acids are produced by successive desaturation and elongation reactions of the two parent fatty acids, 18:3ω3 and 18:2ω6, respectively. In placental mammals, the process of PUFA desaturation and elongation is progressively "magnified" from maternal blood to placenta, fetal liver and fetal brain, so that the latter has the longest PUFA with the highest degree of unsaturation[20]. It seems, however, that the desaturase system is not very active in the human species[21]. Thus, it is important to know the PUFA patterns in the two last steps of this "magnifying" desaturation/elongation chain, trying to correlate liver and brain polyunsaturated fatty acids during perinatal life.

A tissue of special interest within the central nervous system is the retina, for its very high content of 22:6ω3. It seems that docosahexaenoate plays an important role in the visual process[22], and that visual impairment can be produced by ω3 deprivation[12,13]. Therefore, it is also necessary to know the developmental profiles of the main PUFA in retinal tissue, in order to delimit the normal values during maturation.

NORMAL DEVELOPMENTAL PROFILES OF DOCOSAHEXAENOIC AND ARACHIDONIC ACIDS IN THE HUMAN LIVER AND BRAIN

In quantitative terms, ethanolamine phosphoglycerides are the most important unsaturated phospholipids of cell membranes, and in normal brain they contain more than 50% of PUFA. It is, therefore, necessary to have a look at the maturational changes of 22:6ω3 and 20:4ω6 in EPG of the developing human liver and brain[23].

As can be seen in Fig. 1a, the developmental profile of 22:6ω3 in liver EPG is curvilinear, with a clear increase after 30 weeks of gestational age. Arachidonic acid (Fig. 1b), on the other hand, shows a descending parabolic profile in liver EPG, very abrupt after 30 weeks of gestation.

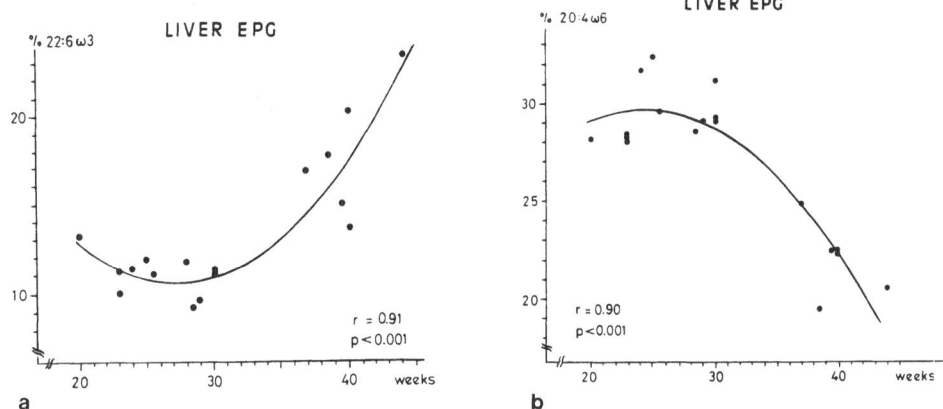

Fig. 1. Docosahexaenoate (a) and arachidonate (b) proportions in ethanolamine phosphoglycerides of the developing human liver.

When looking at these two PUFA in forebrain EPG (Fig. 2), a similar trend can be discerned, docosahexaenoate ascending quadratically towards the end of gestation (Fig. 2a) and 20:4ω6 descending almost linearly during this time (Fig. 2b).

The opposite behaviour of these two fatty acids can be emphasized by plotting one against the other in liver and brain EPG (Fig. 3). This shows a curvilinear negative correlation between the proportions of 20:4ω6 and 22:6ω3 in EPG in the liver (Fig. 3a). In the brain (Fig. 3b), these two fatty acids are only negatively correlated until about 19-20% of 20:4ω6.

DEVELOPMENTAL CHANGES OF RETINAL POLYUNSATURATED FATTY ACIDS

When looking at the developmental profiles of 22:6ω3 and 20:4ω6 in EPG of the human retina[24] (Fig. 4), it can be seen that they are quite similar to those in liver and brain EPG, 20:4ω6 decreasing and 22:6ω3

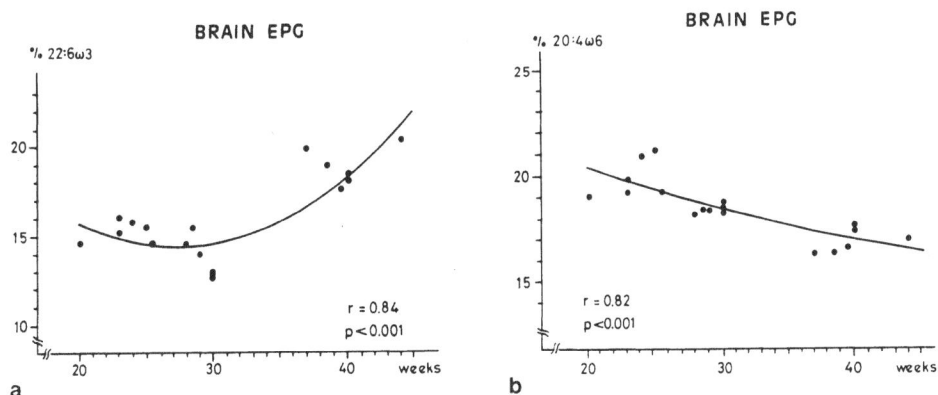

Fig. 2. Docosahexaenoate (a) and arachidonate (b) proportions in ethanolamine phosphoglycerides of the developing human brain.

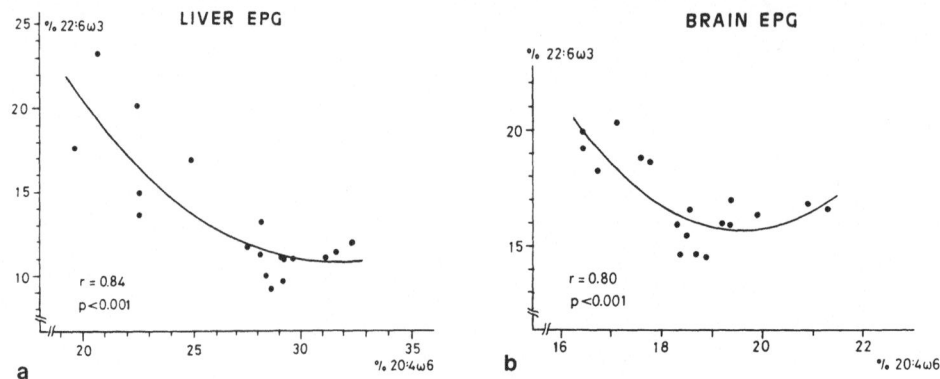

Fig. 3. Docosahexaenoate versus arachidonate percentages
in the developing human liver (a) and brain (b).

increasing in a curvilinear manner during the second half of gestation.
However, when comparing the retinal profile of 22:6ω3 (Fig. 4a) with that
of the brain (Fig. 2a), the shapes of the curves are quite different. In
the brain, the increase in 22:6ω3 is only apparent after 30–31 weeks of
gestation[25], whereas the increase in retinal docosahexaenoate is faster
before 35 weeks of gestational age and seems to slow down towards the end
of gestation. Although the absolute values cannot be directly compared
because they belong to different cases and were obtained by different methods
(see ref. 23 and 24), it is clear that the proportion of 22:6ω3 in retinal
EPG is higher than in brain EPG (as it should be, due to the high content of
22:6ω3 in photoreceptor cells), especially if considering that the values
of the retina represent molar percentages whereas in the case of the brain
they are weight per cent values. Arachidonic acid decreases with maturation
in retinal EPG in a very similar way to that in the brain.

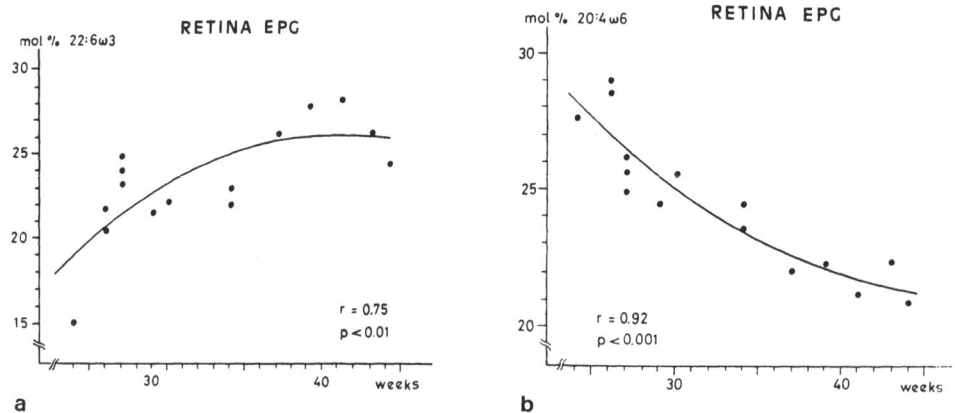

Fig. 4. Molar per cent values of docosahexaenoic (a) and arachidonic (b)
acids in ethanolamine phosphoglycerides of the developing human retina.

126

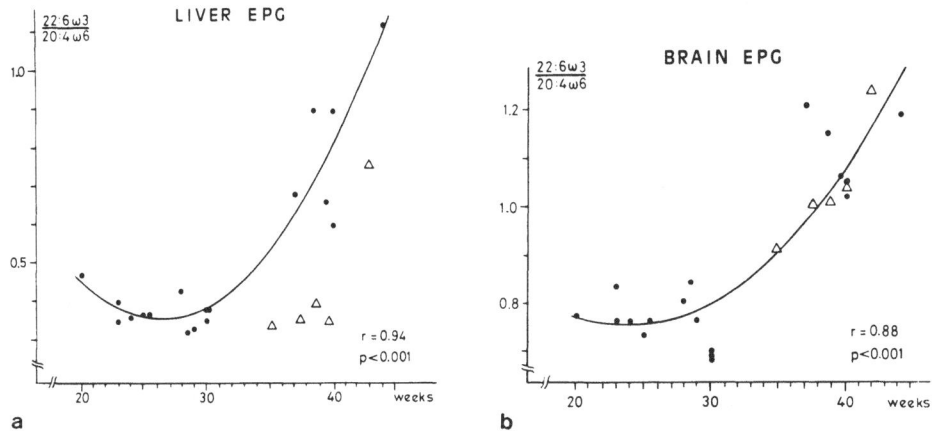

Fig. 5. Ratio docosahexaenoate/arachidonate in ethanolamine phosphoglycerides of the developing human liver (a) and brain (b). ● = controls; Δ = infants receiving TPN with Intralipid.

EFFECTS OF UNBALANCED OMEGA-3/OMEGA-6 DIETS ON THE DEVELOPING HUMAN LIVER, BRAIN AND RETINA

In view of the opposite profiles of 22:6ω3 and 20:4ω6 in the human liver, brain and retina, the ratio 22:6ω3/20:4ω6 can be used in order to emphasize the developmental changes of these PUFA in all three tissues. By so doing, the curvilinear profiles are more abrupt and significant, and any deviation from the normal can be more clearly detected.

When studying the effects of total parenteral nutrition (TPN) with high doses of linoleate (Intralipid), administered to 5 neonates for a short period of 4-12 days, some important alterations could be found in the liver PUFA of these children[23]. As can be seen in Fig. 5a, all the infants receiving TPN had a 22:6ω3/20:4ω6 ratio clearly below the normal profile in liver EPG. In the brain (Fig. 5b), however, the 22:6ω3/20:4ω6 ratio was not altered, probably due to the short duration of the treatment, although the effect of TPN was beginning to be reflected in the brain by a clear increase in the proportion of 18:2ω6 in cerebral choline phosphoglycerides (CPG)[23]. Despite the great increase in 18:2ω6 and the reduction in 22:6ω3, the proportion of 20:4ω6 was not increased in liver EPG (Fig. 6), and it was even clearly decreased in liver CPG[23].

Fig. 8 shows the 22:6ω3/20:4ω6 ratio in human retinal EPG throughout development in 15 neurologically normal, well-nourished infants. An adult case is included just to show that the 22:6ω3/20:4ω6 ratio goes on increasing after infancy. A premature infant undernourished in utero and two postnatally malnourished children are also included in this graph. One was a 9-month-old infant with mucoviscidosis, who died from a hepatic complication, in a state

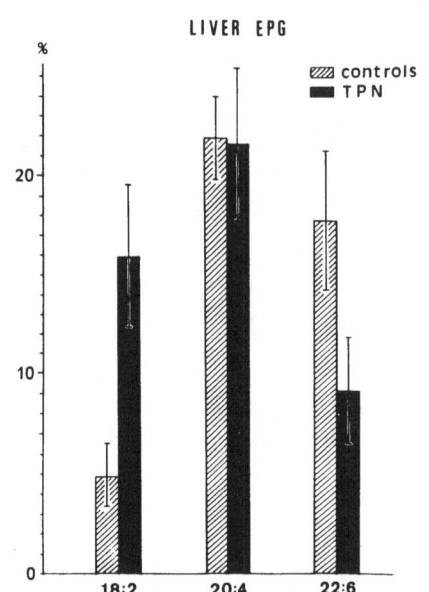

Fig. 6. Weight% of linoleic, arachidonic and docosahexaenoic acids
in liver EPG of controls and TPN.
All bars are mean ± SD (n = 5).

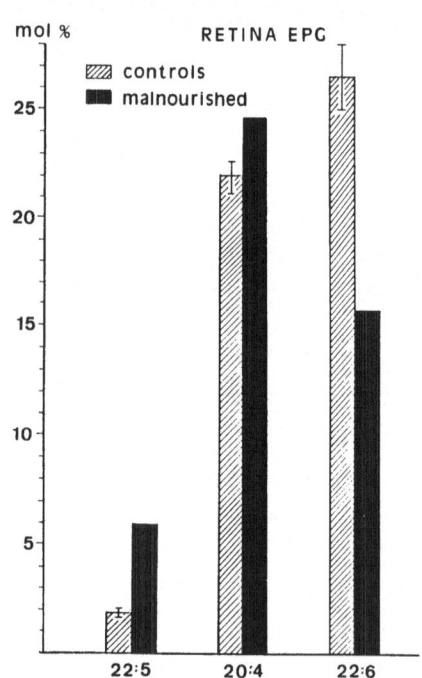

Fig. 7. Mol% of 22:5ω6, 20:4ω6 and
22:6ω3 in retinal EPG of 5 controls
(mean ± SEM) and of a child receiving
milks with very low ω3/ω6 ratios.

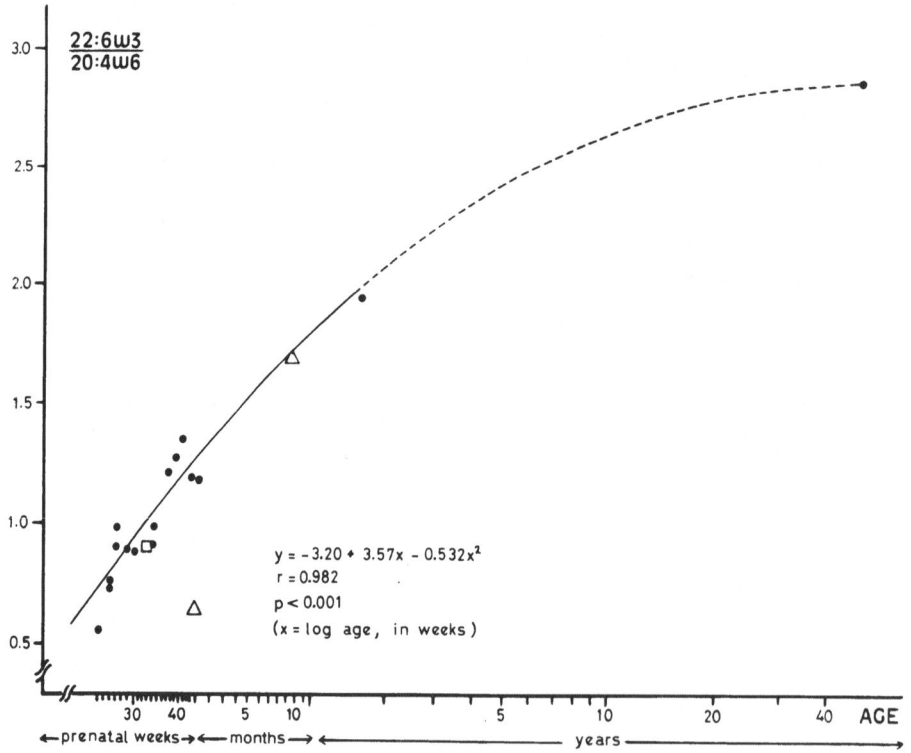

$$y = -3.20 + 3.57x - 0.532x^2$$
$$r = 0.982$$
$$p < 0.001$$
(x = log age, in weeks)

Fig. 8. Docosahexaenoate/arachidonate ratio in retinal ethanolamine
phosphoglycerides during maturation. ● = controls; Δ = postnatally
malnourished (see text); □ = prenatally undernourished (small-for-dates).

of severe marasmus, due to the continual loss of nutrients (body weight,
5,290 g). This child had been born at term (40 weeks of gestational age),
with a body weight of 2,780 g, and had been fed since birth on maternal milk,
and later on a milk with a ω3/ω6 ratio of 1:3.6. The other was a premature
infant, born at 25 weeks of gestation, with a body weight appropriate for
gestational age (880 g). This infant was fed on artificial milk formulas
with ω3/ω6 ratios ranging from 1:18 to 1:66 since birth, until she died
from acute viral pneumonia at 4 months of age (corresponding to a postconceptional age of 44 weeks), with a body weight of 2,460 g. This child had a
docosahexaenoate content of about half the normal value in her retinal EPG
(Fig. 7), and the $22:6\omega3/20:4\omega6$ ratio was greatly reduced, as can be seen
in the graph (Fig. 8). The decrease in the levels of $22:6\omega3$ was not only in
the proportion of this fatty acid in retinal EPG but also in absolute terms.
As Table 1 shows, this 4-month-old child (about 1 month old if considering
postconceptional age) had in her retina only a little more than half the total
$22:6\omega3$ content of a normal neonate. Correspondingly, $22:5\omega6$ was very much
increased in absolute terms, and even more in EPG (fig. 7) The retinal content
of $20:4\omega6$ and $20:3\omega9$ were normal, due to the high linoleate supply. On
the other hand, when looking at the total amount of these two fatty acids
in the retina of the undernourished child with malabsorption problems (Table
1), both $20:3\omega9$ and $22:5\omega6$ were greatly increased, as corresponding to
a loss of essential fatty acids of both $\omega6$ and $\omega3$ series. However, the

Table 1. Main total fatty acids in the developing human retina
of well-nourished and two malnourished children.

	MN	Controls		UN
	4 mo	37-44 wk (n = 5)	18 mo	9 mo
16:0	10533	11689 ± 523	8931	10082
16:1ω7	293	490 ± 72	300	541
18:0	9856	10016 ± 247	9646	11085
18:1ω9	6282	6034 ± 306	5069	6100
18:2ω6	1169	656 ± 129	480	1163
20:3ω9	65	53 ± 11	24	133
20:3ω6	858	562 ± 41	612	919
20:4ω6	5807	5693 ± 352	3970	4772
22:4ω6	1613	1285 ± 103	748	1082
22:5ω6	1142	382 ± 30	184	1276
22:5ω3	206	252 ± 17	489	369
22:6ω3	3655	5548 ± 233	6277	6141

Data are nmol/g wet tissue (mean ± SEM for grouped values). MN = malnourished
by unbalaced diets; UN = undernourished by intestinal losses.

retinal content of 22:6ω3 was totally normal, especially if considering
that the control was somewhat older than the patient. Fig. 8 shows that the
22;6ω3/20:4ω6 ratio was entirely normal in retinal EPG of this child.

CONCLUSIONS

It can be concluded that a diet unbalanced in ω3/ω6 fatty acids can
be quite damaging to the PUFA composition of the developing human central
nervous system. The question of why nutritionists have for so long neglected
the supply of fatty acids of the linolenate family is intriguing. Probably,
α-linolenic acid is so widely distributed in nature, the minimal requirements
are so low and the enzyme systems are so biased in favour of ω3 fatty acids
that it is very difficult, if at all possible, to find a case of ω3 deficiency
spontaneously produced, even in severe undernutrition. This point is very
well illustrated by the child undernourished for intestinal losses who, nevertheless, had a normal level of docosahexaenoic acid in her retina.

However, what occurs with great difficulty in nature can be produced
relatively easily by providing a large excess of ω6 fatty acids in the diet,
especially if this excess is accompanied by a very low intake of ω3 fatty
acids and is given during the vulnerable period of development. This situation is exemplified by the child born very immature and nourished on a diet
extremely poor in ω3 fatty acids and very rich in ω6 fatty acids, in whom
even the retina had been depleted of 22:6ω3.

When given intravenously, it seems that a large excess of 18:2ω6 can
be very damaging, even with a high intake of 18:3ω3. Since Δ6 desaturase
seems not to be very active in the human species, an excessive supply of
precursors, not accompanied by the longer PUFA, can probably be very dangerous. A possible mechanism of substrate inhibition[26], added to the well-known
enzyme competition between families of fatty acids[27,28], cannot be ruled out
in these cases.

In view of the findings reported, it seems advisable to use formulas
of a more physiological nature for the nutrition of infants, especially
prematures. There is no reason why any formula should have as much as 50%
of linoleic acid and/or a ω3/ω6 ratio much higher than 1:4. On the other
hand, special attention should be paid to the presence of long PUFA, mainly
22:6ω3, in all formulas devised for infant nutrition. However, while we do
not have conclusive data on the advisability of adding substantial amounts
of these very unsaturated fatty acids, it is better to be on the safe side
and try to imitate nature as closely as possible. An excessive use of fish
oil may probably be dangerous in pediatrics, due to its extremely high
content of 22:6ω3 and, above all, 20:5ω3, a fatty acid potentially harmful

when given in excess for its competitive role in eicosanoid formation. The fact that there is very little $20:5\omega3$ in human milk and only trace amounts in cell phospholipids should serve as a warning against adding too much of this fatty acid to infant formulas. Perhaps, the most physiological means of enriching an infant nutrient in long PUFA would be to use balanced mixtures of unsaturated fatty acids, such as present in brain phospholipids of the kind of phosphatidyl ethanolamine or serine. In any case, whatever the source of fatty acids chosen, it would be advisable not to provide any excess of precursors, $18:2\omega6$ and $18:3\omega3$, and only a moderate, balanced amount of long polyunsaturated fatty acids, in a proportion not far from that found in the normal composition of human tissues.

REFERENCES

1. Crawford, M.A., Casperd, N.M. and Sinclair, A.J., The long chain metabolites of linoleic and linolenic acids in liver and brain in herbivores and carnivores, Comp. Biochem. Physiol. 1976; 54B, 395-401.

2. Dobbing, J., Vulnerable periods in developing brain. In A.N. Davison and J. Dobbing (Eds.), Applied Neurochemistry, Blackwell, Oxford, 1968: 287-316.

3. Mohrhauer, H. and Holman, R.T., Alteration of the fatty acid composition of brain lipids by varying levels of dietary essential fatty acids, J. Neurochem. 1963; 10: 523-530.

4. Galli, C., Trzeciak, H.I. and Paoletti, R., Effects of dietary fatty acids on the fatty acid composition of brain ethanolamine phosphoglyceride: Reciprocal replacement of n-6 and n-3 polyunsaturated fatty acids, Biochim. Biophys. Acta, 1971; 248: 449-454.

5. Svennerholm, L., Alling, C., Bruce, Å, Karlsson, I. and Sapia, O., Effects on offspring of maternal malnutrition in the rat. In Lipids, Malnutrition and the Developing Brain, A Ciba Fndn. Symposium, Elsevier, Amsterdam, 1972: 141-157.

6. Sinclair, A.J. and Crawford, M.A., The effect of a low-fat diet on neonatal rats, Br. J. Nutr., 1973; 29: 127-137.

7. Sun, G.Y., Go, J. and Sun, A.Y., Induction of essential fatty acid deficiency in mouse brain: effects of fat deficient diet upon acyl group composition of myelin and synaptosome-rich fractions during development and maturation, Lipids, 1974; 9: 450-454.

8. Karlsson, I., Effects of different dietary levels of essential fatty acids on the fatty acid composition of ethanolamine phosphoglycerides in myelin and synaptosomal plasma membranes, J. Neurochem., 1975;25: 101-107.

9. Tinoco, J., Williams, M.A., Hincenbergs, I. and Lyman, R.L., Evidence for nonessentiality of linolenic acid in the diet of the rat, J. Nutr., 1971; 101:937–946.

10. Sanders, T.A.B., Mistry, M. and Naismith, D.J., The influence of a maternal diet rich in linoleic acid on brain and retinal docosahexaenoic acid in the rat, Br. J. Nutr., 1984; 51: 57–66.

11. Bourre, J.M., Pascal, G., Durand, G., Masson, M., Dumont, O. and Piciotti, M., Alterations in the fatty acid composition of rat brain cells (neurons, astrocytes, and oligodendrocytes) and of subcellular fractions (myelin and synaptosomes) induced by a diet devoid of n–3 fatty acids, J. Neurochem., 1984; 43: 342–348.

12. Neuringer, M., Connor, W.E., Lin, D.S., Barstad, L. and Luck, S., Biochemical and functional effects of prenatal and postnatal ω3 fatty acid deficiency on retina and brain in rhesus monkeys, Proc. Natl. Acad. Sci. USA, 1986; 83: 4021–4025.

13. Holman, R.T., Johnson, S.B. and Hatch, T.F., A case of human linolenic acid deficiency involving neurological abnormalities, Am. J. Clin. Nutr., 1982; 35: 617–623.

14. McClead, Jr., R.E., Meng, H.C., Gregory, S.A., Budde, C. and Sloan, H.R., Comparison of the clinical and biochemical effect of increased α–linolenic acid in a safflower oil intravenous fat emulsion, J. Pediatr. Gastroenterol. Nutr., 1985; 4: 234–239.

15. Bjerve, K.S., Mostad, I.L. and Thoresen, L., Alpha–linolenic acid deficiency in patients on long–term gastric–tube feeding: estimation of linolenic acid and long–chain unsaturated n–3 fatty acid requirement in man, Am. J. Clin. Nutr. 1987; 45: 66–77.

16. Martinez, M. and Ballabriga, A., A chemical study on the development of the human forebrain and cerebellum during the brain "growth spurt" period. I. Gangliosides and plasmalogens, Brain Res. 1978; 159:351–362.

17. Martinez, M., Myelin lipids in the developing cerebrum, cerebellum and brain stem of normal and undernourished children, J. Neurochem. 1982; 39: 1684–1692.

18. Clandinin, M.T., Chappell, J.E., Leong, S., Heim, T., Swyer, P.R. and Chance, G.W., Intrauterine fatty acid accretion rates in human brain: implications for fatty acid requirements, Early Hum. Dev. 1980; 4: 121–129.

19. Svennerholm, L., Distribution and fatty acid composition of phosphoglycerides in normal human brain, J. Lipid Res. 1968; 9: 570–579.

20. Crawford, M.A., Hassam, A.G., Williams, G. and Whitehouse, W.L., Essential fatty acids and fetal brain growth, Lancet 1976; i: 452–453.

21. Carlson, S.E., Rhodes, P.G., Rao, V.S. and Goldgar, D.E., Effect of fish oil supplementation on the n-3 fatty acid content of red blood cell membranes in preterm infants, Pediatr. Res. 1987; 21: 507–510.

22. Anderson, R.E., Benolken, R.M., Dudley, P.A., Landis, D.J. and Wheeler, T.G., Polyunsaturated fatty acids of photoreceptor membranes, Exp. Eye Res., 1974; 18: 204–213.

23. Martinez, M. and Ballabriga, A., Effects of parenteral nutrition with high doses of linoleate on the developing human liver and brain, Lipids 1987; 22: 133–138.

24. Martinez, M., Ballabriga, A. and Gil-Gibernau, J.J., Lipids of the developing human retina. I. Total fatty acids, plasmalogens and fatty acid composition of ethanolamine and choline phosphoglycerides, J. Neurosci. Res. 1988 (in press).

25. Martinez, M., Conde, C. and Ballabriga, A., Some chemical aspects of human brain development. II. Phosphoglyceride fatty acids, Pediatr. Res. 1974; 8: 93–102.

26. Brenner, R.R., The oxidative desaturation of unsaturated fatty acids in animals, Mol. Cell. Biochem. 1974; 3: 41–52.

27. Holman, R.T., Nutritional and metabolic interrelationships between fatty acids, Fed. Proc. 1964; 23: 1062–1067.

28. Brenner, R.R. and Peluffo, R.O., Effect of saturated and unsaturated fatty acids on the desaturation in vitro of palmitic, stearic, oleic, linoleic and linolenic acids, J. Biol. Chem. 1966; 241: 5213–5219.

SOURCES OF ω3 FATTY ACIDS IN ARCTIC DIETS AND THEIR EFFECTS

ON RED CELL AND BREAST MILK FATTY ACIDS IN CANADIAN INUIT

Sheila M. Innis

Department of Paediatrics
Faculty of Medicine
University of British Columbia
950 West 28th Avenue
Vancouver, B.C., Canada V5Z 4H4

INTRODUCTION

Considerable interest has been given to the relationship of dietary ω3 fatty acids, particularly eicosapentaenoic acid (20:5ω3), to human disease since the publication of epidaemiological data demonstrating a relationship in Greenland Eskimo between a diet containing large amounts of marine lipid and a low incidence of ischaemic heart disease[1-5]. The ensuing flurry of clinical and research studies have focused almost exclusively on diets high in fish or supplemented with fish oil concentrates, and the biochemistry of 20:5ω3, particularly as it relates to the ω6 fatty acid arachidonic acid (20:4ω6) and eicosanoid metabolism. These studies have shown that incorporation of dietary ω3 fatty acids, largely at the expense of ω6 fatty acids, into membrane structural lipid results in altered membrane physical properties and membrane-dependent biochemical and physiological processes[6]. Substantial knowledge on the function of 20:4ω6 and 20:5ω3 as substrates for eicosanoid synthesis has been gained, although much remains to be learnt on the specific relationship between diet and the tissue fatty acids pools utilized for eicosanoid production in vivo[7]. The epidaemiological association between a high marine lipid intake and a low prevalence of ischaemic heart disease also directed attention to the possible effects of ω3 fatty acids on cholesterol and triacylglycerol metabolism in both normo- and hyper-lipidaemia[8]. Despite considerable research, the efficacy of dietary fish in reducing disease incidence remains uncertain. Both a reduction[9,10] and no reduction[11,12] in ischaemic heart disease has been described in free-living populations with a high dietary fish intake.

Several potentially important points have been largely overlooked in the study of dietary 3 fatty acids. The original epidaemiological work in the Greenland Eskimo did not implicate fish as a major source of dietary lipid[1-5]; rather, these Eskimo obtained the majority of their lipid calories from marine mammal flesh[2]. Although marine mammal lipid contains high quantities of ω3 fatty acids, the relative distribution of the major components, $20:5\omega3$, docosapentaenoic acid ($22:5\omega3$) and docosahexaenoic acid ($22:6\omega3$) is different from that in fish lipid[13-19]. Further, the positional esterification of $20:5 3$, $22:5\omega3$ and $22:6\omega3$ differs between fish and marine mammal lipid; $\omega3$ fatty acids are primarily esterified to glycerol sn1,3 in mammals[20] but to glycerol sn2 in fish[21]. Ackman,[22] has recently drawn attention to this, and speculated that resultant differences in absorption, re-esterification and metabolic routing may lead to a greater incorporation of $\omega3$ fatty acids into hepatic phospholipids following their consumption as fish.

In September 1985 a dietary study with collection of blood samples was undertaken in the Canadian Arctic community of Broughton Island, Northwest Territories. This island lies off the Eastern coast of Baffin Island and with a population of approx 400, largely Inuit, residents was selected for study because of evidence that the diet continues to be based on native foods, of an abundant availability of marine foods, and the high cost of air transportation which remains as a barrier to the use of non-native foods. The composition of red blood membrane fatty acids is known to reflect the composition of the diet fat,[21,22] and may, through their effect on membrane properties, alter red cell physiology[23-25]. Thus, this research undertook to document the red cell membrane fatty acid composition of one of the few remaining free-living populations who continue to practice a diet high in marine mammal lipid and who are potentially similar to the Eskimo in whom the link between dietary $\omega3$ fatty acids and ischaemic heart disease was made[1-5]. In addition, the influence of the Inuit diet on breast milk lipids was studied in mature milk from 5 lactating women. Samples of the hunter's catch, caught during the survey period was also analyzed. Detailed reports have been published[16,28].

RESULTS

Inuit diet

A dietary survey was conducted in the native language (Inuktitut) to establish patterns of traditional food use and quantitative food intake for each household. The per capita intake of the commonly consumed native animal foods for 325 Inuit who participated, and for the adult men is summarized in Tables 1 and 2, respectively. The survey confirmed that the major portion of marine lipid in the Inuit diet was from marine mammal flesh (164g/person/day) rather than fish (13g/person/day). This predominance of marine mammals as the major source diet fat and $\omega3$ fatty acids is similar to the Eskimo studied by Bang and colleagues[1-5] for whom an average number of 5.7 and 6.4 meals/wk containing whale or seal, respectively, and only 1.4 meals/wk containing fish was recorded[2]. Of historical interest, the diet of the Broughton Island Inuit retains much of the traditional pattern described in the 1953 publication by Sinclair,[29] in which he calculated that Greenland Eskimo in 1885 consumed 103 g fat/day from seal and 30 g fat/day from fish.

TABLE 1. Average daily consumption and major dietary sources of meat and fish for Inuit

	g/person/day (mean values for 312 Inuit)
Marine mammals:	
Ringed seal meat	80
Narwhal meat	29
Walrus meat	12
Bearded seal meat	6
Beluga meat	<1
Narwhal matak	15
Walrus matak	2
Ringed seal blubber	8
Narwhal blubber	3
Walrus blubber	3
Bearded seal liver	1
Ringed seal liver	3
Bearded seal, other	2
Ringed seal, other	<1
Total	164
Fish:	
Char	13
Caribou	31

TABLE 2. Average daily consumption of meat and fish by adult male Inuit.

	g/person/day
Marine mammal flesh, total	307
Blubber and oils	7
Fish	26
Caribou	68

The ω3 fatty acid composition of the marine mammals (narwhal, seal and walrus) arctic char, polar bear and caribou caught by the Inuit hunter's during the survey period (Table 3) illustrates the substantial quantities of ω3 fatty acids present in marine mammal and polar bear fat. When compared to arctic char, and previos data on fish fatty acids,[15,17] the proportion of 22:5ω3 is much higher in the mammals than in fish. Whether this is due to further elongation and Δ4 desaturation of 20:5ω3, retroconversion of 22:6ω3 or preferential acylation and retention of 22:5ω3 consumed by these mammals is not known. The presence of large amounts of ω3 fatty acids in the polar bear, but low amount in caribou, appears to reflect the dietary habits of the two species. The polar bear feeds largely on young seals with smaller amounts of shoreline carrion, whilst caribou graze on mosses and lichens[30].

TABLE 3. Content of major ω3 fatty acids in lipid of meat and fish caught by the Inuit

Fatty acids (% total)	Caribou	Polar bear	Narwhal	Mammals Ringed seal	Bearded seal	Walrus	Fish Arctic char
20:5ω3	0.2	2.9	2.2 ± 1.2	9.8 ± 0.6	9.0 ± 1.1	12.2 ± 1.3	9.8 ± 1.0
22:5ω3	0.1	7.5	1.0 ± 0.4	5.9 ± 0.3	5.0 ± 0.3	5.8 ± 0.2	3.1 ± 0.2
22:6ω3	0.2	8.4	2.9 ± 0.3	10.0 ± 0.2	10.2 ± 0.8	9.0 ± 0.6	20.7 ± 1.4

Inuit red cell fatty acids

The fatty acid composition of phosphatidylcholine and phosphatidylethanolamine of red blood cell membranes of 185 Inuit, aged 2mth-85 yr and for reference, of 24 men and women, 21-50 yr, living in Vancouver and consuming a typical North American diet was analyzed. No differences were found due to sex in either the Inuit or Vancouver samples. A summary of the major ω6 and ω3 fatty acids for Inuit children and adult males, and Vancouver males is given in Tables 4 and 5. The highest levels of ω3 fatty acids, and greatest difference between Inuit and Vancouver residents was found in phosphatidylethanolamine. The ω3 fatty acids in the Inuit was higher in the adult over 20 yrs than in the children and adolescents. This may reflect a lower consumption marine foods by the children during their school yrs at Frobisher Bay, rather than progressive or developmental accretion. Dietary information to substantiate this, however, is not yet available.

Compared to Canadians in Vancouver, the Inuit phosphatidylethanolamine had less 20:4, 22:4 and 22:5ω6 and more 18:3, 20:5 and 22:6ω3. Their phosphatidylcholine had similar levels of 18:2 and 20:4, but higher 22:5ω6, and higher 18:3, 20:5, and 22:6ω3. It is particularly interesting that, despite the marked difference in % distribution of individual ω6 and ω3 fatty acids, and in the total ω6 versus ω3 series, the Inuit red cell phospholipids had a similar unsaturation index and mean chain length to the Vancouver red cells. This is of potential significance since, in contrast to these Inuit, short term clinical studies with fish oils have found the increase in red cell phospholipid ω3 fatty acids, largely at the expense of 18:2ω6 and 20:4ω6, (particularly in phosphatidylcholine), is accompanied by an increased unsaturation index. The latter was associated with altered red cell deformability and blood viscosity[25-27]. Thus, the changes in red cell fatty acids and membrane fluidity present in short-term clinical trials with fish appear different from those of a life-time diet high in marine mammal lipid. Whether or not the apparent maintenance of red cell unsaturation in the Inuit is due to long term adaptation to the diet in order to maintain normal membrane physical properties, or if it can be explained by the specific ω3 fatty acid composition and triglyceride structure of marine mammal lipid is not known.

Inuit breast milk fatty acids

The Inuit breast milk contained a similar amount of total lipid to that secreted by other Canadian or North American women. The levels of 20:5ω3 and 22:6ω3, however, were approx 5.5 and 3.5 fold higher, respectively (Table 6). It is noteworthy that, as in the red cell, the increase in 3 fatty acids was not at the expense of 20:4ω6.

TABLE 4. Major ω6 and ω3 fatty acids in red cell phosphatidylethanolamine.

| Fatty acids (% total) | Inuit | | | | | | Vancouver Males |
| | Children | | | | Males | | |
	0-5 yr (13)	6-10 yr (9)	11-15 yr (11)	16-20 yr (9)	21-50 yr (41)	>50 yr (14)	21-50 yr (12)
ω6 series							
18:2	8.0 ± 1.9	8.0 ± 1.2	8.3 ± 0.9	8.2 ± 1.3	6.8 ± 0.4	5.3 ± 0.6	6.8 ± 0.4
20:4	12.7 ± 1.1	15.0 ± 2.8	17.5 ± 1.4	19.2 ± 1.1	15.1 ± 1.0⁺	12.0 ± 0.9	25.7 ± 0.6
22:4	2.2 ± 0.6	3.4 ± 1.4	2.8 ± 0.2	3.8 ± 0.1	1.9 ± 0.2⁺	1.3 ± 0.1	7.0 ± 0.7
22:5	0.4 ± 0.1	0.4 ± 0.2	0.4 ± 0.0	0.6 ± 0.2	0.4 ± 0.1⁺	0.2 ± 0.1	0.9 ± 0.0
ω3 series							
18:3	1.2 ± 0.9	0.8 ± 0.5	0.9 ± 0.3	1.0 ± 0.6	1.4 ± 0.2⁺	1.3 ± 0.3	0.3 ± 0.1
20:5	2.2 ± 0.6	2.0 ± 0.6	3.6 ± 1.0	3.6 ± 1.6	7.1 ± 0.6⁺	9.6 ± 1.2	1.6 ± 0.1
22:5	3.1 ± 0.8	3.2 ± 0.5	5.0 ± 0.4	4.5 ± 0.9	5.2 ± 0.2	5.4 ± 0.5	5.5 ± 0.6
22:6	5.7 ± 1.7	5.7 ± 1.0	9.0 ± 0.6	7.9 ± 2.2	11.3 ± 1.7	10.0 ± 1.0	7.4 ± 0.2
ΣC≥20ω6	19.7 ± 4.2	21.4 ± 5.0	22.3 ± 1.3	24.4 ± 0.9	18.6 ± 1.3⁺	14.6 ± 0.8⁺	36.9 ± 1.0
ΣC≥20ω3	11.1 ± 3.0	11.0 ± 2.0	18.2 ± 1.8	16.0 ± 4.7	23.6 ± 2.1⁺	24.9 ± 2.6⁺	14.8 ± 0.3
X chain Lth	18.5 ± 0.4	18.6 ± 0.3	18.9 ± 0.0	18.9 ± 0.1	18.9 ± 0.1	18.8 ± 0.1	19.3 ± 0.1
U.I.	182 ± 30	189 ± 22	232 ± 8	232 ± 12	214 ± 10	237 ± 14	250 ± 4

Data are means ± SE for Inuit male and female children, and Inuit males over 11 yr, by age (number of Inuit). ⁺indicates value for Inuit male different from Vancouver males, p < 0.01. X Ch. length, U.I. represents mean length and unsaturation index, respectively.

TABLE 5. Major ω6 and ω3 fatty acids in red cell phosphatidylcholine.

Fatty acids (% total)	Inuit						Vancouver
	Children		Males				Males
	0-5 yr (13)	6-10 yr (9)	11-15 yr (11)	16-20 yr (9)	21-50 yr (41)	>50 yr (14)	21-50 yr (12)
ω6 series							
18:2	13.9 ± 3.4	17.1 ± 3.6	16.6 ± 0.8	13.8 ± 3.8	13.9 ± 0.8	10.8 ± 1.8	15.7 ± 3.0
20:4	9.6 ± 2.9	6.9 ± 1.3	8.8 ± 2.1	12.7 ± 2.6	9.6 ± 1.0	7.4 ± 1.3	9.2 ± 1.4
22:4	1.2 ± 0.0	1.7 ± 0.7	0.8 ± 0.2	1.1 ± 0.4	1.0 ± 0.2	1.3 ± 0.6	1.0 ± 0.6
22:5	0.4 ± 0.2	0.2 ± 0.0	0.3 ± 0.1	0.6 ± 0.2	0.4 ± 0.0[+]	0.2 ± 0.1	0.1 ± 0.0
ω3 series							
18:3	1.3 ± 0.4	1.1 ± 0.2	0.8 ± 0.2	1.0 ± 0.1	0.9 ± 0.2[+]	0.8 ± 0.0	0.0 ± 0.0
22:5	0.4 ± 0.0	0.6 ± 0.0	1.2 ± 0.4	1.2 ± 0.3	1.4 ± 0.2	2.1 ± 0.5	1.7 ± 0.3
22:6	2.5 ± 1.0	1.2 ± 0.3	2.3 ± 0.3	3.5 ± 0.8	3.9 ± 0.4[+]	3.1 ± 0.6	1.6 ± 0.4
ΣC≥20 ω6	11.9 ± 2.9	10.4 ± 0.6	12.4 ± 2.4	17.0 ± 3.3	12.7 ± 1.2	10.4 ± 1.9	12.7 ± 2.6
ΣC≥20 ω3	3.4 ± 1.0	4.2 ± 0.6	4.9 ± 0.8	5.8 ± 1.1	8.0 ± 0.7[+]	8.0 ± 1.2	4.4 ± 0.6
X Chain Lth	18.0 ± 0.0	17.8 ± 0.1	18.2 ± 0.1	18.3 ± 0.1	18.0 ± 0.1	18.0 ± 0.1	18.1 ± 0.2
U.I.	102 ± 10	118 ± 12	142 ± 8	161 ± 9	141 ± 5	141 ± 5	140 ± 14

Data are means ± SE for Inuit male and female children, and Inuit males over 11 yr, by age (number of Inuit). [+]indicates value for Inuit male different from Vancouver males, $p < 0.01$. X Ch. length, U.I. represents mean length and unsaturation index, respectively.

TABLE 6. Major fatty acid components of breast milk total lipid.

Fatty acid (% total)	Inuit	Vancouver
Saturates	(n=5)	(n=12)
14:0	5.7 ± 1.0	6.7 ± 0.5
16:0	18.0 ± 0.4	22.1 ± 2.7
18:0	7.1 ± 0.5	8.2 ± 0.8
Monoenes		
16:1	5.0 ± 0.8	3.3 ± 0.6
18:1	38.1 ± 2.4	36.3 ± 2.7
ω6 series		
18:2	11.5 ± 0.7	12.7 ± 1.8
20:4	0.6 ± 0.0	0.7 ± 0.0
ω3 series		
18:3	0.5 ± 0.2	0.6 ± 0.2
20:5	1.1 ± 0.3*	0.2 ± 0.2
22:5	0.8 ± 0.2	0.4 ± 0.1
22:6	1.4 ± 0.4*	0.4 ± 0.1

Values are means ± SE, *indicates value significantly different, $p < 0.05$ from corresponding value for breast milk from women in Vancouver.

CONCLUSIONS

This study of Canadian Inuit emphasizes that the traditional dietary habits of Arctic populations, such as this and those described by Bang and colleagues during the 1970s[1-5] is not high in fish. Rather the primary source of protein and fat is marine mammal flesh. Although the tissues of these mammals contains similar high ω3 and low ω6 fatty acids levels,[16] to that of fish[15,17] the proportion of 22:5ω3 is higher in the mammals, and the positional esterification of 20 and 22 ω3 fatty acids on the glycerol backbone of triacylglycerols is differs from fish[20,21]. Clinical and research studies have focused almost exclusively on ω3 fatty acids from fish. Possibly this is due to the ready availability of fish for North American and European populations in which ischaemic heart disease is prevalent and for whom a diet of seal or whale is neither acceptable nor feasible, as well as to frequent citation that the Greenland Eskimo diet was high in fish.

Comparison of plasma phospholipid 18:2 and 20:4ω6 of different "fish" or "non-fish" eating racial groups such as the Eskimo or Canadian West Coast Indians to their countrymen, has given rise to an hypothesis that different patterns of lipid-related diseases between races is partly related to genetic differences in ω6 fatty acid desaturation[30,31]. No evidence to support this was found in the Baffin Island Inuit. Rather, the composition of the Broughton Island

Inuit red cells appears to be readily explained by their high marine mammal consumption. Despite the presence of higher amounts of ω3 fatty acids in the Inuit red cell lipids, the levels of 18:2ω6 and 20:4ω6 in the phosphatidylcholine and the unsaturation index of both phosphatidylcholine and phosphatidylethanolamine was similar to Canadians in Vancouver eating a typical mixed diet.

The level of 20:5 and 22:6ω6 in the Inuit breast milk is comparable the levels found to following administration of 10g/day x 14 days or 47g/day x 8 days fish oil concentrate to lactating North American women[32]. Concern has been raised that carbon 20 and 22 ω3 fatty acids may be essential nutrients for the human newborn[33]. The analyses of the Inuit milk illustrates the dependance of milk ω3 fatty acids on the maternal diet. Hence, if maternal diet and not the mammary gland regulates the quantity of ω3 fatty acids secreted in the milk, it must be concluded that definitive information on infant ω3 fatty acid requirements cannot be extrapolated from knowledge of breast milk compositions.

RECOMMENDATIONS FOR FUTURE RESEARCH

Much confusion over the effects of dietary ω3 fatty acids is likely to continue until research is undertaken to clarify the importance of ω3 fatty acid source, triacylglycerol structure and effect of specific ω3 fatty acids and unesterified ω3 fatty acid preparations, on tissue lipid composition and metabolic processes. Researchers should be cognisant of the potential limitations of comparing or extrapolating data from studies with ω3 fatty acids from different sources until this information is available. The degree of change in membrane fatty acids as it relates to specific organ and membranes sites, and membrane metabolic and physical properties should be considered specifically, and distinctly, from eicosanoid metabolism, and related to the diet ω3 fatty acid source.

Most of the existing literature on ω3 fatty acids has utilized composite ω3 fatty acid mixtures. More definitive study of the individual components, 20:5, 22:5 and 22:6ω3 is needed. The extent of retroconversion, effects on ω6 series fatty acid metabolism, membrane incorporation and contribution to physiological pools used for eicosanoid synthesis should be considered.

Studies are needed to determine the extent to which adaptations of membrane lipid composition may occur on long-term exposure to fairly constant fatty acid intakes, such as the life-time dietary of the Inuit or Eskimo, in order to maintain membrane properties. This knowledge is important to assess the relevance of changes in tissue lipid compositions and metabolism documented following short-term clinical trials with different dietary oils.

Whether or not the secretion of increased amounts of ω3 fatty acids in breast milk has any significant physiological effect on the nursing infant has yet to be considered. Whether or not the levels of ω3 fatty acids in breast milk could be manipulated by maternal diet to therapeutic advantage in certain groups of infants is a potentially exciting area for research investigation.

REFERENCES

1. H.O. Bang and J. Dyerberg, Plasma lipids and lipoproteins in Greenlandic west coast Eskimos, Acta Med. Scand. 192:85-94 (1972).
2. H.O. Bang and J. Dyerberg, The composition of food consumed by Greenland Eskimos. Acta Med. Scand. 200:69-73 (1976).
3. H.O. Bang, J. Dyerberg, and A.B. Nielsen, A.B. Plasma lipid and lipoprotein pattern in Greenland west-coast Eskimos, Lancet 1:1143-1146 (1971).
4. H.O. Bang, J. Dyerberg, and H.M. Sinclair, The composition of the Eskimo food in north western Greenland, Amer. J. Clin. Nutr. 33:2657-2661 (1980).
5. J. Dyerberg, H.O. Bang, and N. Hjorne, Fatty acid composition of the plasma lipids in Greenland Eskimos, Amer. J. Clin. Nutr. 28:958-966 (1975).
6. S.S. Kantha, Dietary effects of fish oils on human health: A review of recent studies, Yale J. Biol. & Med. 60:37-44 (1987).
7. P. Hoffman and H.J. Mest, What about the effects of dietary lipids on endogenous prostanoid synthesis? A state-of-the-art review, Biomed. Biochem. Acta 46:639-650 (1987).
8. P.M. Herold and J.E. Kinsella, Fish oil consumption and decreased risk of cardiovascular disease: a comparison of findings from animal and human feeding trials. Am. J. Clin. Nutr. 43:566-598 (1986).
9. D. Kromhout, E.B. Bosschieter, and C.L. Coulander, The inverse relation between fish consumption and 20-year mortality from coronary heart disease, New Engl. J. Med. 312:1205-1209 (1985).
10. R.B. Shekelle, L.V. Missel, P. Oglesby, A.M. Shryolk, and J. Stamler, Fish consumption and mortality from coronary heart disease, New Engl. J. Med. 313:820 (1985).
11. J.D. Curb and D.M. Reed, Letter to Editor (Fish consumption and mortality from coronary heart disease), New Engl. J. Med. 313:821-822 (1985).
12. S.E. Norell, A. Ahlbom, M. Feychting, Fish consumption and mortality from coronary heart disease, Brit. Med. J. 293:426 (1986).
13. R.G. Ackman, S. Epstein, and C.A. Eaton, Differences in the fatty acid compositions of blubber fats from northwestern Atlantic finwhales (Balaenoptera Physalus) and harp seals (Pagophilus Groenlandica, Comp. Biochem. Physiol. 40B:683-697 (1971).
14. F.R. Engelhardt and B.L. Walker, Fatty acid composition of the harp seal, Pagophilus Groenlandicus (Phoca Groenlandica), Comp. Biochem. Physiol. 47B:169-179 (1974).
15. J. Exler and J.L. Weihrauch, Comprehensive evaluation of fatty acids in foods, J. Amer. Diet. Assoc. 69:243-248 (1976).
16. S.M. Innis and H.V. Kuhnlein, The fatty acid of Northwestern-Canadian marine and terrestrial mammals, Acta Med. Scand. 222:105-109 (1987).
17. T. Puustinen, K. Punnonen, and P. Uotila, The fatty acid composition of 12 North-European fish species, Acta Med. Scand. 218:59-62 (1985).

18. G.C. West, J.J. Burns, and M. Modafferi, Fatty acid composition of blubber from the four species of Bering Sea phocid seals, Can. J. Zool. 57:189-195 (1979).
19. G.C. West, J.J. Burns, and M. Modafferi, Fatty acid composition of Pacific walrus skin and blubber fats, Can. J. Zool. 57:1249-1255 (1979).
20. H. Brockerhoff, R.J. Hoyle, and P.C. Huang, Positional distribution of fatty acids in fats of a polar bear and a seal, Can. J. Biochem. 44:1519 (1966).
21. H. Brockeroff, R.J. Hoyle, On the structure of the depot fats of marine fish and mammals. Arch. Biochem. Biophys. 102:452 (1963).
22. R.G. Ackman, Some possible effects on lipid biochemistry of differences in the distribution on glycerol of long-chain n-3 fatty acids in the fats of marine fish and marine mammals, Atherosclerosis 70:171-173 (1988).
23. R.M. Dougherty, C. Galli, A. Ferro-Luzzi, and J.M. Iacono, Lipid and phospholipid fatty acid composition of plasma, red blood cells, and platelets and how they are affected by dietary lipids: a study of normal subjects from Italy, Finland, and the USA, Am. J. Clin. Nutr. 45:443-455, (1987).
24. J.W. Farquhar and E.H. Ahrens, Effects of dietary fats on human erythrocyte fatty acid patterns, J. Clin. Invest. 42:675-685 (1963).
25. I.J. Cartwright, A.G. Pockley, J.H. Galloway, M. Greaves, and F.E. Preston, The effects of dietary w-3 polyunsaturated fatty acids on erythrocyte membrane phospholipids, erythrocyte deformability and blood viscosity in healthy volunteers. Atherosclerosis 55:267-281 (1985).
26. C. Popp-Snijders, J.A. Schouten, J. van der Veen, and E.A. van der Veen, Fatty fish-induced changes in membrane lipid composition and viscosity of human erythrocyte suspensions. Scand. J. Clin. Lab. Invest. 46:253-258 (1986).
27. C. Popp-Snijders, J.A. Schouten, W.J. van Blitterswijk, and E.A. van der Veen, Changes in membrane lipid composition of human erythrocytes after dietary sup-plementation of (n-3) polyunsaturated fatty acids. Maintenance of membrane fluidity, Biochim. Biophys. Acta 854:31-37 (1986).
28. S.M. Innis and H.V. Kuhnlein, Long chain n-3 fatty acids in breast milk of Inuit women consuming traditional foods, Early Human Develop. In press (1988).
29. H.M. Sinclair, The diet of Canadian Indians and Eskimos, Br. J. Nutr. 12:69-82 (1952).
30. A.M. Banfield, The Mammals of Canada, Univ. Toronto Press., (1981).
31. C. Bates, C. van Dam., H.F. Horrobin, N. Morse, Y-S. Huang, M.S. Manku, Plasma essential fatty acids in pure and mixed race American Indians on and off a diet exceptionally rich in salmon, Prostaglan. Leukotr. & Med. 17:77-84 (1985).
32. D.F. Horrobin, Low prevalences of coronary heart disease (CHD) psoriasis, asthma and rheumatoid arthritis in Eskimos: are they caused by high dietary intake of eicosapentaenoic acid (EPA), a genetic variation of essential fatty acid (EFA) metabolism or a combination of both? Med. Hypoth. 22:421-428 (1987).

33. W.S. Harris, W.E. Connor, S. Lindsey, Will dietary w-3
    fatty acid change the composition of human milk. Am. J.
    Clin. Nutr. 40:780-785 (1984).
34. M.T. Clandinin, J.E. Chappell, T. Heim, Do low birth
    weight infants require nutrition with chain elongation-
    desaturation products of essential fatty acids? In:
    "Essential faty acids and prostaglandins", R.T. Holman,
    ed., pp. 901-904, Publ. Pergamon Press, N.Y. (1982).

POLYUNSATURATED FATTY ACIDS AND INFANT NUTRITION

Susan E. Carlson

Newborn Center
The University of Tennessee
Memphis, Tennessee

INTRODUCTION

Infants consume n-6 and n-3 fatty acids both as linoleic (18:2n-6) and linolenic (18:3n-3) acids and as their 20 and 22 carbon products of elongation and desaturation. Abnormal elevation of eicosatrienoic acid (20:3n-9) occurs with diets deficient in n-6 and n-3 fatty acids since of these families only the n-9 family can be synthesized de novo by mammals. Human symptoms of deficiency include scaly dermatitis, hair loss, and impaired wound healing. If deficiency occurs during development, growth is limited. When n-3 but not n-6 fatty acids are deficient, animals grow normally but demonstrate subtle differences in retinal physiology, visual acuity, and learning.[1-6] Such an n-3 deficiency can be produced by feeding diets with very high ratios of n-6 to n-3 fatty acids to developing animals as shown by Galli and co-workers.[7] Normally, docosahexaenoic acid (22:6n-3) is a major component of central nervous system synaptosomes and photoreceptor disk membranes.[8-10] In n-3 deficiency, docosahexaenoic acid (22:6n-3) is partially replaced by the equivalent elongation-desaturation product of linoleic acid, docosapentaenoic acid (22:5n-6).

In human infants, most 22:6n-3 present at birth accumulates during the last intrauterine trimester. When infants are born between 26 and 32 weeks' gestation, they have low concentration of 22:6n-3 in gray matter[11,12] and limited hepatic stores.[13] Furthermore, infants appear to be limited in their ability to elongate and desaturate 18:3n-3 to 22:6n-3. Unlike animals made n-3 deficient, preterm infants fed formula generally receive good amounts of 18:3n-3. Nevertheless, infants fed formula in which 18:3n-3 accounts for total n-3 intake have only half the membrane phospholipid 22:6n-3 of infants fed human milk.[14,15] Furthermore, 22:6n-3 declines in red blood cell and plasma phospholipids following early delivery.[15] Studies which demonstrate that fish oil 20 and 22 carbon n-3 fatty acids can prevent declines in membrane phospholipid n-3 fatty acids in premature infants[16,17] have paved the way for evaluation of the contribution of dietary 18:3n-3 metabolites to functional development of prematurely born infants.

This manuscript deals exclusively with the role of premature delivery and the qualitative nature of subsequent dietary lipids on the fatty acid status of the human infant. Since our studies have focused almost exclusively on the effects of these variables on n-3 status, the subject will be narrowed even further. To place the infant data in perspective, it is important to understand how n-3 fatty acids are accumulated by the developing mammal, the role of dietary n-3 fatty acids, and the evidence that n-3 fatty acids have a unique role in brain and retina. These topics were addressed in this conference by investigators responsible for primary work in the area and are included in this proceedings. They will not be reviewed in this article. The factors known to influence neural and retinal accumulation of the n-3 metabolite docosahexaenoic acid (22:6n-3) and documented functional alterations which result from failure to accumulate 22:6n-3 are listed in Fig. 1. They also have been reviewed briefly in a recent proceedings.[18]

Studies of the Term Infant

Our interest in the 22:6n-3 status of infants originated with the finding that infants born at term and subsequently fed mother's milk had approximately twice as much 22:6n-3 in red blood cell phospholipids as infants receiving formulas with 18:3n-3 but not 22:6n-3.[14] The infants studied ranged in age from 4.5 to 6 months and received human milk or one of two commercial formulas for at least 80% of energy

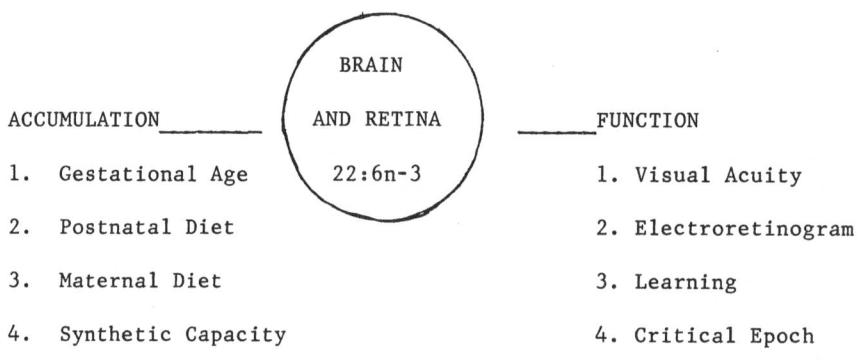

ACCUMULATION

BRAIN AND RETINA 22:6n-3

FUNCTION

1. Gestational Age

2. Postnatal Diet

3. Maternal Diet

4. Synthetic Capacity

1. Visual Acuity

2. Electroretinogram

3. Learning

4. Critical Epoch

Fig. 1. Docosahexaenoic Acid (DHA) accumulation/function

prior to this time (Table 1). Fortuitously, human milk and Formula A had equivalent amounts (wt %) of oleic (18:1n-9) and linoleic (18:2n-6) acids but differed in their provision of very long chain n-6 and n-3 metabolites. Only human milk contained n-6 and n-3 fatty acids of 20 and 22 carbons. Infants fed human milk compared to Formula A accumulated significantly more phospholipid arachidonic acid (20:4n-6) and 22:6n-3.

Formula A was equivalent to Formula B in that neither contained metabolites of n-6 and n-3 fatty acids, but the 18 carbon fatty acids differed in these formulas (Table 1): in Formula B 18:2n-6 substituted for much of 18:1n-9 found in Formula A so that the P/S ratio increased

Table 1. Selected Fatty Acids in Term Human Milk
Formula A and Formula B (wt %)
(See Reference 14)

Fatty Acid	Human Milk	Formula A	Formula B
18:1	37.6	43.3	19.9
18:2n-6	15.8	14.0	45.1
18:3n-3	0.8	1.2	5.0
20:2 to 4n-6	1.4	--	--
20:5n-3	0.1	--	--
22:4 and 5n-6	0.3	--	--
22:5 and 6n-3	0.2	--	--

fivefold in B compared to A although the ratio of 18:2n-6 to 18:3n-3 was comparable. Infants fed Formula A had a higher percent of 18:1n-9 and a lower percent of 18:2n-6 than those fed Formula B. Despite the absolute increases in 18:2n-6 and 18:3n-3 fed Formula B infants, no increase in long chain metabolites was observed in their phospholipids compared to those fed Formula A. Only 22:5n-6 differed significantly between the two formula-fed groups, and it declined in infants receiving the greater amount of 18:2n-6. We concluded that dietary 20 and 22 carbon n-6 and n-3 metabolites increased their incorporation into red blood cell membranes but that 18 carbon n-6 and n-3 fatty acids alone did not.

Sinclair and Crawford[19,20] had demonstrated earlier the more efficient neural accumulation of preformed 22:6n-3 compared to that synthesized from 18:3n-3. The red blood cell data fit very nicely with theirs, but it was not clear if dietary 20 and 22 carbon fatty acids were solely responsible for the larger amounts of n-6 and n-3 metabolites present in phospholipids of human milk-fed infants. A lower ratio of 18:2n-6 to 18:3n-3 also could have resulted in greater amounts of n-3 metabolites. We were able to rule out this possibility in this study since infants fed human milk actually received the highest ratio of 18:2n-6 to 18:3n-3. Nevertheless, it seems reasonable that the preterm infant like other species would demonstrate altered accumulation of n-6 and n-3 metabolites as a result of diets differing dramatically in their ratio of 18:2n-6 to 18:3n-3. This hypothesis has never been tested directly.

Only small amounts of 20:4n-6 and 22:6n-3 are found in human milk in comparison to total n-6 and n-3 fatty acids and yet the longer chain metabolites accounted for much of those found in the infant's red blood cell phospholipids. The relative contribution of milk with an average of 0.6% 20:4n-6 to total phospholipid 20:4n-6 in 4.5 to 6 month old infants was 37% in phosphatidylcholine (PC) and 16% in phosphatidylethanolamine (PE). Inclusion of only 0.1% 22:6n-3 as 22:6n-3 accounted for 58% of PC and 53% of PE 22:6n-3 in these same infants.

Studies of the Preterm Infant

The human infant is a special case among mammals in that medical advances have made ex-utero survival of the fetus possible, and alternatives to mother's milk are available for nutritional support. The developing premature infant may be at risk for inadequate accumulation of 22:6n-3 in the brain and retina due to a combination of factors including: 1) delivery prior to the accumulation of 22:6n-3 from the mother in the last intrauterine trimester; 2) failure to provide a dietary source of 22:6n-3; e.g., human milk; and 3) poor ability to convert 18:3n-3 to 22:6n-3. Following early delivery, infants fed formulas in the United States do not receive dietary 22:6n-3 although they receive large amounts of its substrate 18:3n-3.

All of the above suggest that the preterm infant is at risk for inadequate accumulation of n-3 metabolites, especially 22:6n-3 in the brain and retina. At the same time, evidence has been accumulating for an important role of 22:6n-3 in retinal physiology, retinal function, and learning since these are impaired in animals failing to accumulate the usual amounts of DHA during development.[1-6] These observations provide further incentive for investigations of 22:6n-3 status in preterm infants and are reinforced by indications that there may be only a finite period during development when repletion of membrane 22:6n-3 by dietary intervention can reverse deficits in retinal function.[2]

The study of the effects of human milk and formula feeding done in term infants[14] was repeated in preterm infants (<32 weeks' gestation).[15] Blood samples were analyzed at birth (cord blood) when infants were fed only human milk or formula and after a mean of seven weeks of feeding. The additional time points permitted a comparison of the effects of both diets on red blood cell n-6 and n-3 metabolites. The normal range in membrane n-6 and n-3 metabolites at birth, the effects of less than optimal nutrition in the early weeks following delivery, and the effect of time after delivery also were observed.

Cord blood levels of red blood cell phosphatidylethanolamine changed between early delivery and the institution of solely enteral feeding. Linoleic acid increased significantly while 22:6n-3 and 20:4n-6 declined (Table 2). After an average of seven weeks of feeding either human milk or formula, infants fed human milk had significantly lower 18:2n-6 and higher 20:4n-6 and 22:6n-3 compared to those fed a formula designed for preterm infants. In comparison with prefeeding phospholipid composition, it could be seen that 22:6n-3 continued to decline with formula-feeding and increased with human milk-feeding although it did not reach levels as high as observed in cord blood in the relatively short course of feeding. Infants fed human milk achieved cord blood levels of total 22 carbon n-3 fatty acids (22:6n-3 and 22:5n-3) whereas infants fed formula had levels lower than cord blood. Arachidonic acid (20:4n-6) was maintained with human milk and declined in formula-fed infants. Not all of these changes were significant.

One somewhat surprising finding was that the wt % of 22:6n-3 in both phosphatidylcholine and phosphatidylethanolamine was very similar after feeding human milk to term and preterm infants fed human milk. Since infants born early in the last trimester are presumed to have a greater need for 22:6n-3, it might have been anticipated that their 22:6n-3 would be lower indicating a poorer status. On the other hand, preterm infants completed the protocol at approximately the time of expected term delivery, while term infants were studied in the middle of infancy. Additional declines in 20:4n-6 and 22:6n-3 might have been expected to occur with time on diets devoid of n-6 and n-3 metabolites. It was clear more information was needed about the effects of long term feeding of diets with and without n-6 and n-3 metabolites. A study now in progress should provide this information throughout infancy.

Table 2. Red Blood Cell Phosphatidylethanolamine
Fatty Acids (wt %)

	Cord Blood	Preliminary	After Feeding Human Milk	Formula
18:2n-6	2.3	5.8 →	6.8	→ 9.1*
20:4n-6	27.1	24.0 →	24.6	→ 22.6**
22:6n-3	8.0	4.8 →	6.2	→ 4.0**
22:5n-3 + 22:6n-3	8.8	6.7 →	8.4	→ 6.0**

Differs from human milk fed infants: *$p < 0.05$; **$p < 0.001$

When membrane 20 and 22 carbon n-6 and n-3 fatty acids four-six months after delivery and feeding of human milk are compared with cord blood values, there is a dramatic decline in PC 20:4n-6 with time from 15.7% to 8.8% (Table 3). On the other hand, infants maintained on human milk for long periods of time have identical amounts of PE products of Δ5- and Δ4-desaturation and equivalent amounts of Δ4-desaturation products in PC (Tables 3 and 4). These data strongly suggest that there is a normal physiological decline in PC 20:4n-6 with time. As noted previously, formula-fed infants experience much more dramatic declines in n-3 metabolites compared to n-6 metabolites. For this reason and because functional deficits have been associated with failure to accumulate normal proportions of n-3 metabolites in development, our subsequent focus has been on the n-3 status of the preterm infant. (This may have been a fortuitous choice since Dr. Martinez has reported data at the conference which suggests that 22:6n-3 but not 20:4n-6 accumulation of preterm infants is limited by malnutrition following early delivery.)

Thus it has been demonstrated that 1) products of Δ4- and Δ5-desaturase in human milk account for significant proportions of these fatty acids in red blood cell phospholipids and 2) that there is an apparently normal physiological decline in phosphatidylcholine 20:4n-6 beginning in the last intrauterine trimester and occurring through at least the first several months of infancy. A major question remaining is to what degree the red blood cell fatty acid pattern reflects neural status of n-6 and n-3 metabolites. Before proceeding with studies of infants, these membranes were compared in developing rats exposed to n-3 deficient and sufficient diets and found to be related.[21] Data from Neuringer et al.[22] show a relationship between retinal and plasma phospholipid 22:6n-3. For the precise degree of relationship, we have few clues since the growing preterm infant receiving 18:3n-3 differs from animal models deprived of this fatty acid.

Additional studies of n-3 status in preterm infants have focused on determining if fish oil could substitute for 22:6n-3 found in human milk to maintain phospholipid 22:6n-3. In the first study,[16] it was determined that a bolus once per day of 71 mg/kg of 22:6n-3 in 750 mg/kg/d of MaxEPA (R. P. Scherer, Troy, MI) maintained red blood cell phospholipid 22:6n-3 at enrollment levels during a six-week study period while 22:6n-3 declined significantly in phospholipids of infants who did not receive the supplement. The large dose necessary for maintenance of membrane 22:6n-3 suggested, however, that the fish oil was very poorly absorbed.

Table 3. Total RBC PC Δ5- and Δ4-Desaturase Products: Effect of Time in Infants Receiving Only 18 C PUFA

	Δ5-Desaturase	Δ4-Desaturase
Birth (PT)	15.7%	2.7%
4 Wks (PT)	10.0%	1.2%
11 Wks (PT)	6.3%	0.9%
4-6 Mos (T)	5.2%	1.0%

Reference: Human milk-fed term infant at 4-6 mos
8.8%   2.3%

Table 4. Total RBC PE Δ5- and Δ4-Desaturase Products: Effect of Time in Infants Receiving Only 18 C PUFA

	Δ5-Desaturase	Δ4-Desaturase
Birth (PT)	27.0%	10.4%
4 Wks (PT)	25.0%	6.9%
11 Wks (PT)	23.0%	5.6%
4-6 Mos (T)	23.1%	5.0%

Reference: Human milk-fed term infant at 4-6 mos
27.0%   8.8%

An additional study was undertaken in which fish oil was mixed into formula using an ultrasonicator. Ultrasonication clearly improved uptake as two infants receiving 81 mg/kg/d of 22:6n-3 (an amount equivalent to the bolus dose used earlier) resulted in dramatic elevations in plasma and red blood cell 20:5n-3 and 22:6n-3 over a five-day period. A study subsequently was undertaken in which infants fed formula containing only 18:3n-3 were compared with infants receiving one of two levels of 11 or 27 mg/kg/d of 22:6n-3 or 0.2% and 0.5% of total fatty acids, respectively.[17] Infants fed the smaller amount of fish oil demonstrated significant elevations in plasma phospholipid 22:6n-3 during the two weeks of administration. Plasma phospholipid 22:6n-3 was quite similar (i.e., not dose responsive) at the higher intake, but the 20:5n-3 level was much higher. When we compared the increase in plasma phospholipid 22:6n-3 of infants receiving a bolus of MaxEPA one time per day with those receiving sonicated MaxEPA in formula in eight daily feedings, it was clear that the latter mode of administration resulted in a dramatic improvement in uptake.

Subsequently, preterm infants receiving formula have been supplemented with 22:6n-3 as 0.2% and 0.4% of total fatty acids. After four weeks of supplementation, supplemented groups show a significant increase in PE 22:6n-3 (wt %) at the time that 22:6n-3 is declining in unsupplemented infants (Fig. 2). Red blood cell PE 20:4n-6 was not affected by fish oil supplementation after four weeks of feeding (Fig. 3). Total plasma 22:6n-3 was twofold greater in supplemented compared to unsupplemented (11 µg/ml vs 21 µg/ml) infants (Fig. 4) while plasma PC 20:4n-6 declined with time regardless of dietary supplementation (Fig. 5). The apparent rise in plasma PC 20:4n-6 between the third and fourth week of feeding does not appear to be maintained judging from additional samples obtained at 2 months of age.

Most fish oil in addition to containing 22:6n-3 provides considerable amounts of eicosapentaenoic acid (20:5n-3). The feeding of 0.4% 22:6n-3 with its accompanying 0.6% 20:5n-3 resulted in a linear increase in red blood cell PE 20:5n-3 over four weeks of study. There was no suggestion that 20:5n-3 would plateau (Fig. 6). In order to minimize this increase, subsequent feeding of 22:6n-3 has been limited to 0.2% of total dietary fatty acids. Fortuitously, this amount of 22:6n-3 is what we find in average samples of preterm human milk[15] and appears to provide equivalent amounts of 22:6n-3 to plasma and red blood cell phospholipids.

The goals of this work have been 1) to determine the concentration of dietary 22:6n-3 which will retain red blood cell phospholipid 22:6n-3 in the range found in cord blood and in human milk-fed infants, 2) to maintain 22:6n-3 concentrations at this level throughout infancy in supplemented infants, 3) to document 22:6n-3 in plasma and red blood cell phospholipids of supplemented and unsupplemented infants, and 4) to measure visual acuity/cognitive development in infancy as a function of 22:6n-3 status. A protocol currently underway is addressing these goals.

CONCLUSION

Human preterm infants may be at risk for inadequate accumulation of neural and retinal 22:6n-3 because of a combination of factors including separation from the maternal supply at a critical period of accumulation, poor ability to form 22:6n-3 from precursor 18:3n-3, and diets which do not contain 22:6n-3. Since dietary 18:3n-3 is provided to formula-fed infants, however, it is unlikely that their neural membranes become as depleted of 22:6n-3 as in animal models of n-3 deficiency. Nevertheless, studies of infants fed 22:6n-3 compared to those receiving 18:3n-3 alone clearly show that 01%-0.2% of total fatty acid as 22:6n-3 provides more than 50% of red blood cell phospholipid 22:6n-3. It is also clear that formula-fed infants experience a deterioration of 22:6n-3 status. Whether this decline is associated with functional deficits such as have been demonstrated in the animal models remains to be seen. The pattern of 22:6n-3 accumulation in brain and retina of the normally growing preterm infant is not likely to be determined. However, it is possible to test the hypothesis that provision of supplemental 22:6n-3 to preterm infants can improve functions such as visual acuity. Studies of this nature are currently underway in several laboratories.

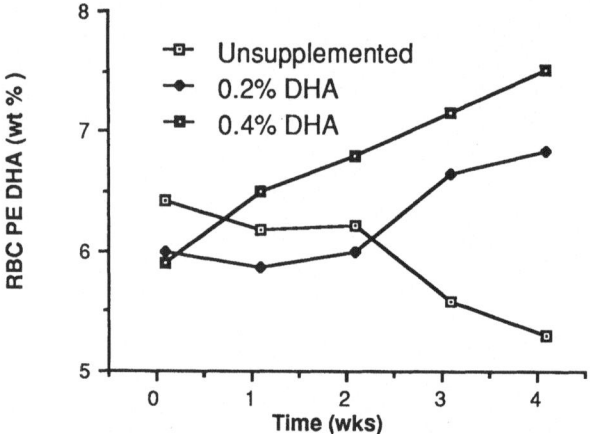

Fig. 2. Red blood cell PE DHA (22:6n-3): Effect
of time and n-3 supplementation

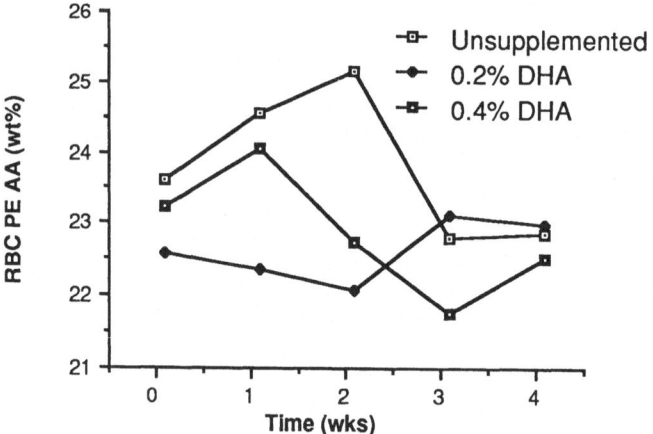

Fig. 3. Red blood cell PE AA (20:4n-6): Effect
of time and n-3 supplementation

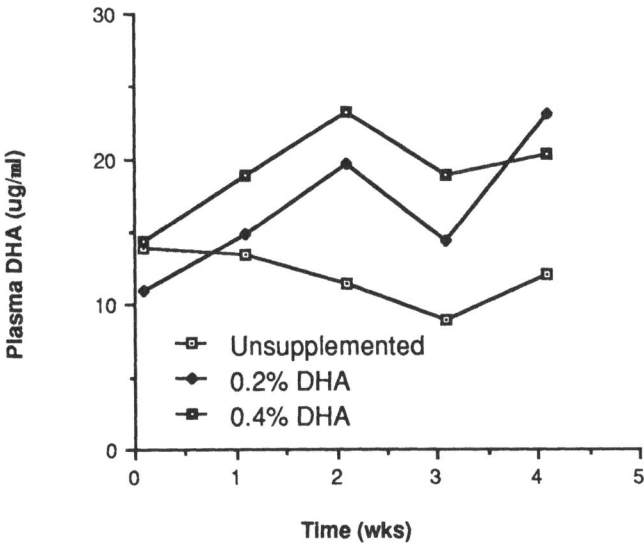

Fig. 4. Plasma DHA (22:6n-3) concentration: Effect
of time and n-3 supplementation

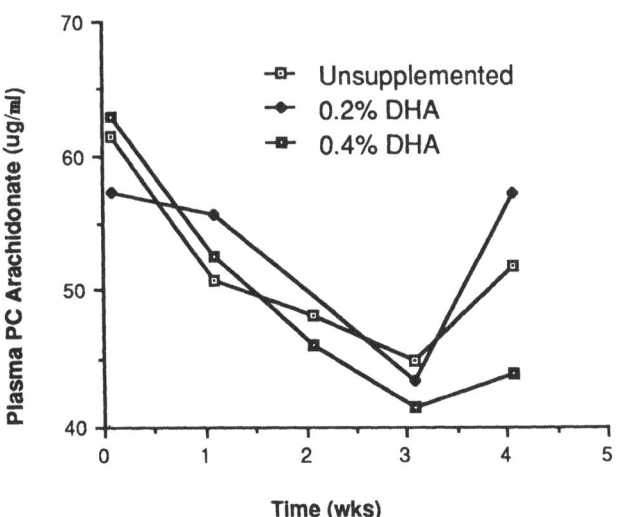

Fig. 5. Plasma PC AA (20:4n-6) concentration: Effect
of time and n-3 supplementation

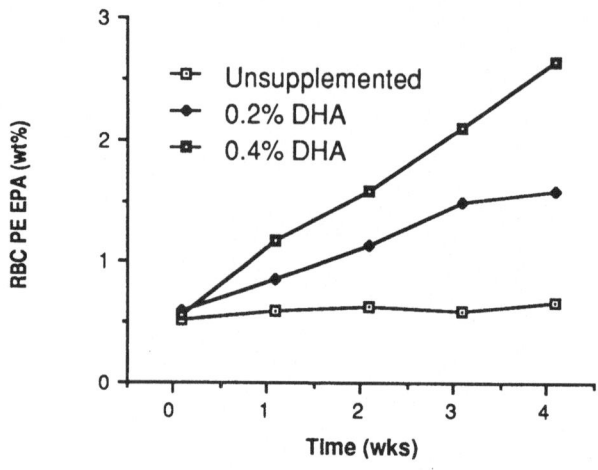

Fig. 6. Red blood cell PE EPA (20:5n-3): Effect of time and n-3 supplementation

REFERENCES

1. R. M. Benolken, R. E. Anderson, and T. G. Wheeler, Membrane fatty acids associated with the electrical response in visual excitation, Science 182:1253 (1973).

2. M. Neuringer, W. E. Connor, and S. L. Luck, Omega-3 fatty acid deficiency in rhesus monekys: Depletion of retinal docosahexaenoic acid and abnormal electroreginograms, Am. J. Clin. Nutr. 43:706 (1985).

3. M. Neuringer, W. E. Connor, C. Van Petten, and L. Barstad, Dietary omega-3 fatty acid deficiency and visual loss in infant rhesus monkeys, J. Clin. Invest. 73:272 (1984).

4. N. Yamamoto, M. Saitoh, A. Moriuchi, M. Nomura, and H. Okuyama, Effect of dietary alpha linolenate/linoleate balance on brain lipid compositions and learning ability of rats, J. Lipid Res. 28:144 (1987).

5. M. Neuringer and W. E. Connor, The importance of dietary n-3 fatty acids in the development of the retina and nervous system, in: "Proceedings of the AOCS Short Course on Polyunsaturated Fatty Acids and Eicosanoids," W. E. M. Lands (ed.), American Oil Chemists Society, Champaign, IL (1987).

6. H. Okuyama, M. Saitoh, Y. Naito, T. Hori, A. Hashimoto, A. Moriuchi, and N. Yamamoto, Re-evaluation of the essentiality of alpha-linolenic acid in rats, in: "Proceedings of the AOCS Short Course on Polyunsaturated Fatty Acids and Eicosanoids," W. E. M. Lands (ed.), American Oil Chemists Society, Champaign, IL (1987).

7. C. Galli, H. I. Trzeciak, and R. Paoletti, Effect of dietary fatty acids on the fatty acid composition of brain ethanolamine phosphoglyceride: Reciprocal replacement of n-6 and n-3 polyunsaturated fatty acids, Biochim. Biophys. Acta 248:449 (1971).

8. J. S. O'Brien, D. L. Fillerup, and J. F. Mean, Quantification of fatty acid and fatty aldehyde composition of ethanolamine choline and serine phosphoglycerides in human cerebral gray and white matter, J. Lipid Res. 5:329 (1964).

9. R. E. Anderson, M. B. Maude, and W. Zimmerman, Lipids of ocular tissues. X. Lipid composition of subcellular fractions of bovine retina, Vision Res. 15:1087 (1975).

10. S. J. Fliesler and R. E. Anderson, Prog. Lipid Res. 22:79 (1983).

11. M. T. Clandinin, J. E. Chappell, S. Leong, T. Heim, P. R. Swyer, and G. W. Chance, Intrauterine fatty acid accretion rates in human brain: Implications for fatty acid requirements, Early Hum. Dev. 4:121 (1980).

12. L. Svennerholm, Distribution and fatty acid composition of phosphoglycerides in normal human brain, J. Lipid Res. 9: 570 (1968).

13. M. T. Clandinin, J. E. Chappell, T. Heim, P. R. Swyer, and G. W. Chance, Fatty acid accretion in fetal and neonatal liver: Implications for fatty acid requirements, Early Hum. Dev. 5:1 (1981).

14. J. C. Putnam, S. E. Carlson, P. W. DeVoe, and L. A. Barness, The effect of variations in dietary fatty acids on the fatty acid composition of erythrocyte phosphatidylcholine and phosphatidylethanolamine in human infants, Am. J. Clin. Nutr. 36:106 (1982).

15. S. E. Carlson, P. G. Rhodes, and M. G. Ferguson, Docosahexaenoic acid status of preterm infants at birth and following feeding with human milk or formula, Am. J. Clin. Nutr. 45:798 (1986).

16. S. E. Carlson, P. G. Rhodes, V. S. Rao, and D. E. Goldgar, Effect of fish oil supplementation on the omega-3 fatty acid content of red blood cell membranes in preterm infants, Pediatr. Res. 21:507 (1987).

17. C-C. F. Liu, S. E. Carlson, P. G. Rhodes, V. S. Rao, and E. F. Meydrech, Increase in plasma phospholipid docosahexaenoic acids as a reflection of their intake and mode of administration, Pediatr. Res. 22:292 (1987).

18. S. E. Carlson, Docosahexaenoic acid in mammalian development, in: "31st Perinatal and Developmental Symposium, Mead Johnson Symposium on Infant Nutrition," Mead Johnson and Co., Evansville, IN (1987) (in press).

19. M. A. Crawford, N. M. Casperd, and A. J. Sinclair, The long chain metabolites of linoleic and linolenic acids in liver and brain in herbivores, Comp. Biochem. Physiol. 54B:395 (1976).

20. A. J. Sinclair and M. A. Crawford, The accumulation of arachidonate and docosahexaenoate in the developing rat brain, J. Neurochem. 19:1753 (1972).

21. S. E. Carlson, J. D. Carver, and S. G. House, High fat diets varying in ratios of polyunsaturated to saturated fatty acid and linoleic to linolenic acid: A comparison of rat neural and red blood cell membrane phospholipids, J. Nutr. 116:718 (1986).

22. M. Neuringer, W. E. Connor, C. Van Petten, and L. Barstad, Dietary omega-3 fatty acid deficiency and visual loss in infant rhesus monkeys, J. Clin. Invest. 73:272 (1984).

POLYUNSATURATED FATTY ACIDS OF THE n-3 SERIE AND NERVOUS SYSTEM DEVELOPMENT

Jean-Marie Bourre[1], Odile Dumont[1], Michèle Piciotti[1], Gérard Pascal[2], and Georges Durand[2]

[1] INSERM Unité 26, Hôpital Fernand Widal, 200 rue du Faubourg St-Denis, 75475 Paris Cedex 10 (France)
[2] INRA-CNRZ, 78350 Jouy-en-Josas (France)

INTRODUCTION

It is necessary to ensure that brain cells receive adequate supplies, especially of lipids, during their differentiation and multiplication. A lipidic anomaly could result in altered function of the membranes and a greater susceptibility of the membranes to aggression, particularly toxic.

The nervous system is the organ with the greatest concentration of lipids, immediately after adipose tissue. These lipids are practically all structural and not energetic ; they participate directly in the functioning of cerebral membranes. Cerebral development is genetically programmed : if one stage is missed or perturbed, the chances of recuperation are greatly reduced. Moreover, the renewal of neurons and oligodendrocytes is nil (a cell that disappears is not replaced), the renewal of membranes is often very slow. The study of structural role of polyunsaturated essential fatty acids in nerve membranes, qualitatively and quantitatively very important, has been neglected[1]. The dietary polyunsaturated fatty acids, linoleic and linolenic acid, are the necessary precursors of longer chains. The latter control the composition of membranes, hence their fluidity, and in consequence the enzymatic activities, the binding between molecules and receptors, the cellular interactions, and the transport of nutrients. As far as the nervous system is concerned, these fatty acids can control certain electrophysiologic parameters as well as the functions of learning. It is well known that dietary polyunsaturated fatty acids control fatty acids in the membranes[2] and that they are particularly important for harmonious cerebral development[3]. Results demonstrating the influence of polyunsaturated fatty acids on the structure and function of the nervous system are thus numerous[4-16]. However, these authors have generally used diets simultaneously deficient in fatty acids of the n-6 and n-3 series. Deficiency in all the essential fatty acids changes the course of cerebral development, period when nutrient requirements are particularly important. Nerve tissue compensates partially for this deficit by synthesizing polyunsaturated acids of the n-9 series which it incorporates in all its membranes. Polyunsaturated fatty acids of the n-3 series have a very specific role in the membranes, especially in the nervous system : all the cells and cerebral organelles are extremely rich in them. The very long chain cerebral polyunsaturated fatty acids are derived from precursors that have a dietary origin.

MATERIALS AND METHODS

Two groups of wistar rats were used. One group was fed for four generations with a semi-synthetic diet containing 1.5% sunflower seed oil (1000 mg of n-6 fatty acids and 6 mg of n-3 fatty acids). The other group was fed a soybean oil diet (1000 mg n-6 fatty acids, and 135 mg per 100 g). At 60 days after birth, animals receiving the "sunflower" diet were changed to a "soybean" diet in order to study the rate of recuperation as from the 60th day. The animals were killed at 67, 74, 81, 88, 95, 102 and 130 days (7, 14, 21, 28, 35, 42 and 70 days after changing the diet). Diets intermediate in linolenic acid content were obtained by adding variable and increasing amounts of linolenic acid by addition of soybean oil to the sunflower diet. Diets intermediate in linoleic acid content were obtained by adding variable amount of linseed oil and hydrogenated palm oil to rapeseed oil or sunflower oil. Thus in these experiments linolenic acid content varied from 6 to 681 mg/100 g diet and linoleic acid from 150 to 6200 mg/100 g diet. Four weeks before mating these diets were given to females (previously fed with sunflower oil). Their pups were sacrificed when 21 and 60 days old.

Separation of neurons, oligodendrocytes, astrocytes, myelin, nerve endings (synaptosomes), mitochondria, endoplasmic reticulum (microsomes) has been previously described[17]. The purity of fractions was evaluated by phase-contrast microscopy, enzyme marker assay, specific protein analysis (radioimmunoassay), electrophoresis, and lipid analysis[17]. Brain capillaries were prepared according to Goldstein[18] and their purity checked[19]. Extraction methods for lipid fractions, their transmethylation and the analysis of methyl esters by capillary column gas chromatography have also been previously described[17].

Electroretinogram recordings were performed as previously published[20].

Learning behavior and motor activity were measured using open field and shuttle box tests in 60-day-old animals.

Motor activity measurement was carried out in an individual box equipped with photoelectric cells that automatically recorded the movements of the animal (counts were made in 5 min periods for one hour).

For measurement of resistance to poisons, Triethyltin was made up in physiologic solution and administered at a dosage of 0.5 ml/100 g body weight. The product was administered by the intraperitoneal route in one single dose to animals that had been fasted for 16 hours and had been under observation for 8 days. Feeding was resumed 4 hours after administration of the product. Behavior, appearance of the animals, side effects, and mortality were noted daily for 14 days after the injection of triethyltin.

RESULTS AND DISCUSSION

1/ Deficiency in linolenic acid produced anomalies in the composition of cells and organelles in the nervous system as well as in other organs

In animals fed the sunflower diet, cells and organelles show a very marked deficit in cervonic acid (22:6 n-3) that is compensated by an excess of 22:5 n-6. If 60-day-old animals fed either the sunflower or soybean diet are compared, the total n-3/n-6 ratio is 16 times less in oligodendrocytes, 12 times less in myelin, 2 times less in neurons, 6 times less in synaptic vesicles, 3 times less in astrocytes, 7 times less in mitochondria, and 5 times less in microsomes. Cells and intracellular organelles in all dietary groups are found to have an identical total amount of polyunsaturated fatty acids with the two diets. The saturated and monounsaturated fatty acids are practically unchanged.

Faced with a deficiency in linolenic acid the membranes of cerebral cells and organelles are just as deficient, if not more so, as those of other organs. However, there is marked preservation of dietary linolenic acid (and a reutilization of its very long chain derivatives) because a 25-fold decrease in the diet only results in, at the worst, a 10-fold decrease in the fractions we have examined. The importance of fatty acids of the n-3 series can similarly be well demonstrated by specifically studying certain phospholipids such as phosphatidyl-ethanolamine in animals fed diets based on peanut or rapeseed oils[21,22].

After changing from the "sunflower" diet to the "soybean diet", the rate of recuperation is remarkably slow : several months are necessary before the cerebral cells and organelles[23] recover their normal levels of cervonic acid and lose the excess 22:5 n-6 in contrast with other organs. This recuperation in brain is slow regardless of the cell or organelle. One might have foreseen that recuperation would not be rapid in myelin, a membrane with a slow rate of renewal. But it is very unexpected to find that nerve endings also recuperate very slowly, although the renewal of molecules comprising their membranes is supposed to be rapid. It can be proposed that the regulation of recuperation is situated either at the level of the hepatic production of chain terminals (cervonic and arachidonic acid), or at the level of the enzymatic activities of desaturation and elongation that are known to be very low in the brain after birth[9,24], or at the level of transport across the blood-brain barrier.

It is interesting to note that the cerebral microvessels and capillaries also have a very slow rate of recuperation even though they are in contact with normal plasmatic lipoproteins[25].

In the sciatic nerve, the rate of recuperation (after changing from a diet deficient in linolenic acid to a normal diet) is also very slow[26].

2/ Enzymatic activities

Na-K-ATPase is decreased by half in the nerve endings of animals fed a "sunflower" diet as compared to those fed a "soybean" diet, who show normal levels of enzyme activity. Interestingly, a specific deficiency in linolenic acid produces a decrease in this enzymatic activity, while a simultaneous deficiency in linoleic and linolenic acids results in an increase in this same activity[16]. This enzyme, Na-K-ATPase, controls the ionic flow resulting from nerve transmission. It consumes half the energy used by the brain (in an adult man this organ only represents 2% of body weight but consumes 20% of energy).

The activity of 5'-nucleotidase is decreased by 30% in total brain, but not in myelin nor in nerve endings, which suggests that the enzyme level in cellular membranes is probably very altered. These results are in agreement with those of Bernsohn et al.[27] who showed that the decrease in the activity of the enzyme produced by a simultaneous deficiency of linoleic and linolenic acids is corrected only by the addition of linolenic acid to the diet.

CNPase, specific of myelin, is decreased by a deficiency in linolenic acid, even though this membrane is considered to be very rigid and metabolically little active. Another enzyme, acetylcholinesterase, also has its activity modulated by dietary lipids[28].

3/ Electroretinogram

The retina is one of the tissues that is richest in n-3 polyunsaturated fatty acids[5]. Overall deficiencies in polyunsaturated fatty acids induce, in the long term, modifications in the distribution of membrane fatty acids in the retina[29] which are related to perturbations in the amplitude of the a and b waves in the electroretinogram[30]. But the direct influence of dietary fatty acids on the electroretinogram has been reported to depend on the species, age, and the duration of the deficiency, and it has even been denied[31].

In 4-week-old sunflower animals the a and b waves are only detectable at stimulation intensities that are ten times those of soybean-fed animals (Fig. 1). These perturbations begin to decline at the age of 6 weeks and disappear in adults (16 to 18 weeks), except for the a wave whose amplitude, at maximum stimulation intensity only, has a value 10% lower than that found in the control group, perhaps indicating an irreversible change.

Similarly in young animals (21 days) fed a "peanut" diet the amplitude of the ERG is three times less than that obtained with animals fed a "rapeseed" diet[32].

4/ Learning test results

A deficiency in linoleic and linolenic acids alters the learning capacity of animals[15,33,34].

Results for the motor activity and open field tests were practically normal for "sunflower" animals. On the other hand, the learning behavior of these animals was very perturbed, as shown by the shuttle box test. At the first test period it is clear that "soybean" animals made a quicker association between the light signal and the electric shock, since on average they avoid 7 shocks out of 30, while the "sunflower" animals only avoided 2. The number of passages with shock is the same in both groups, but the number of non-passages is much higher in the "sunflower" group. This means that "sunflower" animals undergo electric shock for the whole duration of the stimulus without attempting, or without being able, to escape. The "soybean" animals total 2 non-passages out of 30 stimulations while "sunflower"animals total 10 (Fig. 2). As a function of the number of test periods, differences tend to decline and disappear at the fourth test period. Learning impairment in "sunflower" animals could be related to a decrease in learning capacity, but also to more intense emotivity. These results are in agreement with those of Lamptey and Walker[35].

5/ Less rapid mortality in animals not deficient in linolenic acid when tested with a neurotoxic agent, triethyltin

The LD 50 of animals given the "soybean" or "sunflower" diets does not differ significantly (6.18 versus 6.02 µl/kg). But the animals given the "sunflower" diet die more rapidly than those given the "soybean" diet (Fig. 3).

6/ Definition of the minimum necessary amount of n-3 fatty acids to assure normal composition of cerebral membranes

Diets intermediate in linolenic acid content (obtained by adding variable and increasing amounts of linolenic acid by addition of soybean

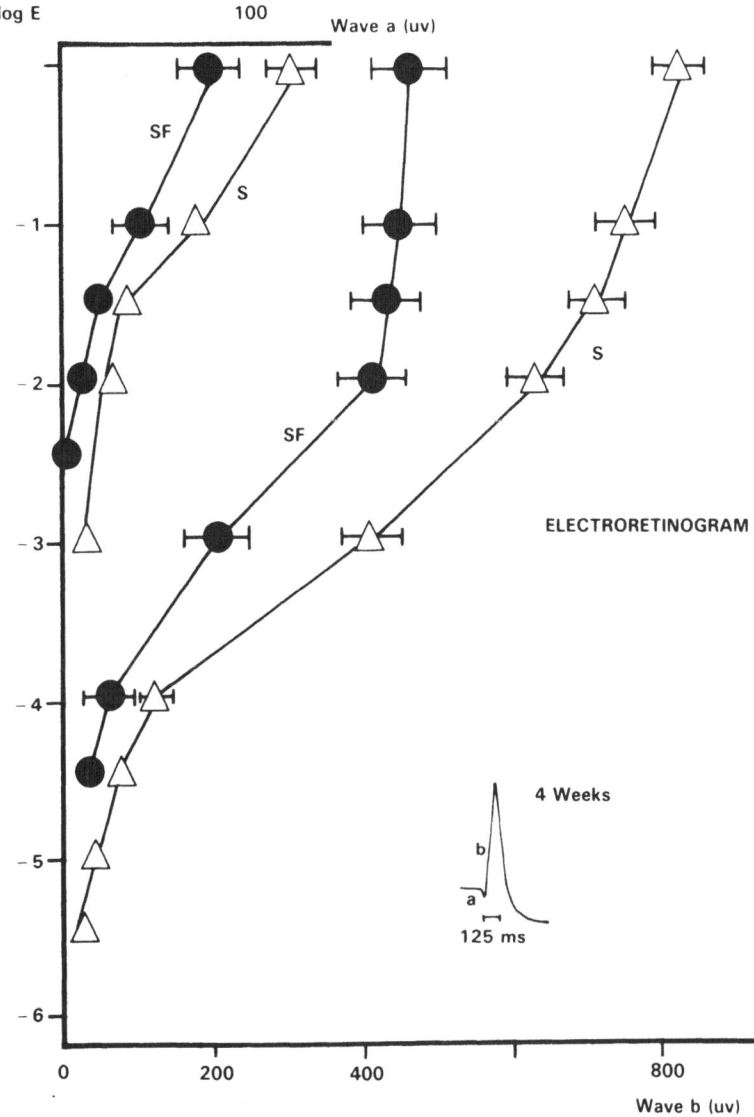

Fig. 1. Electroretinogram a and b waves are altered in young animals
deficient in linolenic acid

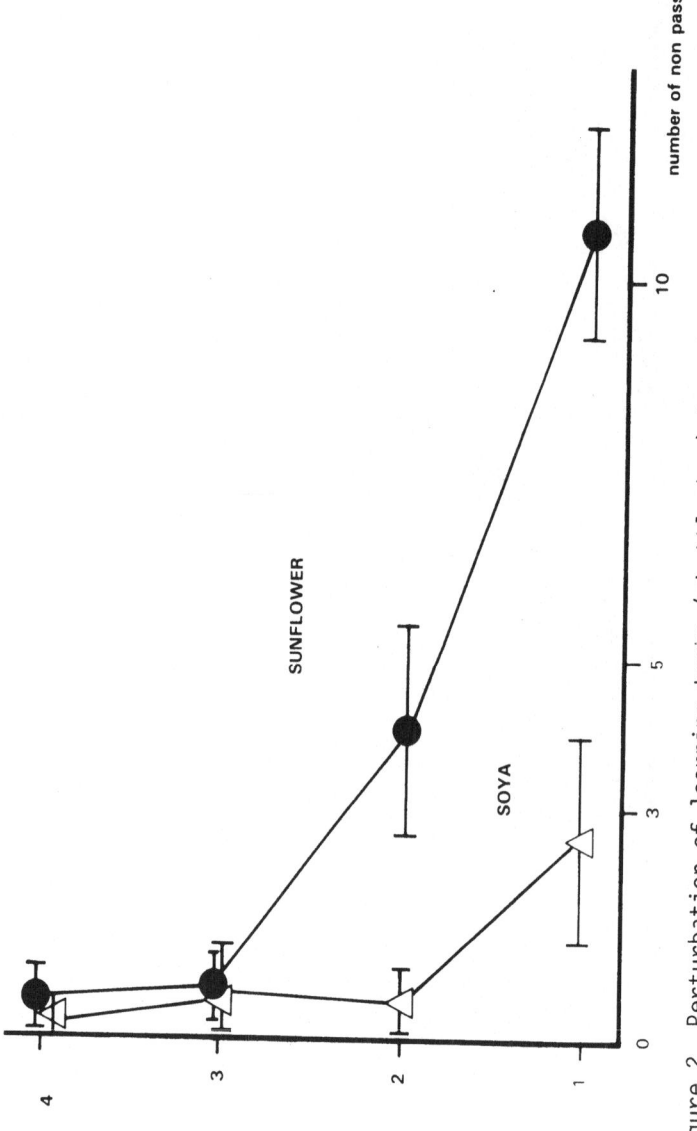

Figure 2. Perturbation of learning tests (shuttle box) in animals deficient in linolenic acid

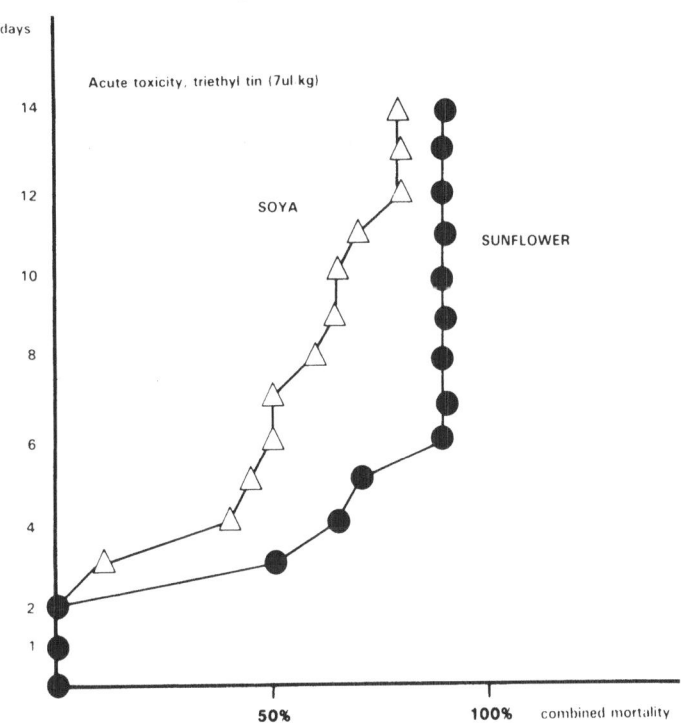

Figure 3. **A neurotoxic agent (acute triethyltin) is more rapidly fatal in "sunflower" animals than in "soybean" animals.**
The curves represent combined mortality (total number of deaths on a given day) as a function of delay after acute intoxication with 7 µl/kg triethyltin. The difference are significant up to the tenth day.

oil to the sunflower diet) were given. Increasing quantities of 18:3 n-3 results in all tissues studied in an overall increase of 22:6 n-3 (Fig. 4) and inversely a decrease 22:5 n-6. In whole brain, in myelin and nerve endings the level of 22:6 n-3 increases linearly with a 18:3 n-3 intake varying from 0 to 200-250 mg/100 g of diet, it then reaches a plateau (the inverse is observed for 22:5 n-6). In the liver the response is rapid up to 300 mg per 100 g chow, beyond that point there is a slower increase. Very interestingly, all organs need the same level of dietary linolenic acid (approx. 0.4% calories) (data not shown).

There is a direct relationship between dietary and gastric in n-3 fatty acid content. In serum, only the HDL (but not the LDL nor the VLDL) n-3 levels increase with increased n-3 dietary intake. Study of the composition of lipoprotein polyunsaturated fatty acids demonstrates the importance of intrahepatic metabolism in the supply of polyunsaturated fatty acids to the brain ; direct capture by the brain of 18:2 n-6 and 18:3 n-3 precursors is probably small. These precursors have to be desaturated and elongated in the liver to longer chains which are, in fact, the essential cerebral fatty acids[32], as appears to have been established by cell cultures[36]. The brain contains practically no linoleic or linolenic acid ; cultured nerve cells can not synthesize measurable amounts of docosahexanoic acid (22:6 n-3). Only the addition of 20:4 n-6 and 22:6 n-3 to nerve cell culture medium results in, on the one hand, a better functioning of the neurons (measured, for example, by the release of neurotransmitters)[37], and on the other, the multiplication and differentiation of oligodendrocytes thanks to a normal fatty acid composition of their membranes.

The linolenic acid requirements increase slightly with the dietary content of linoleic acid. For the developing brain in the rat, for 1000, 2500 and 5000 mg/100 g 18:2 n-6, requirements in 18:3 n-3 are approx. 175, 200 and 250 mg/100 g respectively (Fig. 5).

Increasing amount of 18:3 n-3 in the diet do not change the level of arachidonic acid in the membrane (Fig. 6) 22:4 n-6 is poorly affected (not shown).

Interestingly, increasing the amount of 18:2 n-6 in the diet from 150 mg to 6200 mg/100 g diet had little effect on the level of 22:6 n-3 in brain, sciatic nerve, myelin, and nerve endings or in liver (Fig 7). Though, increasing the amount of 18:2 n-6 from 300 - 1400 mg/100 g diet increased 20:4 n-6 content ; after 1200 mg it reached a plateau in the liver (Fig. 8). Higher quantities of 18:2 n-6 in the diet only altered 22:5 n-6 and to a lesser extent 22:4 n-6 (not shown). Thus the levels of 22:6 n-3 and 20:4 n-6 are tightly controlled by some unknown mechanism: deficiency in n-3 fatty acids or excess of 18:2 n-6 both provoke only accumulation of 22:5 n-6.

Excess dietary n-3 polyunsaturated fatty acids does not lead to an accumulation in the nervous system. In the brains of animals whose diet is enriched in n-3 fatty acid (5% lipid and 1% menhaden oil), the vitamin E, conjugated dienes, and malonaldehyde contents are not changed, nor are the glutathione peroxidase, catalase, and superoxide dismutase activities[38]. Slight excess fats are unlikely to be implicated in attacks by free radicals or to be broken down into toxic derivatives by peroxidation in the brain. However, feeding animals a large excess of fish oil (15% cod liver) alters C22:6 and C20:5 n-3 brain contents[39].

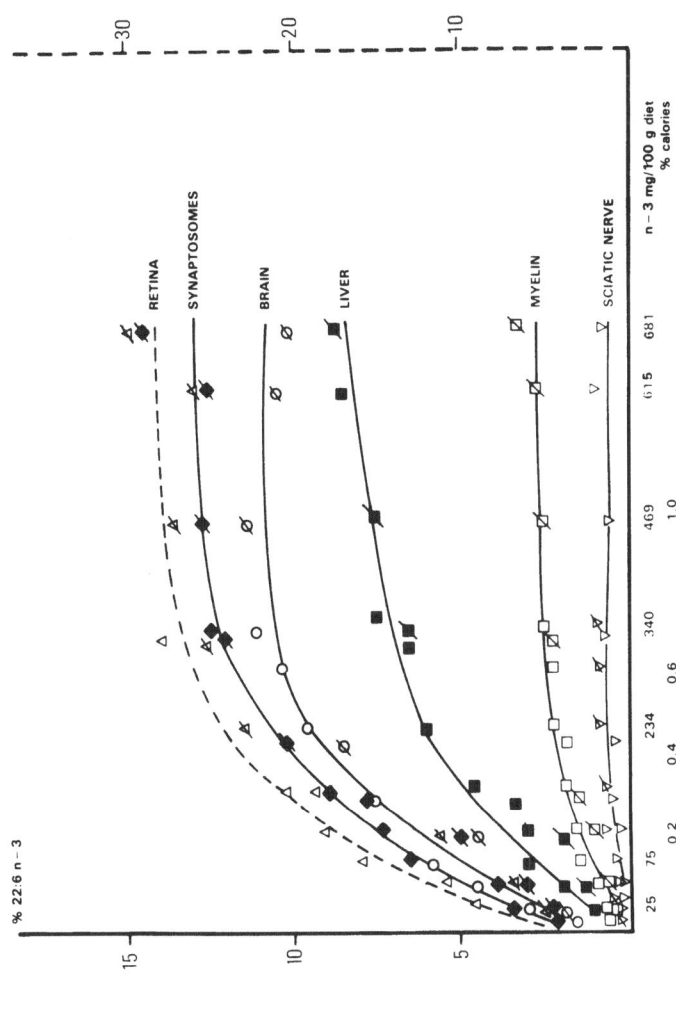

Figure 4. Relationship between dietary linolenic acid content and 22:6 n-3 levels.
Animals were fed a diet containing 5 or 10% lipids (thus providing two levels or 18:2 n-6).
3200 mg/100 g diet: normal symbols; 6400 mg/100 g diet: symbols with a diagonal bar. Intermediate
linolenic acid contents were obtained by adding increasing quantities of sunflower oil to soya oil.

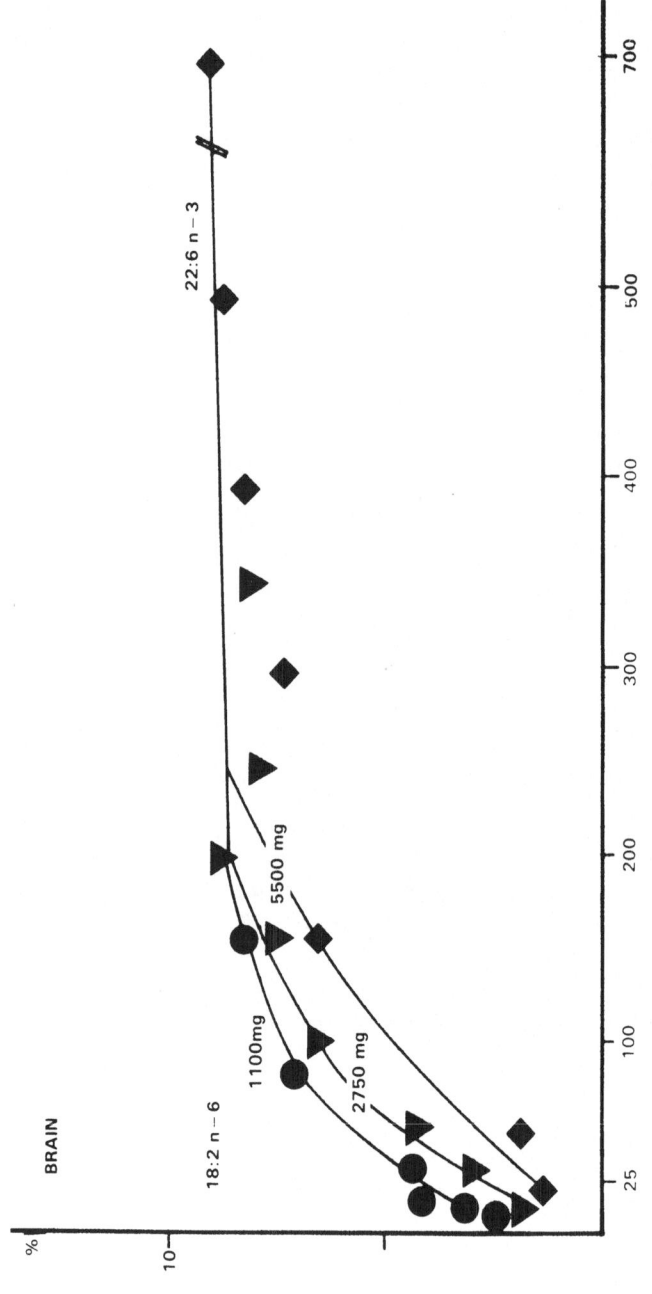

Figure 5. Brain linolenic acid requirement increases slightly with the dietary content of linoleic acid.

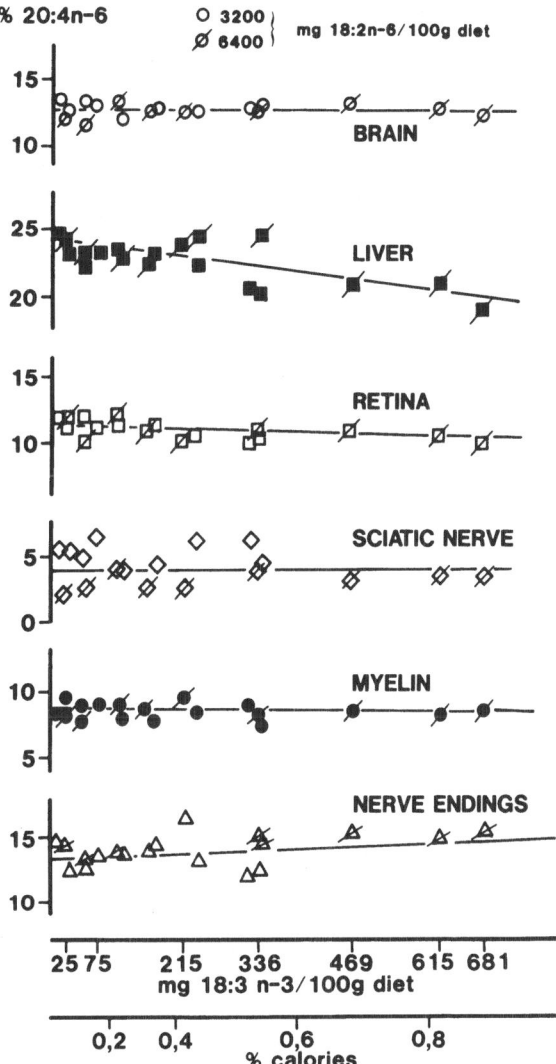

Figure 6. Level of arachidonic in membrane does not depend on dietary linolenic acid.
Animals were fed a diet containing 5 or 10% lipids (thus providing 2 levels of 18:2 n-6). 3200 mg/100 g diet : normal symbols ; 6400 mg/100 g diet : symbols with a diagonal bar. Intermediate linolenic acid contents were obtained by adding increasing quantities of sunflower oil to soya oil.

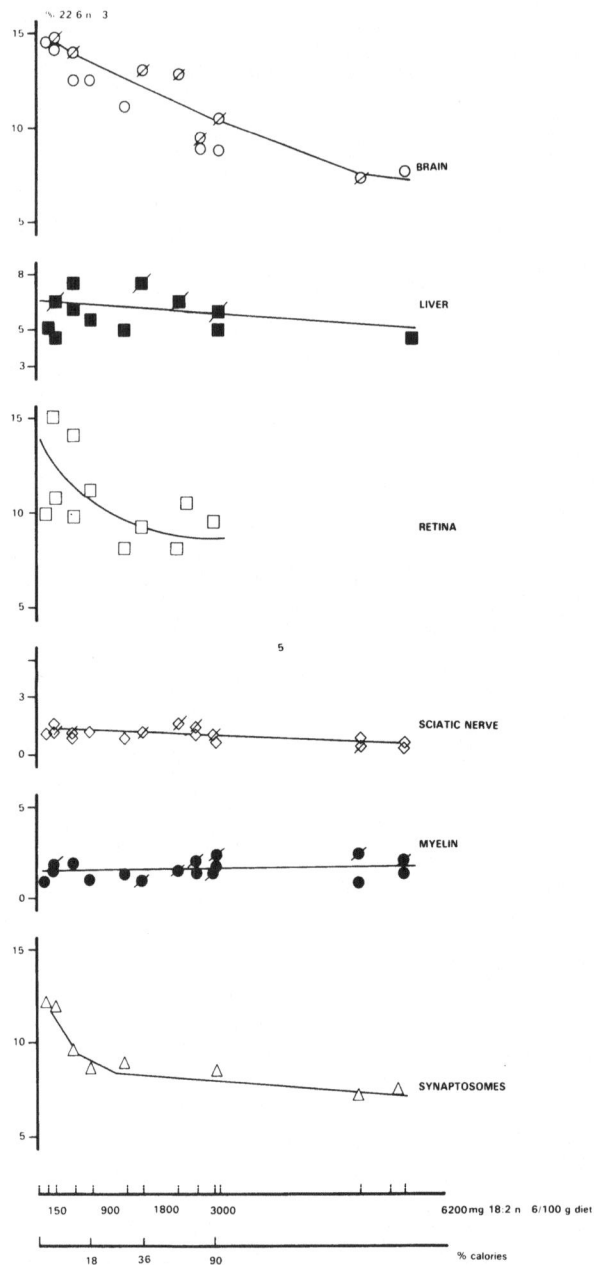

Figure 7. 22:6 n-3 in membranes is poorly affected by dietary linolenic acid 18:3 n-3: 150 mg/100 g diet: normal symbols; 300 mg/100g diet: symbols with a diagonal bar.

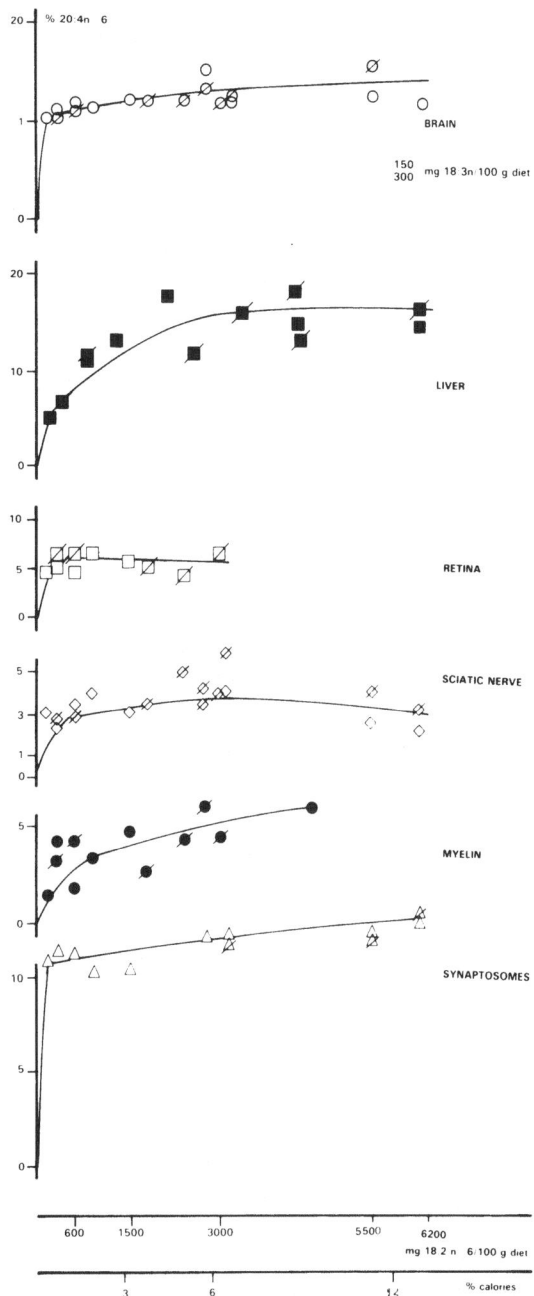

Figure 8. Arachidonic acid level in membranes depends on the dietary
linoleic acid. 18:2 n-6: 3200 mg/100g diet: normal symbols;
6400 mg/100 g diet: symbols with a diagonal bar.

CONCLUSIONS

A pathogenesis of linolenic acid deficiency has been described in the monkey[40], and in man[14,41-43]. A syndrome of modern society has been proposed as being a deficiency in series n-3 fatty acids[48]. It is, therefore, certainly very important to verify exactly the amount of n-3 acids in the diet : a minimum must be supplied to allow cerebral membranes to have a normal composition and function : 200 mg/100 g food intake is a minimum in the developing animal, hence also in man. Thus, this work recommends that alpha-linolenic acid represent 0.4% of the total dietary calories, in agreement with studies in animals[45] as well as in man[14,46,47].

The origin of brain saturated and unsaturated fatty acids (in situ synthesis and nutritional origin) is well documented[48] ; in contrast, polyunsaturated fatty acid metabolism in nerve tissue needs further study.

This study shows that the dietary linolenic requirement for membrane synthesis is the same whatever the organ (200 mg/100 g diet). In contrast, linoleic requirements vary between 150 and 1200 mg/100 g diet according to the organ (data to be published). Dietary n-6/n-3 ratio is between 1 and 6 according to the organ. Thus we suggest that minimum linolenic and linoleic acid intake should be 200 mg and 1200 mg/100 g diet respectively, i.e. 0.4% and 2.4% calories. If no pharmacological effect is required the n-6/n-3 ratio must be approx. 1/6.

Acknowledgements

This work was supported by INSERM,INRA, CETIOM and ONIDOL. The authors thank Michelle Bonneil for secretarial services.

REFERENCES

1. J.F. Mead, The non-eicosanoid functions of the essential fatty acids. J. Lipid Res. 25:1517-1521 (1984).
2. R.T. Holman, Control of polyunsaturated acids in tissue lipids. J. Amer. Col. Nutr. 5:183-211 (1986).
3. N.K. Menon, and G.A. Dhopeshwarkar, Essential fatty acid deficiency and brain development. Prog. Lipid Res. 21:309-326, 1982.
4. C. Alling, A. Bruce, I. Karlsson, O. Sapia, and L.J. Svennerholm, Effect of maternal essential fatty acid supply on fatty acid composition of brain, liver, muscle and serum in 21-day-old rats. J. Nutr. 102:773-782 (1971).
5. N. Bazan, S. Di Fazio De Escalante, M. Careaga, H.E.P. Bazan, and N.M. Giusto, High content of 22:6 (docosahexaenoate) and active (2-^{3}H) glycerol metabolism of phosphatidic acid from photoreceptor membranes. Biochim. Biophys. Acta 712:702-706 (1982).
6. R.R. Brenner, Effect of unsaturated acids on membrane structure and enzyme kinetics. Prog. Lipid Res. 23:69-96 (1984).
7. M.T. Clandinin, J.E. Chappell, S. Leong, T. Heim, P.R. Swyer, and G.W. Chance, Intrauterine fatty acid accretion rates in human brain : implications for fatty acid requirements. Early Human Development, 4/2:121-129 (1980).
8. M.T.Clandinin, J.E. Chappell, S. Leong, T. Heim, P.R. Swyer, and G.W. Chance, Extrauterine fatty acid accretion in infant brain : implication for fatty acid requirements. Early Human Development 4/2:131-138 (1980).
9. H.W.Cook, "In vitro" formation of polyunsaturated fatty acids by desaturation in rat brain : some properties of the enzyme in developing brain and comparison with liver. J. Neurochem. 30:1327-1334 (1978).

10. M.A. Crawford, and A.J. Sinclair, Nutritional influences in the evolution of mammalian brain. Ciba Foundation Symposium, Published by ASP (Elsevier. Excerpta Medical, North-Holland), Amsterdam. pp. 267-292 (1971).

11. M.A. Crawford, A.G. Hassam, and P.A. Stevens, Essential fatty acid requirements in pregnancy and lactation with special reference to brain development. Prog. Lipid Res. 20:31-40 (1981).

12. G.A. Dhopeshwarkar, and J.F. Mead, Uptake and transport of fatty acids into the brain and the role of the blood-brain barrier system. Adv. Lipid Res. 11:109-142 (1973).

13. R.T. Holman, R.T. Essential fatty acid deficiency. In : progress in the Chemistry of Fats and Other Lipids. 9, Part. 2. pp. 279, ed. R.T. Holman. Pergamon Press, Oxford (England) (1968).

14. R.T. Holman, S.B. Johnson, and T.F. Hatch, A case of human linolenic acid deficiency involving neurological abnormalities. Am. J. Clin. Nutr. 35:617-623 (1982).

15. R. Paoletti, and C. Galli, Effect of essential fatty acid deficiency on the central nervous system in the growing rat. In Lipid. Malnutrition and the Developing Brain. Ciba Foundation Symposium, pp. 121-140 (1972).

16. G.Y. Sun, and A.Y. Sun, Synaptosomal plasma membranes : acyl group composition of phosphoglycerides and $(Na^+ + K^+)$-ATPase activity during fatty acid deficiency. J. Neurochem. 22:15-18 (1974).

17. J.M. Bourre, G. Pascal, G. Durand, M. Masson, O. Dumont, and M. Piciotti, Alterations in the fatty acid composition of rat brain cells (neurons, astrocytes and oligodendrocytes) and of subcellular fractions (myelin and synaptosomes) induced by a diet devoid of n-3 fatty acids. J. Neurochem. 43:342-348, (1984).

18. G.W. Goldstein, Relation of potassium transport to oxidative metabolism in isolated brain capillaries. J. Physiol. 286:185-195 (1979).

19. P. Homayoun, F. Roux, E. Niel, and J.M. Bourre, The synthesis of lipids from $(1-^{14}C)$ acetate by isolated rat brain capillaries. Neurosci. Lett. 62:143-147 (1985).

20. C. Weidner, The presence of an albino ERG in the pigmented rat : genetic implications. J. Physiol. 77:813-821 (1981).

21. A. Nouvelot, J.M. Bourre, G. Sezille, P. Dewailly, and J. Jaillard, Changes in the fatty acid patterns of brain phospholipids during development of rats fed peanut or rapeseed oil, taking into account differences between milk and maternal food. Ann. Nutr. Metabol. 27:233-241 (1983).

22. A. Nouvelot, C. Delbart, and J.M. Bourre, Hepatic metabolism of dietary alpha-linolenic acid in suckling rats, and its posible importance in polyunsaturated fatty acid uptake by the brain. Ann. Nutr. Metab. 30:316-323 (1986).

23. A. Youyou, G. Durand, G. Pascal, M. Piciotti, O. Dumont, and J.M. Bourre, Recovery of altered fatty acid composition induced by a diet devoid or n-3 fatty acids in myelin, synaptosomes, mitochondria and microsomes of developing rat brain. J. Neurochem. 46:224-228 (1986).

24. Strouve-Vallet and M. ascaud, Désaturation de l'acide linoléique par les microsomes du foie et du cerveau du rat en développement. Biochimie 53:699-703 (1971).

25. P. Homayoun, G. Durand, G. Pascal, and J.M. Bourre, Alteration in fatty acid composition of adult rat brain capillaries and choroid plexus induced by a diet deficient in (n-3) fatty acids. Slow recovery by substitution with a non deficient diet. J. Neurochem. (in press).

26. J.M. Bourre, A. Youyou, G. Durand, and G. Pascal, Slow recovery of the fatty acid composition of sciatic nerve in rats fed a diet initially low in n-3 fatty acids. Lipids 22:535-537 (1987).

27. J. Bernsohn, and F.J. Spitz, Linoleic and linolenic acid dependency of some brain membrane-bound enzymes after lipid deprivation in rats. Biochem. Biophys. Res. Com. 57:293-298 (1974).

28. M. Foot, T.F. Cruz, and M.T. Clandinin, Influence of dietary fat on the lipid composition of rat brain synaptosomal and microsomal membranes. Biochem. J. 208:631-640 (1982).

29. J. Tinoco, P. Miljanich, and B. Medwadowski, Depletion of docosahexaenoic acid in retinal lipids of rats fed a linolenic acid-deficient, linoleic acid-containing diet. Biochim. Biophys. Acta 486:575-578 (1977).

30. M. Neuringer, and W.E. Connor n-3 fatty acids in the brain and retina : evidence for their essentiality. Nutr. Rev. 44, 289 (1986).

31. W.M.F. Leat, R. Curtis, N.J. Millichamp, and R.W. Cox, Retinal function in rats and guinea-pigs. Reared on diets low in essential fatty acids and supplemented with linoleic or linolenic acids. Ann. Nutr. Metab. 30:166-174 (1986).

32. A. Nouvelot, E. Dedonder, Ph. Dewailly, and J.M. Bourre, Influence des n-3 exogènes sur la composition en acides gras polyinsaturés de la rétine. Aspects structural et physiologique. Cah. Nutr. Diét. XX, 2:123-125 (1985).

33. J.A. Caldwell, and J.A. Churchill, Learning impairment in rats administered a lipid free diet during pregnancy. Psychol. Rep. 19:99-102 (1966).

34. M.S. Lamptey and B.L. Walker, Learning behaviour and brain lipid composition in rats subjected to essential fatty acid deficiency during gestation. Lactation and growth. J. Nutr. 108:358-367 (1978).

35. M. Lamptey and B.L. Walker, A possible essential role for dietary linolenic acid in the development of the young rat. J. Nutr. 106:86-93 (1976).

36. J.M. Bourre, A. Faivre, O. Dumont, A. Nouvelot, C. Loudes, J. Puymirat, and A. Tixier-Vidal, Effect of polyunsaturated fatty acids on fetal mouse brain cells in culture in a chemically defined medium. J. Neurochem. 41:1234-1242 (1983).

37. C. Loudes, A. Faivre, A. Barret, D. Grouselle, J. Puymirat, and M. Tixier-Vidal, Release of immunoreactive TRH in serum free culture of mouse hypothalamic cells. Dev. Brain Res. 9:231-234 (1983).

38. J. Chaudière, M. Clément, F. Driss, and J.M. Bourre, Unaltered brain membranes after prolonged intake of highly oxidizable long-chain fatty acids of the (n-3) series. Neurosci. Lett. 82, 233-239 (1987).

39. J.M. Bourre, M. Bonneil, O. Dumont, M. Piciotti, G. Nalbone, and H. Lafont, High dietary fish oil alters the brain polyunsaturated fatty acid composition. BBA 960:458-461 (1988).

40. R.N.T.Fiennes, A.J. Sinclair, and M.A. Crawford, Essential fatty acid studies in primates linolenic acid requirements of capuchins. J. Med. Prim. 2:155-169 (1973).

41. Anonymous, Combined EFA deficiency in a patient on long term TPN. Nutrition Reviews 44:301-305 (1986).

42. K.S. Bjerne, I.L. Mostad, and L. Thoresen, Alpha-linolenic acid deficiency in patients on long term gastric tube feeding : estimation of linolenic acid and long-chain unsaturated n-3 fatty acid requirement in man. Scand. Am. J. Clin. Nutr. 45:66-77 (1987).

43. K.S. Bjerve, S. Fisher, and K. Alme, Alpha-linolenic acid deficiency in man : effect of ethyl linolenate on plasma and erythrocyte fatty acid composition and biosynthesis of prostanoids. Am. J. Clin. Nutr. 46:570-576 (1987).

44. D. Rudin, The dominant diseases of modernized societies as omega-3 essential fatty acid deficiency syndrome : substrate beriberi. Med. Hypotheses. 8:17-47 (1982).

45. C. Pudelkewicz, J. Seufert, and R.T. Holman, Requirements of the female rat for linoleic and linolenic acids. J. Nutr. 94, 138-146 (1968).

46. M. Lasserre, F. Mendy, D. Spielmann, and B. Jacotot, Effects of different dietary intake of essential fatty acids on C20:3 w6 and C20:4 w6 serum levels in human adult. Lipids 4:227-233 (1985).

47. J.E. Kinsella, Food components with potential therapeutic benefits : the n-3 polyunsaturated fatty acids of fish oils. Food Technology pp. 89-97 (1986).

48. J.M. Bourre, Origin of aliphatic chains in brain. Dans "Neurological Mutations Affecting Myelination" (N. Baumann ed.) INSERM Symposium n°14. Elsevier/North Holland Biomedical Press pp. 187-206 (1980).

49. J.M. Bourre, G. Durand, G. Pascal, and A. Youyou, Recovery of altered polyunsaturated fatty acid composition induced by a diet deficient in n-3 fatty acids in brain cells (neurons, astrocytes and oligodendrocytes). Comparison with other organs. J. Nutr. (accepted).

OMEGA-3 FATTY ACIDS IN THE RETINA

Martha Neuringer and William E. Connor

Section of Clinical Nutrition, Department of Medicine
Oregon Health Sciences University
Portland, Oregon; and
Division of Neuroscience
Oregon Regional Primate Research Center
Beaverton, Oregon, USA

The retina and brain contain exceptionally high concentrations of docosahexaenoic acid (DHA, or 22:6 omega-3).[1-3] This fatty acid is a component of the phospholipids, especially phosphatidylethanolamine and phosphatidylserine, which are basic structural constituents of cell membranes. Particularly rich in DHA are specialized neural membranes, such as those of synaptic endings[4,5] and photoreceptor outer segments.[6,7]

The outer segment of the photoreceptor cell (Fig. 1) is a highly structured organelle designed for absorbing light and transducing its energy, via a series of amplifying biochemical reactions, into electrical signals which are communicated to other cells of the retina and then to the brain. The outer segment consists of stacks of hundreds of flattened sacs or disks of membrane. These membranes are very fluid because of their high content of phospholipid (80-90%) and low content of cholesterol (8-10%),[8] as well as their high level of polyunsaturated fatty acids.

DHA accounts for 35-60% of the total fatty acids in the phosphatidylethanolamine and phosphatidylserine of photoreceptor outer segment membranes, the highest percentage in any tissue.[6,7] In the sn-2 position of these phospholipids, which generally contains a high proportion of polyunsaturated fatty acids, DHA accounts for 75-100% of all fatty acids.[9,10] It is even more unusual that the sn-1 position, which typically contains saturates or monoenes, also is occupied by DHA 5-25% of the time.[9,11] Consequently, many of these phospholipid molecules contain DHA in both positions.[10,12] Indeed, rod outer segment membranes have been found to contain a number of phospholipid molecular species, termed "supraenes", with combinations of polyunsaturated fatty acids in the two positions,[12] and these species appear to have a particularly active metabolism.[13,14]

The reasons for the unique lipid composition of outer segment disk membranes, and the role of DHA in the visual process, remain unclear. The fact that DHA is selectively incorporated and retained by outer segment membrane lipids[15] suggests an important functional role, especially given the metabolic costs and biological hazards which are imposed by its presence. Because of its high degree of polyunsaturation, DHA is more

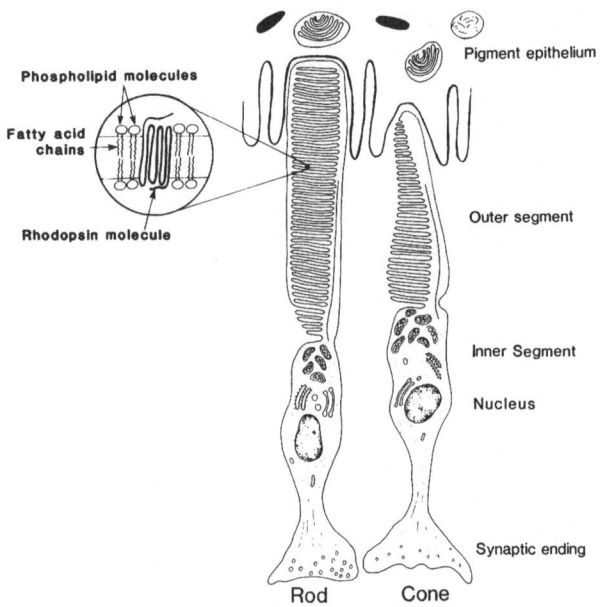

Fig.1. Schematic drawings of rod and cone photoreceptor cells, showing the stacks of DHA-rich membranous disks which form the outer segments. Packets of disk membranes in the pigment epithelium represent old disks which have been shed and phagocytized. The magnified section of disk membrane on the left represents a visual pigment molecule imbedded in the phospholipid membrane bilayer.

susceptible to lipid peroxidation that any other major tissue fatty acid, and therefore is selectively lost during oxidative stress. In the retina, this potential for damage is exacerbated by exposure to light and high oxygen tension. Damaging levels of light and oxidants such as ferrous sulfate are thought to precipitate retinal degeneration through their peroxidation of DHA.[16,17] A battery of protective mechanisms, including vitamins E and C and the glutathione enzymes, is needed to prevent or limit lipid peroxidation in the retina. In addition, the fragile membranes of outer segment disks must be continually renewed.[18] New disks form at the base of the outer segment, are displaced upward over a period of 9-14 days, and then separate in small packets to be phagocytized and digested by the retinal pigment epithelium (Fig. 1). Failure of this disk shedding and renewal process, as in some congenital retinal degenerations, leads to loss of the outer segment and eventually the entire photoreceptor cell.[19] Given the disadvantages of the unstable nature of membranes with high levels of DHA, these levels must convey substantial counterbalancing benefits, such as increased functionality of membrane proteins or increased speed of neurotransmission.[20]

The specialized lipid environment of the outer segment disk membrane contains the visual pigment as an integral membrane protein. In the case of vertebrate rod photoreceptors, this pigment is rhodopsin, a complex of the protein opsin and the chromophore 11-cis-retinal. Cone photoreceptors contain similar pigments with other chromophore groups. These pigments account for 80-90% of the protein in photoreceptor outer segments.[8] The rhodopsin molecule has alternating hydrophilic and hydophobic segments which are believed to form seven loops spanning the outer segment disk membrane (Fig. 1),[21] and it is tightly bound to the surrounding phospholipid molecules.

Rhodopsin initiates the process of vision when it is activated by absorption of a photon of light and is transformed through a series of intermediate steps to opsin and all-trans-retinal or free retinol (vitamin A). In the process, it undergoes an unfolding or conformational change which exposes G-protein binding sites, thus initiating a cascade of biochemical events culminating in generation of a neural signal.[21] The conformational change is thought to occur at the transition between two particular intermediate forms, metarhodopsin I and metarhodopsin II.[21,22] This step is thought to be critical for generation of electrical potentials within the photoreceptor.[23] During the excitation process, rhodopsin also rotates and moves laterally within the membrane.[24,25] Both the conformational change and this movement of rhodopsin appear to require a high degree of membrane fluidity and flexibility.[20]

Although outer segment disk membranes are very fluid, DHA does not appear to specifically enhance this property when compared to other polyunsaturated fatty acids. Indeed, in synthetic phosphatidylcholine membranes, fluidity appears to decrease slightly when DHA replaces linoleic (18:2 omega-6) or arachidonic (20:4 omega-6) acids in the sn-2 position.[26,27] Thus, membrane fluidity is not monotonically related to the number of double bonds in the constituent fatty acids of membrane phospholipids.

Biophysical properties other than fluidity may be more specifically related to high levels of DHA. It recently has been reported that the conversion of metarhodopsin I to metarhodopsin II is optimized in recombinant or synthetic outer segment membranes when the proportion of DHA in the membrane phospholipids is approximately 50%, as in normal outer segment membranes.[28] This conversion is accompanied by a change in conformation of the rhodopsin molecule, as mentioned above, together with an increase in the thickness of the outer segment disk membrane.[22] Applying hydrostatic pressure to the membrane inhibits both its thickening and the conversion to metarhodopsin II.[22] It has been proposed that the DHA chain has an unusual ability to coil and uncoil, giving the membrane greater capacity to compress and expand and thereby accommodate the conformational change.[29] In addition, DHA-containing phospholipids have been found to reduce the deformability of the membrane and the order of the fatty acid chain in the sn-1 position.[30] Finally, DHA shows substantial conformational motion of the methylene units making up the fatty acid chain,[31] and greater mobility of the terminal methyl group than any other fatty acid.[32]

In addition to its effects on the biophysical properties of the membrane, DHA may also have functionally important specific interactions with membrane proteins. Changes in the fatty acid composition of membrane phospholipids have been shown to affect the characteristics of membrane-associated enzymes, receptors, and transport systems.[27,33,34] Activities of brain 5'-nucleotidase, lysosomal β-N-acetyl-D-glucosaminidase, and Na+,K+-ATPase appeared to increase in the presence of omega-3 fatty acids,[35-37] whereas synaptosomal acetylcholinesterase activity decreased.[38] Salem and others have presented evidence that DHA-containing phosphatidylethanolamine and phosphatidylserine are selectively associated with membrane proteins in erythrocyte and synaptic plasma membranes,[39,40] and DHA-containing phospholipids appear to interact most strongly with rhodopsin.[41]

Another possible role for DHA in the retina is as a precursor of functionally important lipoxygenase products, either directly or via retroconversion to EPA.[42] Mono- and dihydroxy derivatives of DHA have been detected in both the retina and the brain.[39,43]

One way to assess the importance of DHA in the visual process is to examine the effects of DHA depletion, produced by dietary omega-3 fatty acid deprivation, on retinal function. The electroretinogram (ERG), a standard measure of retinal function, is an evoked response elicited by brief flashes of light. It is useful in the diagnosis of human retinal diseases such as retinitis pigmentosa.[44] Different stimulating conditions and states of light or dark adaptation can be used to differentiate responses of the rod system, which is responsible for night vision, and the cone system, which mediates daylight vision, color vision, and high visual acuity. Furthermore, activity of the photoreceptors is specifically seen in the A-wave, an early component of the ERG response, whereas the later B-wave reflects electrical currents in other, downstream retinal cells.

In 1973, Benolken et al.[45] first demonstrated a reduction in the amplitude of the ERG after feeding rats a completely fat-free diet. The A-wave was diminished more than the B-wave. A subsequent study showed that dietary omega-3 fatty acids, in the form of α-linolenic acid, fully prevented this abnormality, whereas linoleic (18:2 omega-6) or oleic (18:1 omega-9) acids did not.[46] However, the duration of feeding in the second study was only 25-40 days, too short a time to produce measurable changes in the fatty acid composition of the retina. More recently, Nouvelot et al.[47] and Watanabe et al.[48] found similar changes in ERG amplitude in rats deprived of omega-3 fatty acids for two generations, together with reductions of 30-50% in retinal DHA levels.

Two other studies have reported effects of omega-3 fatty acid deficiency on learning in rats.[49,50] The learning tasks were visual discriminations, in one case between the black and white alleys of a maze and in the other study between a bright and a dim stimulus light. The use of these tasks leaves open the possibility that slower learning in the deficient rats was due to an effect on vision rather than learning ability.

The visual system of rats, as well as their lipid metabolism, differs in many ways from that of the human. We therefore chose to examine the effects of omega-3 fatty acid deficiency in a higher primate species which closely resembles the human in its retinal anatomy, visual capacities, and nutritional requirements.

METHODS

Rhesus monkeys were studied after a combination of maternal and postnatal deprivation of dietary omega-3 fatty acids. Throughout pregnancy, adult females received a semipurified diet with safflower oil as the only fat. Their infants were hand-reared and received a similar liquid diet from birth. Both diets contained less than 0.3% of total fatty acids as α-linolenic acid (18:3 omega-3), and the ratio of omega-6 to omega-3 acids was approximately 250:1.[51,52] Mothers and infants in the control group received diets with the same composition except that the fat source was soybean oil, providing 7.7% of total fatty acids as α-linolenic acid and an omega-6:omega 3 ratio of 7:1. Neither diet contained detectable amounts of any longer-chain omega-3 fatty acids.

To test the reversibility of the deficiency, five of the deficient offspring were repleted with long-chain omega-3 fatty acids beginning at 10-24 months of age. In the repletion diet, fish oil (either San Omega® Nippon Oil & Fat Co., or MaxEPA®) replaced 80% of the safflower oil; the remaining safflower oil provided ample linoleic acid (4.5 percent of calories).

Plasma, erythrocyte, and tissue samples, including whole retina and cerebral gray matter, were analyzed for fatty acid composition by capillary column gas-liquid chromatography after separation of total phospholipids or individual phospholipid classes by thin layer chromatography[51,52]. In the repletion study, serial biopsies of prefrontal cortical gray matter were obtained by craniotomies under thiamylal anesthesia.

Visual function was assessed behaviorally by measuring visual acuity at 4, 8, and 12 weeks of age with the preferential-looking method, as described previously.[51] In addition, the specific physiological effects of omega-3 fatty acid deprivation on the function of the retina were examined by electroretinogram recordings. The ERG response of the cone system was isolated with flashes presented at 31 Hz in the light-adapted state. Rod responses were recorded to a series of flash intensities

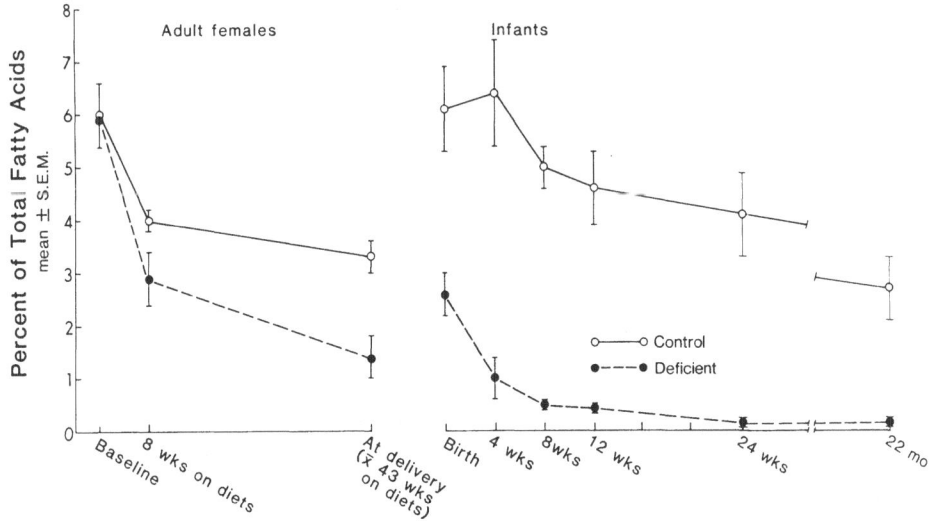

Fig. 2. Total omega-3 fatty acids (weight percent of total fatty acids) in plasma phospholipids of adult female rhesus monkeys and their offspring fed control (soybean oil) and omega-3 acid deficient (safflower oil) diets.

presented at 3.2-second intervals after 30 minutes of dark adaptation. The recovery function of the rod ERG was tested by repeating a relatively bright flash (1.75 cd/m^2-sec), bright enough to elicit a maximal ERG response, at intervals from 1 to 30 seconds. The amplitude of the second and subsequent responses was reduced at the shorter intervals, and percent recovery was calculated as the amplitude at these intervals divided by the maximal amplitude evoked at the longest interval.

RESULTS

Fatty Acid Composition

Dietary deprivation of omega-3 fatty acids resulted in low plasma levels of all omega-3 fatty acids (Fig. 2).[51] By 24 weeks of age, DHA was barely detectable in the plasma phospholipids of deficient infants.

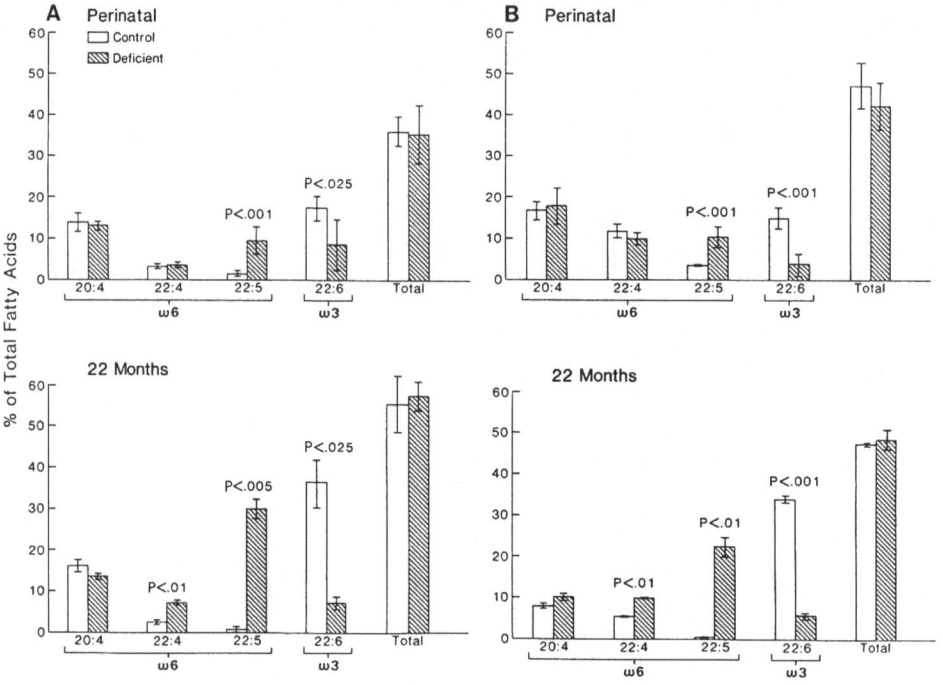

Fig. 3. Major omega-6 and omega-3 fatty acids (weight percent of total fatty acids, mean ± S.D.) of phosphatidylethanolamine in the retina (A) and occipital cortex (B) of rhesus monkeys in the control and omega-3 fatty acid deficient groups during the perinatal period and at 22 months of age.

Tissues were also depleted of DHA.[52] In deficient animals at or near birth, DHA levels in phosphatidylethanolamine were reduced by 50% in the retina and by 75% in cerebral cortex, relative to control values (Fig. 3). The proportion of DHA in both tissues doubled between birth and 22 months of age in control monkeys but failed to increase in the deficient group, so that by 22 months DHA levels in deficient monkeys were reduced by 80-85%. For each tissue at each age, increases in 22:5 omega-6 almost completely compensated for the reduced DHA, so that the degree of polyunsaturation was preserved as much as possible. This omega-6 fatty acid, which comprises less than 1% of total fatty acids in normal tissue phospholipids, rose to approximately 20% in phosphatidylethanolamine of the cerebral cortex and to nearly 30% in the retina of the 22-month-old deficient animals.

After dietary repletion with fish oil, these changes in fatty acid composition were rapidly reversed.[53] Indeed, levels of DHA and eicosapentaenoic acid (EPA, or 20:5 omega-3) increased beyond control values. Omega-3 fatty acids in plasma phospholipids increased from 0.1 to 33.6 percent of total fatty acids, with EPA accounting for 22% and DHA for 8.3% (Fig. 4). Omega-6 fatty acid levels fell reciprocally. Similar changes occurred in erythrocyte phospholipids and reached a new stable level after approximately 12 weeks of repletion. Changes in frontal cortical gray matter also occurred rapidly, as early as one week after fish oil supplementation. By 24-28 weeks, DHA in phosphatidylethanolamine increased from 4.2 to 29.3 percent of total fatty acids (Fig. 5), compared with 22.3 percent in soybean oil-fed control animals, and EPA and 22:5 omega-3 both increased from zero to approximately 3 percent. Meanwhile, levels of 22:5 omega-6 and other longer-chain omega-6 fatty acids decreased.

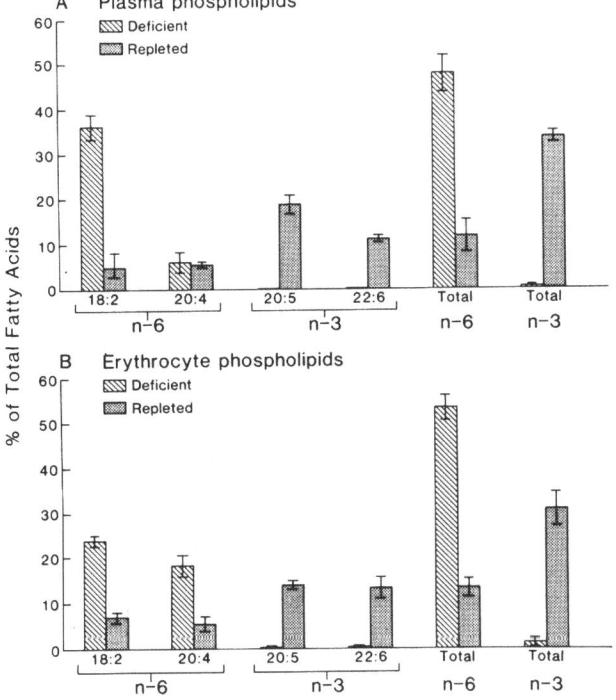

Fig. 4. Major omega-6 and omega-3 fatty acids (weight percent of total fatty acids, mean ± S.D.) in the plasma phospholipids (A) and erythrocyte phospholipids (B) of omega-3 fatty acid deficient juvenile monkeys (n=5) and the same monkeys after 12-28 weeks of dietary fish oil supplementation.

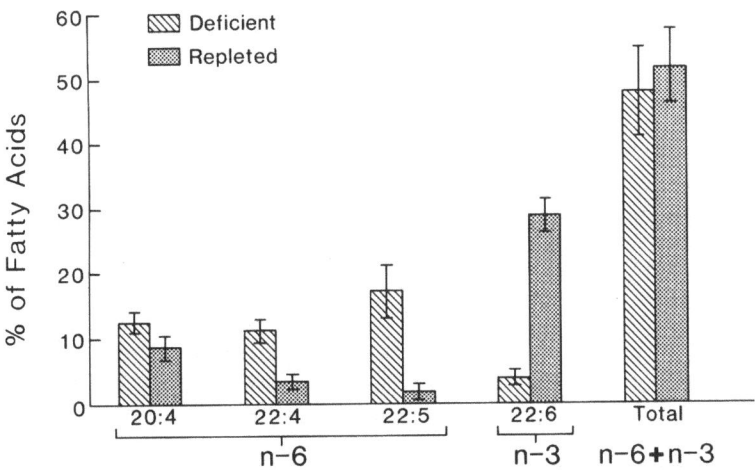

Fig. 5. Major omega-6 and omega-3 fatty acids (weight percent of total fatty acids, mean ± S.D.) in the frontal cortical gray matter of omega-3 fatty acid deficient monkeys (n=5) and the same monkeys after 24-28 weeks of dietary fish oil supplementation.

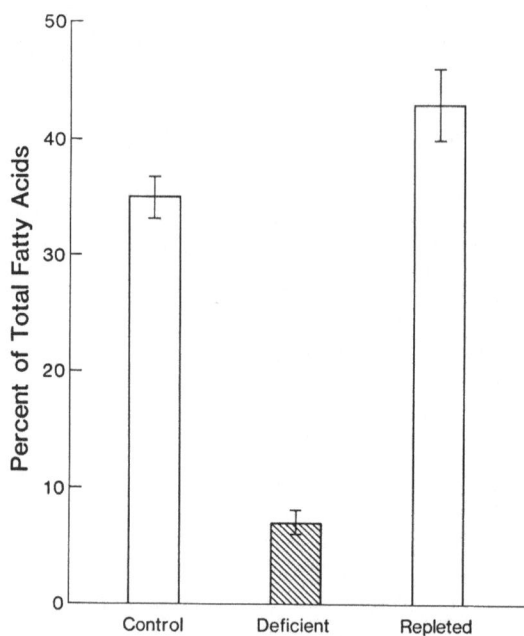

Fig. 6. DHA in phosphatidylethanolamine of the retina in control and omega-3 fatty acid deficient rhesus monkeys and in deficient monkeys after dietary fish oil repletion.

Omega-3 fatty acids stabilized at their new high level within about 12 weeks, while the proportion of 22:5 omega-6 took slightly longer to decline to the low level found in control cerebral cortex. The half-lives of DHA and 22:5 omega-6 in phosphatidylethanolamine were estimated at 21 and 32 days, respectively. The retinas of these animals, examined after durations of fish oil supplementation from 40 to 130 weeks, showed similar compositional changes (Fig. 6).

Visual Function

The visual acuity of omega-3 fatty acid deficient infants was reduced by half at 8 and 12 weeks of age (Fig. 7), as previously reported.[51] Furthermore, deficient monkeys showed a number of abnormalities in the electroretinogram. The timing of the ERG response was altered, with significant delays in the peak latency (time to the B-wave peak) of both cone and rod responses (Fig. 8). In contrast to previous studies in omega-3 fatty acid deficient rats, differences in response amplitudes were not detected at 7-24 months of age. However, more recent recordings of younger infants at 3-4 months of age have demonstrated clear differences in the A-wave amplitudes of both rod and cone responses. The reason for the transient nature of this effect is unknown. Deficient animals also showed a specific abnormality in the rate of recovery of the ERG response after an initial bright flash.[52] This effect was present at 3 months but increased in magnitude with age. With an interval of 3.2 seconds between flashes, response amplitude in the deficient animals was reduced nearly twice as much as in controls, relative to the maximal amplitude seen to the first flash or to flashes presented at long ($>$ 20 sec) intervals (Fig. 9). Thus, recovery of the capacity to generate a full ERG response was significantly slowed.

In deficient monkeys repleted with fish oil, ERGs were recorded at 3, 6, and 9 months after the beginning of the repletion phase. Despite the increase in omega-3 fatty acid levels in tissues, no improvement was seen in either peak latencies or in the ERG recovery function.

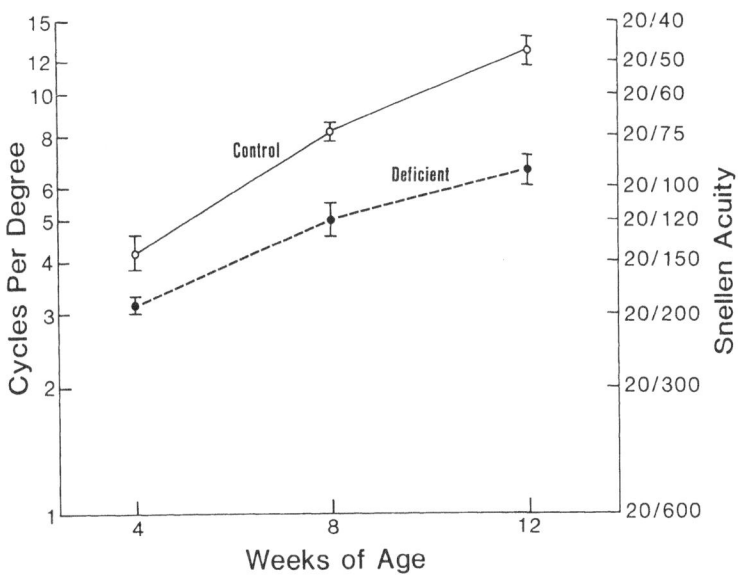

Fig. 7. Visual acuity thresholds (mean ± S.E.M.) of control and omega-3 fatty acid deficient monkeys at 4, 8, and 12 weeks of age. Acuity thresholds are expressed in cycles per degree of visual angle and in the equivalent Snellen acuity values.

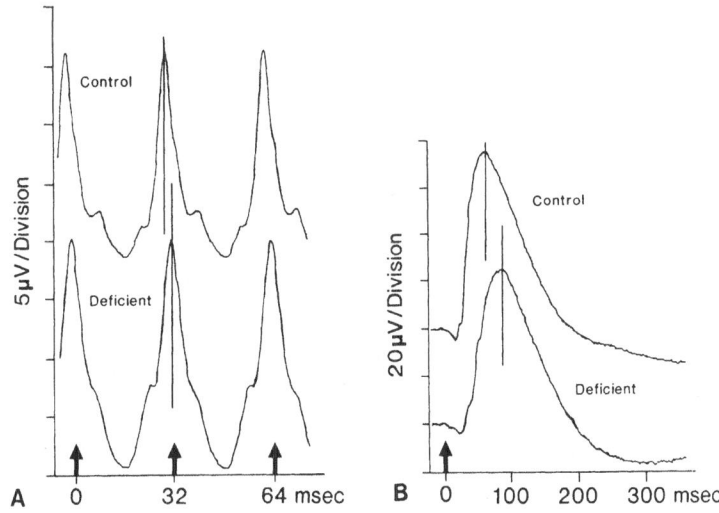

Fig. 8. Representative ERG waveforms from control and omega-3 fatty acid deficient monkeys under conditions selectively stimulating the cone system (A) and the rod system (B). Vertical arrows indicate time of flash. B-wave peak latencies were delayed in the deficient group; vertical lines have been drawn through the peaks to aid comparison.

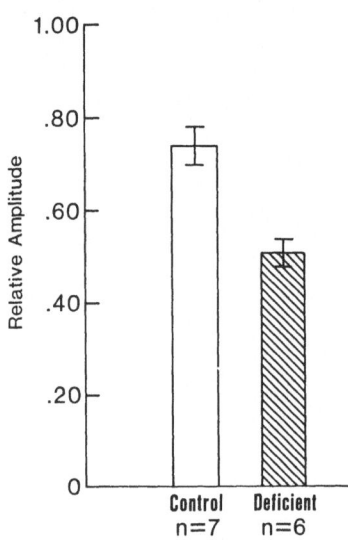

Fig. 9. Relative B-wave amplitude (mean ± S.E.M.) of the ERG elicited by flashes at 3.2-second intervals, as a percent of the maximal amplitude produced at intervals of 20 seconds or more. Relative amplitude is significantly reduced in the deficient group (p <.01).

DISCUSSION

In both rats and monkeys, omega-3 fatty acid deficiency leads to abnormalities in the electroretinogram, the retinal potential evoked by light stimulation. We do not yet understand the mechanisms involved in these changes, and their elucidation will require a far better understanding of how the presence of DHA in membrane phospholipids helps to create optimal functional properties of the membrane, including the proper environment for the activity of membrane proteins such as rhodopsin. However, it appears that the high concentrations of DHA normally found in photoreceptor membranes are necessary for the retina's ability to produce a normal ERG response and to recover this ability rapidly after light stimulation. The failure of fish oil supplementation to reverse these effects suggests that they are an irreversible result of omega-3 fatty acid deficiency during retinal development, or that the fatty acid composition of repleted retinas, although high in omega-3 fatty acids, remained abnormal in a functionally important way.

In monkey infants, we have also documented an effect on visual acuity development. This effect may be due to abnormalities in the function of the central retinal cone photoreceptors which mediate normal high levels of acuity, or to changes in the visual pathways of the central nervous system. Beyond effects on the visual system, studies of the effects of omega-3 fatty acid deprivation on brain function and behavior represent an extremely interesting area for future research, one which is only beginning to be explored.

Abnormalities in visual function provide the strongest evidence presently available for the essentiality of dietary omega-3 fatty acids. In all but one such study, the induction of the deficiency involved a combination of prenatal and postnatal dietary deprivation, with dietary omega-6:omega-3 ratios of 150:1 or greater. Such extreme and long-term deficiency is unlikely to be encountered in humans consuming natural diets. However, the effects of shorter-term deficiency have not been carefully explored. In one study in rats, deprivation for only 25-40 days produced significant changes in the electroretinogram.[46] Our own preliminary studies of monkeys deprived only postnatally also suggest that shorter-term deprivation may have a reduced but still significant impact.[54]

The data now available, although incomplete, suggest that the inclusion of omega-3 fatty acids in the human diet, particularly the diet of infants and of pregnant or lactating women, is the prudent course. For infants, a level of omega-3 acids approximating human milk is a reasonable goal. Most infant formulas contain ample α-linolenic acid, but those with corn oil as the only source of polyunsaturates have relatively low amounts, with a high ratio of omega-6 to omega-3 fatty acids (50 or 60:1). Corn oil diets in rats lead to reductions of omega-3 fatty acids in tissues,[55] and may also produce some functional alterations.[49] Furthermore, no commercial infant formula provides longer-chain polyunsaturates such as DHA and arachidonic acid, both of which are present in human milk. The capacity for desaturation needed to biosynthesize these fatty acids may be low in infants, as indicated by low levels in the erythrocytes of infants fed standard formulas versus human milk or DHA-supplemented formulas.[56-58] Animal studies have not yet adequately investigated the importance of preformed dietary DHA for the development of visual function.

ACKNOWLEDGEMENTS

We thank Betty Lou Hoeppner for preparation of the manuscript. This work was supported by grants DK-29930, HL-07295, and RR-00163 from the National Institutes of Health and by a grant from the Medical Research Foundation of Oregon. This is publication number 1624 of the Oregon Regional Primate Research Center.

REFERENCES

1. R.E. Anderson. Lipids of ocular tissues. IV. A comparison of the phospholipids from the retina of six mammalian species. Exp. Eye Res. 10:339-344 (1970).
2. J.S. O'Brien and E.L. Sampson. Fatty acid and fatty aldehyde composition of the major brain lipids in normal human gray matter, white matter, and myelin. J. Lipid Res. 6:545-551 (1965).
3. L. Svennerholm. Distribution and fatty acid composition of phosphoglycerides in normal human brain. J. Lipid Res. 9:570-579 (1968).
4. W.C. Breckenridge, I.G. Morgan, J.P. Zanetta, and G. Vincendon. Adult rat brain synaptic vesicles. II. Lipid composition. Biochim. Biophys. Acta 320:681-686 (1973).
5. C. Cotman, M.L. Blank, A. Moehl, and F. Snyder. Lipid composition of synaptic plasma membranes isolated from rat brain by zonal ultracentrifugation. Biochemistry 8:4606-4612 (1969).
6. R.E. Anderson, R.M. Benolken, P.A. Dudley, D.J. Landis, and T.G. Wheeler. Polyunsaturated fatty acids of photoreceptor membranes. Exp. Eye Res. 18:205-213 (1974).
7. W.L. Stone, C.C. Farnsworth, and E.A. Dratz. A reinvestigation of the fatty acid content of bovine, rat and frog retinal rod outer segments. Exp. Eye Res. 28:387-397 (1979).

8. F.J.M. Daemen. Vertebrate rod outer segment membranes. Biochim. Biophys. Acta 300:255-288 (1973).

9. R.E. Anderson and L. Sperling. Lipids of ocular tissues. VII. Positional distribution of the fatty acids in the phospholipids of bovine retina rod outer segments. Arch. Biochem. Biophys. 144:673-677 (1971).

10. R.D. Wiegand and R.E. Anderson. Phospholipid molecular species of frog rod outer segment membranes. Exp. Eye Res. 37:159-173 (1983).

11. R.E. Anderson and L.D. Andrews. Biochemistry of retinal photoreceptor membranes in vertebrates and invertebrates, in: "Visual Cells in Evolution," J. Westfall, ed., Raven Press, New York (1982).

12. M.I. Aveldano and N.G. Bazan. Molecular species of phosphatidylcholine, -ethanolamine, -serine, and -inositol in microsomal and photoreceptor membranes of bovine retina. J. Lipid Res. 24:620-627 (1983).

13. N.M. Giusto, M.I. De Boschero, H. Sprecher and M.I. Aveldano. Active labeling of phosphatidylcholines by [1-^{14}C]docosahexaenoate in isolated photoreceptor membranes. Biochim. Biophys. Acta 860:137-148 (1986).

14. N.P. Rotstein and M.I. Aveldano. Labeling of lipids of retina subcellular fractions by [1-^{14}C]eicosatetraenoate (20:4(n-6)), docosapentaenoate (22:5(n-3)) and docosahexaenoate (22:6(n-3)). Biochim. Biophys. Acta 921:221-234 (1987).

15. J. Tinoco, P. Miljanich and B. Medwadowski. Depletion of docosahexaenoic acid in retinal lipids of rats fed a linolenic acid-deficient, linoleic acid-containing diet. Biochim. Biophys. Acta 486:575-578 (1977).

16. R.D. Wiegand, C.D. Joel, L.M. Rapp, J.C. Nielsen, M.B. Maude and R.E. Anderson. Polyunsaturated fatty acids and vitamin E in rat rod outer segments during light damage. Invest. Ophthalmol. Vis. Sci. 27:727-733 (1986).

17. R.D. Wiegand, L.M. Rapp and R.E. Anderson. Ferrous ion-induced retinal degeneration: Biochemical changes in photoreceptor membranes. Invest. Ophthalmol. Vis. Sci. 26(Suppl.3):65 (1985).

18. R.W. Young. The renewal of photoreceptor cell outer segments. J. Cell Biol. 33:61-72 (1967).

19. R.J. Mullen and M.M. LaVail. Inherited retinal dystrophy: Primary defect in pigment epithelium determined with experimental rat chimeras. Science 192:799-801 (1976).

20. S.J. Fliesler and R.E. Anderson. Chemistry and metabolism of lipids in the vertebrate retina. Prog. Lipid Res. 22:79-131 (1983).

21. M.L. Applebury and P.A. Hargrave. Molecular biology of the visual pigments. Vision Res. 26:1881-1896 (1986).

22. A.A. Lamola, T. Yamane and A. Zipp. Effects of detergents and high pressures upon the metarhodopsin I to metarhodopsin II equilibrium. Biochemistry 15:738-745 (1974).

23. R.A. Cone and W.H. Cobbs. Rhodopsin cycle in the living eye of the rat. Nature 221:820-822 (1969).

24. P.K. Brown. Rhodopsin rotates in the visual receptor membrane. Nature New Biology 236:35-38 (1972).

25. R.A. Cone. Rotational diffusion of rhodopsin in the visual receptor membrane. Nature New Biology 236:39-43 (1972).

26. K.P. Coolbear, C.B. Bearde, and K.M.W. Keough. Gel to liquid-crystalline phase transitions of aqueous dispersions of polyunsaturated mixed acid phosphatidylcholines. Biochemistry 22:1466-1473 (1983).

27. E.A. Dratz and A.J. Deese. The role of docosahexaenoic acid (22:6n-3) in biological membranes: Examples from photoreceptors and model membrane bilayers, in: "Health Effects of Polyunsaturated Fatty Acids in Seafoods," A.P. Simopoulos, ed., Academic Press, New York (1986).

28. T.S. Weidmann, R.D Pates, J.M. Beach, A. Salmon, and M.F. Brown. Lipid-protein interactions mediate the photochemical function of rhodopsin. Biochemistry 27:6469-6474 (1988).

29. E.A. Dratz, N. Ryba, A. Watts, and A.J. Deese. Studies of the essential role of docosahexaenoic acid (DHA), 22:6 omega-3, in visual excitation. Invest. Ophthalmol. Vis. Sci. 28(Suppl.3):96 (1987).

30. M.R. Paddy and F.W. Dahlquist. Simultaneous observation of order and dynamics at several defined positions in the single acyl chain using ^{2}H NMR of single acyl chain perdeuterated phosphatidylcholines. Biochemistry 24:5988-5995 (1985).

31. F. Millett, P.A. Hargrave, and M.A. Raftery. Natural abundance ^{13}C nuclear magnetic resonance spectra of the lipid in intact bovine retinal rod outer segment membranes. Biochemistry 12:3591-3592 (1973).

32. C. Arus, W.M. Westler, M. Barany, and J.L. Markley. Observation of the terminal methyl group in fatty acids of the linolenic series by a new ^{1}H NMR pulse sequence providing spectral editing and solvent suppression. Application to excised frog muscle and rat brain. Biochemistry 25:3346-3351 (1986).

33. C.D. Stubbs and A.D. Smith. The modification of mammalian membrane polyunsaturated fatty acid composition in relation to membrane fluidity and function. Biochim. Biophys. Acta 779:87-137 (1984).

34. A.A. Spector and M.A. Yorek. Membrane lipid composition and cellular function. J. Lipid Res. 26:1015-1035 (1985).

35. J. Bernsohn and F.J. Spitz. Linoleic and linolenic acid dependency of some brain membrane-bound enzymes after lipid deprivation in rats. Biochem. Biophys. Res. Commun. 57:293-298 (1974).

36. A. Orlacchio, C. Maffei, L. Binaglia, and G. Porcellati. The effect of membrane phospholipid acyl-chain composition on the activity of brain ß-N-acetyl-D-glucosaminidase. Biochem. J. 195:383-388 (1981).

37. R. Tanaka. Comparison of lipid effects on K^+-Mg^{2+} activated p-nitrophenyl phosphatase and Na^+-K^+-Mg^{2+} activated adenosine triphosphatase of membrane. J. Neurochem. 16:1301-1307 (1969).

38. M. Foot, T.F. Cruz, and M.T. Clandinin. Effect of dietary lipid on synaptosomal acetylcholinesterase activity. Biochem. J. 211:507-509 (1983).

39. N. Salem, H.-Y. Kim, and J.A. Yergey. Docosahexaenoic acid: Membrane function and metabolism, in: "Health Effects of Polyunsaturated Fatty Acids in Seafoods," A.P. Simopoulos, ed., Academic Press, New York (1986).

40. V.A. Tyurin and N.V. Gorbunov. Fatty acid composition of aminophospholipids in protein microenvironment of plasmatic synaptic membranes of the brain in rat (in Russian). J. Evol. Biochem. Physiol., 591-594 (1983).

41. A.J. Deese, E.A. Dratz, F.W. Dahlquist, and M.R. Paddy. Interaction of rhodopsin with two unsaturated phosphatidylcholines: A deuterium nuclear magnetic resonance study. Biochemistry 20:6420-6427 (1981).

42. M.A. Yorek, R.R. Bohnker, D.T. Dudley, and A.A. Spector. Comparative utilization of n-3 polyunsaturated fatty acids by cultured human Y-79 retinoblastoma cells. Biochim. Biophys. Acta 795:277-285, (1984).

43. N.G. Bazan, D.L. Birkle, and T.J. Reddy. Docosahexaenoic acid (22:6n-3) is metabolized to lipoxygenase reaction products in the retina. Biochem. Biophys. Res. Commun. 125:741-747 (1984).

44. E.L. Berson. Electroretinographic testing as an aid in determining visual prognosis in families with hereditery retinal degenerations, in: "Retina Congress," R.C. Pruett and C.D.J. Regan, eds., Appleton-Century-Crofts, New York (1974).

45. R.M. Benolken, R.E. Anderson, and T.G. Wheeler. Membrane fatty acids associated with the electrical response in visual excitation. Science 182:1253-1254 (1973).

46. T.G. Wheeler, R.M. Benolken, and R.E. Anderson. Visual membranes: Specificity of fatty acid precursors for the electrical response to illumination. Science 188:1312-1314 (1975).

47. A. Nouvelot, E. Dedonder, P. Dewailly, and J.M. Bourre. Influence des n-3 exogenes sur la composition en acides gras polyinsatures de la retine, aspects structural et physiologique. Cah. Nutr. Diet. 20:123-125 (1985).

48. I. Watanabe, M. Kato, H. Aonuma, A. Hasimoto, Y. Naito, A. Moriuchi, and H. Okuyama. Effect of dietary alpha-linolenate/linoleate balance on the lipid composition and electroretinographic responses in rats. Adv. Biosciences 62:563-570 (1987).

49. M.A. Lamptey and B.L. Walker. A possible essential role for dietary linolenic acid in the development of the young rat. J. Nutr. 106:86-93 (1976).

50. N. Yamamoto, M. Saitoh, A. Moriuchi, M. Nomura, and H. Okuyama. Effect of dietary alpha-linoleate/linoleate balance on brain lipid compositions and learning ability of rats. J. Lipid Res. 28:144-151 (1987).

51. M. Neuringer, W.E. Connor, C. Van Petten, and L. Barstad. Dietary omega-3 fatty acid deficiency and visual loss in infant rhesus monkeys. J. Clin. Invest. 73:272-276 (1984).

52. M. Neuringer, W.E. Connor, D.S. Lin, L. Barstad, and S. Luck. Biochemical and functional effects of prenatal and postnatal omega-3 fatty acid deficiency on retina and brain in rhesus monkeys. Proc. Natl. Acad. Sci. USA. 83:4021-4025 (1986).

53. W.E. Connor, M. Neuringer, and D. Lin. The incorporation of docosahexaenoic acid into the brain of monkeys deficient in omega-3 essential fatty acids. Clin. Res. 33:598A (1985).

54. M. Neuringer, W.E. Connor, D. Daigle, and L. Barstad. Electroretinogram abnormalities in young infant rhesus monkeys deprived of omega-3 fatty acids during gestation and postnatal development or only postnatally. Invest. Ophthalmol. Vis. Sci. 29(Suppl. 3):145 (1988).

55. B.L. Walker. Maternal diet and brain fatty acids in young rats. Lipids 2:497-500 (1967).

56. S.E. Carlson, P.G. Rhodes, and M.G. Ferguson. Docosahexaenoic acid status of preterm infants at birth and following feeding with human milk or formula. Am. J. Clin. Nutr. 44:798-804 (1986)

57. S.E. Carlson, P.G. Rhodes, V. S. Rao, and D.E. Goldgar. Effect of fish oil supplementation on the n-3 fatty acid content of red blood cell membranes in preterm infants. Pediatr. Res. 21:507-510 (1987).

58. M.A. Crawford, A.G. Hassam, and B.M. Hall. Metabolism of essential fatty acids in the human fetus and neonate. Nutr. Metab. 21:187-188 (1977).

POLYUNSATURATED FATTY ACID METABOLISM IN ENDOTHELIAL CELLS:

RELATIONSHIP TO EICOSANOID FORMATION

Arthur A. Spector and Terry L. Kaduce
Department of Biochemistry
University of Iowa
Iowa City, IA 52242 USA

INTRODUCTION

Arachidonic acid, a member of the omega-6 (n-6) class of polyunsaturated fatty acids, plays a very important role in endothelial function. This is illustrated in Fig. 1. When the endothelium is stimulated by agonists such as thrombin or histamine, arachidonic acid (20:4) is hydrolyzed from phospholipids and converted to several vasoactive compounds. One of the major products is prostacyclin (PGI_2), a potent platelet anti-aggregatory agent and vasodilator[1]. Many other prostaglandin (PG) and lipoxygenase (LX) products are produced by various kinds of endothelial cells[2], among the major ones being prostaglandin E_2 (PGE_2) and 15-hydroxyeicosatetraenoic acid (15-HETE).[3-5] Sizable amounts

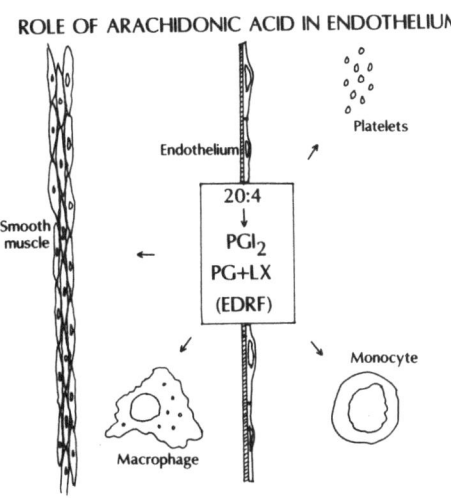

ROLE OF ARACHIDONIC ACID IN ENDOTHELIUM

Fig 1. Schematic representation of the role of arachidonic acid and its products in vascular function. The abbreviations are: 20:4, arachidonic acid; PGI_2, prostacyclin; PG, prostaglandins; LX, lipoxygenase products; EDRF, endothelium-derived relaxing factor. EDRF is listed in brackets because of uncertainty as to whether any of this material is derived from arachidonic acid.

of unmodified arachidonic acid also are released when endothelial cells are activated.[5,6] In addition, endothelium derived relaxing factor (EDRF), a powerful vasodilator, is released by endothelial cells.[7,8] Initially, this substance was thought to be an arachidonic acid metabolite. Although recent studies indicate that much of the EDRF activity is due to a form of nitrite, it is possible that some EDRF component still may be related to arachidonic acid oxidation. To indicate this possibility, EDRF is listed in brackets in Fig 1.

Essentially all of the eicosanoids that are formed by the endothelium are released into the extracellular fluid. Studies with endothelial monolayers grown on micropore filters indicate that PGI$_2$, several other metabolites, and unmodified arachidonic acid are released from both the apical and basolateral surfaces.[6] As shown in Fig 1, release into the apical fluid is thought to prevent platelet aggregation.[1] It may also affect the adherence of monocytes which bind to the endothelial surface.[9] The basolateral release is likely to affect the underlying smooth muscle, thereby causing vasodilation, and possibly influence the properties of macrophages that enter the arterial intima.

In order to better understand the factors that regulate the formation of eicosanoids in endothelium, we have investigated the metabolism of arachidonic acid and related polyunsaturated fatty acids in endothelial cultures.

LINOLEIC ACID UTILIZATION

Endothelial cultures readily incorporate linoleic acid (18:2). As shown in Fig. 2, most of the linoleic acid remains as 18:2 under the usual culture conditions, where fetal bovine serum and moderate amounts of 18:2 are present in the medium.[10,11] Under these conditions, there is a sizable increase in the cellular 18:2 content, as measured by gas-liquid chromatography (GLC).[12]

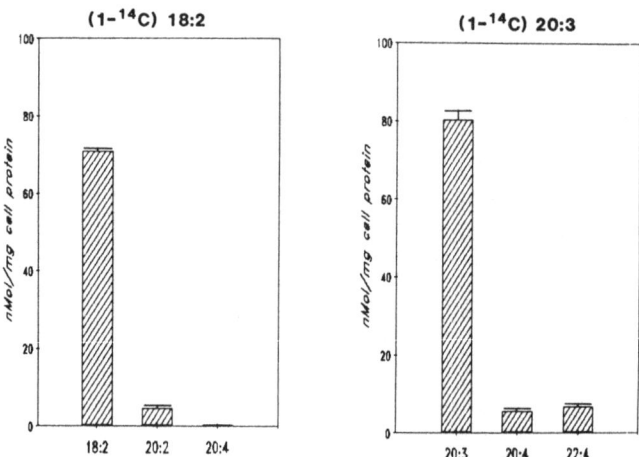

Fig. 2. Elongation and desaturation of fatty acid by human umbilical vein endothelial cells. Subconfluent cultures were incubated for 24 h in 20% fetal bovine serum with either [1-^{14}C]18:2 or [1-^{14}C]20:3n-6, and the radioactive products separated by GLC. A small amount of linoleic acid is elongated to docosadienoic acid (20:2) but very little is converted to 20:4.

By contrast, these endothelial cultures convert considerably more of eicosatrienoic acid (20:3n-6) to 20:4 as seen on the right side of Fig. 2. Some of the 20:4 that is formed subsequently is converted to docosa-tetraenoic acid (22:4). Further studies indicate that endothelial cells can retroconvert 22:4 to 20:4.[13]

Taken together, these findings indicate that the 6-desaturase activity is suppressed when endothelial cells derived from large blood vessels are grown in culture. By contrast the 5-desaturase that converts 20:3 to 20:4 and the elongation steps operate at a higher level. Suppression of the 6-desaturase is produced by the high content of available extracellular lipid, especially polyunsaturated fatty acids.[14] Under physiological conditions, the endothelium is bathed by the plasma, which ordinarily also is quite rich in lipids, including arachidonic acid. Therefore, it is likely that the 6-desaturase activity operates at a relatively low level _in vivo_ and that endothelium derives much of its 20:4 preformed from the plasma. Since there is relatively little 20:3n-6 in plasma, it is unlikely that 20:4 is obtained primarily by uptake and desaturation of 20:3.

This interpretation is supported by the finding that endothelial cells readily take up large quantitites of 20:4, even when it is available in low concentrations.[10] An excess of the commonly occurring saturated and monounsaturated fatty acids does not suppress 20:4 uptake, only the 20- and 22-carbon polyunsaturates compete effectively with 20:4.[15] The 20:4 is rapidly incorporated into all of the endothelial phospho-glycerides[10]. If an excess of 20:4 is available to the endothelium, it is stored in triglycerides or elongated to 22:4.[13,16] Subsequently, the triglyceride can be hydrolyzed, and the 20:4 utilized or even released from the endothelial cells as free fatty acid.[17,18]

When exposed to 20:4, endothelial cells can directly convert some 20:4 into PGI_2 and other eicosanoids, without the need for any added

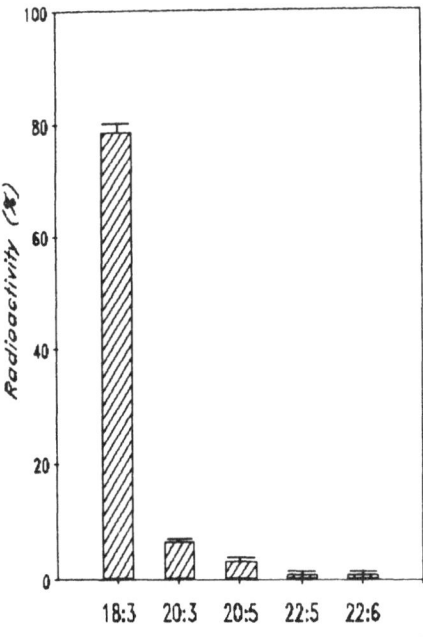

Fig 3. Elongation and desaturation of linolenic acid (18:3n-3). Human umbilical vein endothelial cells were incubated for 48 h with 50 μM [1-^{14}C]18:3 in 2.5% fetal bovine serum.

stimulus and, apparently, without the 20:4 passing through a cellular phospholipid pool.[5,12,19] Thus, endothelial cells exhibit a full range of metabolic pathways for arachidonic acid, except that synthesis from 18:2 is suppressed when an adequate supply of 20:4 is available extra-cellularly.

OMEGA-3 POLYUNSATURATED FATTY ACIDS

As shown in Fig 3, endothelial cells readily take up linolenic acid (18:3n-3). Most of the incorporated radioactivity remains as 18:3. The main metabolite of 18:3 is the elongation product, 20:3n-3, and very little is further desaturated. This is consistent with suppression of the 6-desaturase activity when extracellular lipids are available, a conclusion drawn from the results with linoleic acid.[10,11,14]. A higher level of desaturation and conversion to eicosapentaenoic acid (EPA or 20:5n-3) and docosapentaenoic acid (22:5n-3) occurs when the serum or 18:3 content of the medium is low,[20] but the results shown in Fig 3 probably more closely reflect the physiologic condition. Therefore, it is likely that much of the EPA and 22:6 present in endothelium, like 20:4, is obtained preformed from the plasma rather than through synthesis from 18:3.

Consistent with this interpretation, endothelial cells readily take up preformed EPA and docosahexaenoic acid (22:6n-3).[15,20,21] The EPA is elongated to 22:5, but very little is further desaturated to 22:6. Both EPA and 22:6 are readily incorporated into endothelial phospholipids. Most of the incorporated EPA is present initially in phosphatidylcholine (PC), whereas more of the 22:6 is recovered in the ethanolamine phospholipids (PE) than in PC.[15,21] About 35% of the 22:6 incorporated into the PE fraction is contained in plasmalogens.[21] Furthermore, some of the incorporated 22:6 can be released into the extracellular fluid as free fatty acid, or retroconverted to EPA.[21] Thus, the endothelium may serve as a site for processing 22:6 and under certain conditions may be able to supply surrounding tissues with this fatty acid.

The availability of saturated and monounsaturated fatty acid does not interfere with the ability of endothelial cells to take up EPA and 22:6, even when they are present in low concentrations. Only polyun-

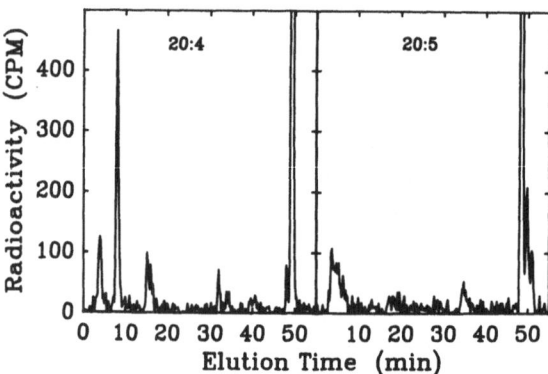

Fig. 4. Eicosanoid formation by endothelial cells. Human umbilical vein endothelial cultures were incubated for 20 min with 7.5 µM of either [1-^{14}C]20:4 or [1-^{14}C]20:5. The radioactive products were separated by HPLC and assayed with an on-line flow scintillation detector.[21]

saturated fatty acids, especially 20:4, compete effectively with EPA and 22:6 for uptake. Likewise, EPA and 22:6 effectively compete with 20:4 for incorporation into endothelial lipids.[15,21] This indicates that the content of EPA, 22:6 and 20:4 in endothelial lipids is determined in large part by the relative availability of each of these polyunsaturated fatty acids in the extracellular fluid.

EICOSANOID FORMATION

The radioactive products formed by human umbilical vein endothelial cultures during incubation with [1-^{14}C]arachidonic acid are shown on the left side of Fig. 4. PGI_2, the main product, has a retention time of 7.5 min in this high performance liquid chromatography (HPLC) system. PGE_2 has a retention time of 16 min, several hydroxylated metabolites elute between 30 and 35 min, and a small amount of two unidentified metabolites elute between 39 and 42 min. The unmodified 20:4 elutes at 49.5 min.

A similar HPLC tracing obtained following incubation with [1-^{14}C]EPA is shown on the right side of Fig. 4. Although a sizable amount of polar radioactivity elutes between 3 and 8 min, there is no prominant peak that corresponds to the expected retention time of PGI_3. Additional experiments with these cultures indicate that at most, the amount of PGI_3 produced is only 8% as much as PGI_2 under comparable conditions.[20] Likewise, there are no prominent peaks corresponding to other prostaglandins in the incubation with [1-^{14}C]EPA. There is, however, a comparable amount of hydroxylated metabolites formed, eluting between 33 and 37 min, as in the incubation with [1-^{14}C]20:4. The unmodified 20:5 elutes at 49 min in this chromatogram. Bovine aortic endothelial cultures also produce little or no PGI_3 from EPA.[21] Therefore, under the conditions that are ordinarily employed where sizable amounts of 20:4 are converted to PGI_2, we find little conversion of 20:5 to PGI_3 or any other prostaglandin.

As shown in Fig. 5, we find that EPA, and to some extent 22:6, inhibit PGI_2 formation. Very little inhibition occurred when the cultures

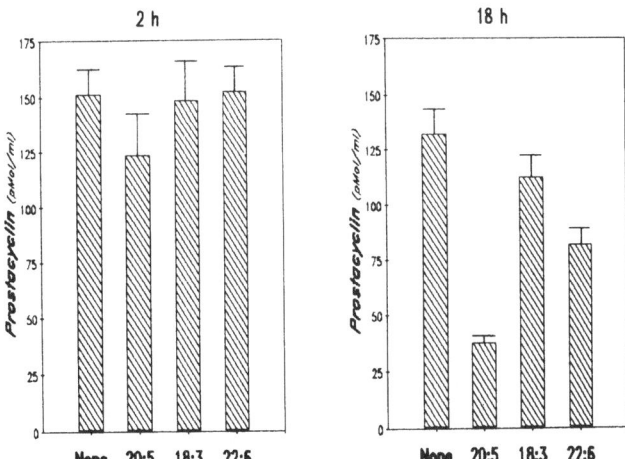

Fig. 5. Inhibition of PGI_2 formation by omega-3 fatty acids. Human umbilical vein endothelial cells were incubated for either 2 h (left side), or 18 h (right side) with either 90 µM albumin alone, or this protein containing 100 µM 18:3, 20:5 or 22:6. After these media were removed and the cultures washed, they were incubated for 5 min with 1 unit thrombin in 1 ml of a buffered salt solution. PGI_2 release was measured by radioimmunoassay for 6-keto-PGF$_{1\alpha}$.[20]

were exposed to these fatty acids for only 2 h. After 18 h of exposure, there still was very little inhibition produced by 18:3. By contrast, PGI_2 formation was reduced by 70% when the cultures were exposed to EPA, and by about 40% after exposure to 22:6. Since a small amount of 22:6 is retroconverted to EPA,[21] it is possible that EPA formation may account for the inhibitory effect of 22:6, rather than a direct action of 22:6 on eicosanoid synthesis. However, additional experiments in which mixtures of arachidonic acid and 22:6 were added directly to the cultures indicate that 22:6 itself can reduce endothelial PGI_2 formation.[21]

Additional studies indicate that EPA reduces PGI_2 formation when it is added together with 20:4 during a 20 min incubation, without any need for either preincubation or the addition of an agonist.[15] Concentrations of EPA as low as 5 to 15 µM are effective in this assay, with 5 µM EPA reducing PGI_2 formation from 7.5 µM 20:4 by 50%.[15] Likewise, EPA reduces PGI_2 formation when it is added as a component of a fatty acid mixture, the EPA accounting for only 5 to 15% of the total added fatty acids.[15] These findings suggest that 20:5, and to a lesser extent 22:6, reduce the capacity of the endothelium to produce PGI_2.

Other studies indicate that the content of 20:4 in endothelial lipids can vary somewhat, depending on the availability of 20:4 in the extracellular fluid.[19] The amount of PGI_2 that the cultures release when they are exposed to thrombin depends to some extent on the cellular 20:4 content.[19]

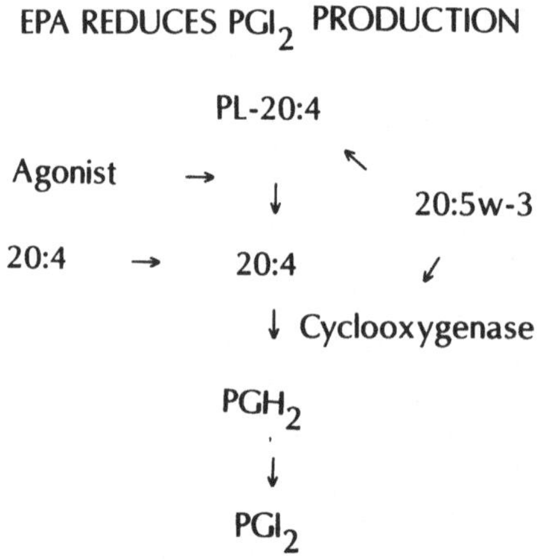

Fig. 6. Proposed effect of eicosapentaenoic acid on endothelial prostacyclin formation. The abbreviations are: EPA, eicosapentaenoic acid or 20:5n-3; PL, phospholipid; 20:4, arachidonic acid.

DISCUSSION AND CONCLUSIONS

A number of important conclusions regarding the metabolism and interactions of n-6 and n-3 polyunsaturated fatty acids in the vascular endothelium are suggested by these studies. When the endothelium is exposed to plasma lipids, the 6-desaturase activity is likely to be suppressed. Under these conditions, most of the arachidonic acid is obtained preformed from the plasma, rather than through desaturation and elongation of 18:2. Likewise, EPA and 22:6 also appear to be taken up preformed instead of being synthesized from 18:3. There is competition between 20:4 and EPA for incorporation into endothelial phospholipids.[15] Thus, the endothelial content of these polyunsaturates depends primarily on the relative amount of each present in the plasma. Since 22:6 can be taken up and retroconverted by endothelial cells,[21] it is a secondary source of EPA.

The most important conclusion is that EPA and 22:6, possibly through retroconversion to EPA, reduce the capacity of the endothelium to produce PGI_2. As illustrated in Fig. 6, this is thought to occur in two ways. First, EPA competes with 20:4 for access to cyclooxygenase. Second, EPA also competes with 20:4 for incorporation into endothelial phospholipids, thereby lowering the 20:4 content of the intracellular storage pools.[15,20] The amount of 20:4 in these pools influences the quantity of PGI_2 formed by endothelial cells in response to a stimulus.[19] Through a combination of both effects, less 20:4[15] available for release from phospholipids because of replacement with EPA, and competition between the available 20:4 and EPA for cyclooxygenase, PGI_2 formation is reduced. The amount of PGI_3 that is formed from EPA does not compensate for the reduction in PGI_2 output (Fig. 4). This may have pathological significance, for it suggests that the vasodilator and platelet anti-aggregatory potential of the endothelium may be reduced when excessive amounts of EPA are available.

Additional studies with a variety of cultured cells, including smooth muscle, agree with the conclusion that less prostaglandins are formed from 20:4 when EPA is available.[22,23]

As opposed to these results with cell cultures, clinical studies indicate the feeding a diet rich in EPA and 22:6 does not reduce PGI_2 formation.[24,25] Furthermore, these clinical studies indicate that sizable amounts of PGI_3 are produced from EPA. It is important to resolve the apparent discrepancy between the clinical and cell culture results. A commonly held advantage of the omega-3 fatty acids is based on the finding that they do not compromise the vasodilator and anti-aggregatory properties of the endothelium, as indicated by the clinical measurements of PGI_2 and PGI_3 excretion in the form of urinary metabolites.[24,25]

It is possible that the results obtained with cultured cells do not properly reflect the physiologic situation. Excessive EPA concentrations may have been employed, autooxidation may have occurred during long-term culture, or free fatty acid may not be the form in which EPA is available to cells in vivo. Alternatively, the clinical data may not adequately reflect endothelial prostaglandin formation. It is possible that the urinary metabolites are not entirely derived from the endothelium, and urinary collections may not indicate the extent to which the endothelium can respond to an acute stimulus. If the capacity of endothelial cells to produce dienoic prostaglandins is truly compromised by EPA, some caution in the use of omega-3 fatty acid supplements may be indicated until it can be demonstrated that such a reduction is not harmful.

ACKNOWLEDGMENT

These studies were supported by Arteriosclerosis Specialized Center of Research grant HL 14230 from the National Heart, Lung and Blood Institute, National Institutes of Health.

REFERENCES

1. S. Moncada, Prostacyclin and arterial wall biology, Arteriosclerosis 2:193 (1982).
2. A. R. Johnson, G. Revtyak, and W. B. Campbell, Arachidonic acid metabolites and endothelial injury: Studies with cultures of human endothelial cells, Fed. Proc. 44:19 (1985).
3. B. Mayer, R. Moser, H. Gleispach, and W. R. Kukovetz, Possible inhibitory function of endogenous 15-hydroperoxyeicosatetraenoic acid on prostacyclin formation in bovine aortic endothelial cells, Biochim. Biophys. Acta 875:641 (1986).
4. N. K. Hopkins, T. D. Oglesby, G. L. Bundy, and R.R. Gorman, Biosynthesis and metabolism of 15-hydroperoxy-5,18,11,13-eicosatetraenoic acid by human umbilical vein endothelial cells, J. Biol. Chem. 259:140 (1984).
5. S. A. Moore, A. A. Spector, and M. N. Hart, Eicosanoid metabolism in cerebromicrovascular endothelium, Am. J. Physiol. 254:C37 (1988).
6. D. M. Shasby, L. L. Stoll, and A. A. Spector, Polarity of arachidonic acid metabolism by bovine aortic endothelial cell monolayers, Am. J. Physiol. 253:H1177 (1987).
7. R. F. Furchgott, Role of endothelium in responses of vascular smooth muscle, Circ. Res. 53:557 (1983).
8. M. J. Peach, H. A. Singer, and A. L. Loeb, Mechanisms of endothelium-dependent vascular smooth muscle relaxation, Biochem. Pharmacol. 34:1867 (1985).
9. P. E. DiCorleto and C.A. de la Motte, Characterization of the adhesion of the human monocytic cell line U937 to cultured endothelial cells, J. Clin. Invest. 75:1153 (1985).
10. A. A. Spector, T. L. Kaduce, J. C. Hoak, and G. L. Fry, Utilization of arachidonic and linoleic acids by cultured human endothelial cells, J. Clin. Invest. 68:1003 (1981).
11. T. L. Kaduce, A. A. Spector, and R. S. Bar, Linoleic acid metabolism and prostaglandin production by cultured bovine pulmonary artery endothelial cells, Arteriosclerosis 22:380 (1982).
12. A. A. Spector, J. C. Hoak, G. L. Fry, G. M. Denning, L. L. Stoll, and J. B. Smith, Effect of fatty acid modification on prostacyclin production by cultured human endothelial cells, J. Clin. Invest. 65:1003 (1980).
13. C. J. Mann, T. L. Kaduce, P. H. Figard, and A. A. Spector, Docosatetraenoic acid in endothelial cells: Formation, retroconversion to arachidonic acid, and effect on prostaglandin production, Arch. Biochem. Biophys. 244:813 (1986).
14. M. D. Rosenthal, and C. Whitehurst, Fatty acid $\Delta 6$ desaturation activity of cultured human endothelial cells. Modulation by fetal bovine serum, Biochim. Biophys. Acta 750:490 (1983).
15. C. Hadjiagapiou, T. L. Kaduce, and A. A. Spector, Eicosapentaenoic acid utilization by bovine aortic endothelial cells: Effects on prostacyclin production, Biochim. Biophys. Acta 875:369 (1986).
16. G. M. Denning, P. H. Figard, T. L. Kaduce, and A. A. Spector, Role of triglycerides in endothelial cell arachidonic acid metabolism, J. Lipid Res. 24:993 (1983).
17. P. H. Figard, D. P. Hejlik, T. L. Kaduce, L. L. Stoll, and A. A. Spector, Free fatty acid release from endothelial cells. J. Lipid Res. 27:771 (1986).
18. L. L. Stoll, and A. A Spector, Lipid transfer between endothelial and smooth muscle cells in coculture, J. Cell. Physiol. 133:103 (1987).

19. A. A. Spector, T. L. Kaduce, J. C. Hoak, and R. L. Czervionke, Arachidonic acid availability and prostacyclin production by cultured human endothelial cells, Arteriosclerosis 3:323 (1983).

20. A. A. Spector, T. L. Kaduce, P. H. Figard, K. C. Norton, J. C. Hoak, and R. L. Czervionke, Eicosapentaenoic acid and prostacyclin production by cultured human endothelial cells, J. Lipid Res. 24:1595 (1983).

21. C. Hadjiagapiou, and A. A. Spector, Docosahexaenoic acid metabolism and effect on prostacyclin production in endothelial cells, Arch. Biochem. Biophys. 253:1 (1987).

22. I. Morita, Y. Saito, W. C. Chang, and S. Murota, Effects of purified eicosapentaenoic acid on arachidonic acid metabolism in cultured murine aortic smooth muscle cells, vessel walls and platelets, Lipids 18:42 (1983).

23. L. Levine, and N. Worth, Eicosapentaenoic acid: Its effects on arachidonic acid metabolism by cells in culture, J. Allergy Clin. Immunol. 74:430 (1984).

24. C. V. Shacky, W. Seiss, S. Fischer, and P. C. Weber, A comparative study of eicosapentaenoic acid metabolism by human platelets in vivo and in vitro, J. Lipid Res. 26:457 (1985).

25. S. Fischer, and P. C. Weber, Prostaglandin I_3 is formed in vivo in man after dietary eicosapentaenoic acid, Nature 307:165 (1984).

MODIFICATION OF THE ARACHIDONIC ACID CASCADE BY LONG-CHAIN W-3 FATTY ACIDS

Peter C. Weber

Institut fur die Prophylaxe der Kreislaufkrankheiten
b.d. Universitat Munchen, Pettenkoferstr. 9, 8000
Munchen 2, FRG

INTRODUCTION

Physiological and pathophysiological reactions like
vascular resistance, thrombosis, wound healing, inflammation
and allergy are modulated by oxygenated metabolites of
arachidonic acid (AA; 20:4w-6 or 20:4n-6) and related
polyunsaturated fatty acids that are collectively termed
eicosanoids. These compounds include prostaglandins,
prostacyclin, thromboxane, leukotrienes and other oxygenated
derivatives of arachidonic acid. The eicosanoids are all
derived from essential fatty acids that must be provided in
the diet, and production of eicosanoids is controlled by
cellular mechanisms for the uptake, release and oxygenation
of the eicosanoid precursor fatty acids.

Mammalian cell membranes consist of a lipid bilayer
composed primarily of phospholipids and cholesterol.
Proteins that have important cellular functions such as
receptors, transporters, and enzymes are embedded in the
lipid bilayer. Although the following review will be
focused on the modification of the eicosanoid system by
dietary fatty acids, it is clear that some of the resulting
cellular functional changes might be related more directly
to the modifications of membrane phospholipid fatty acid
composition and their effects on those proteins embedded in
the lipid bilayer.

EICOSANOIDS AS MODULATORS OF CELL FUNCTION

Eicosanoids and related lipid mediators, such as 1,2
diacylglycerol (DAG) or platelet activating factor (PAF),
function as a modulatory device in cell stimulus-response

coupling and for cell to cell communication. Some eicosanoids, like thromboxane A2 and leukotriene B4, might amplify an initial (Ca++-related) signal for cell activation by stimulating specific membrane receptors coupled to phospholipase C, thereby further increasing i.c. Ca++ concentrations. Other eicosanoids, such as PGI2 or PGD2 -via an increase of cAMP - might, on the contrary, blunt an initial signal for cell activation by decreasing i.c. Ca++ release.

NUTRITION AND THE EICOSANOID SYSTEM

Interference with eicosanoid synthesis is characteristic for many of the therapeutic agents in use in developed countries, including antiinflammatory drugs and antithrombotic agents, antihypertensives and diuretics, suggesting that eicosanoids are involved in a broad spectrum of disease processes prevailing in these countries. A change of eicosanoid production and eicosanoid-dependent cellular functions may, however, also be achieved by altering eicosanoid precursor availability.

Under our "Western" dietary conditions, arachidonic acid is by far the dominant precursor fatty acid of biologically highly active eicosanoids of the two-series. The major primary source of AA in our food chain is linoleic acid (18:2w-6). At variance to the fatty acids of the linoleic- or w-6-family in terrestrial animals and in most plant seeds - which are the major sources of our eicosanoid precursor fatty acids - the fatty acids of the linolenic- or w-3-family predominate in green leaves and the long-chain w-3 fatty acids in marine lipids . Here, eico sapentaenoic acid (EPA; 20:5w-3) and docosahexaenoic acid (DHA; 22:6w-3) are the major polyunsaturated fatty acids. The desaturation step from the w-6 fatty acids to the w-3 fatty acids seems to be carried out exclusively in green leaves and phytoplankton to form linolenic acid (18:3w-3) from linoleic acid (18:2w-6).

Fatty acids belonging to different families such as w-3 and w-6 fatty acids cannot be interconverted in the mammalian organism (28). Therefore, nutritional intake determines the fatty acid composition of phospholipids in plasma and in cell membranes to a great extent. Several independent lines of evidence suggest that changes in the natural history of hypertensive, atherothrombotic and inflammatory disorders may be achieved by altering the eicosanoid precursor availability. Native Greenland Eskimos (3,12) but also Japanese (9,15,29) have a high dietary intake of long-chain w-3 polyunsaturated fatty acids from seafood and a low incidence of myocardial infarction (Table 1). Diets containing w-3 polyunsaturated fatty acids have also been found to reduce the severity of experimental cerebral and myocardial infarction, and retard autoimmune nephritis and prolong survival in NZB x NZW F mice (18).

TABLE 1

AA, EPA and frequency of death from coronary artery
disease (CAD) in three study groups (A).

	Platelet phospholipid fatty acids		Ratio $\frac{w-6}{w-3}$	CAD (% death)
	AA (C20:4w-6)	EPA (C20:5w-3)		
Europe, USA	20-26	0.1-0.7	50	40
Japan	18-22	0.6-1.6	12	12
Greenland Eskimo	8.3-9.0	6.4-8.0	1.2	7

(A) All values are approximate.
(Composite data from ref.3,4,5,8,14,15,20)

Both the short- and the long-chain w-6 and w-3 fatty acids in the diet are rapidly absorbed and incorporated into plasma and cellular lipids (21). It seems also that in vivo in the adult organism the formation of the eicosanoid precursor fatty acids from their respective short-chain parent fatty acids, C18:2w-6 or C18:3w-3, is a slow process (25). This implies that the long-chain eicosanoid precursor fatty acids in the diet might have a more direct and actual effect on the eicosanoid precursor pool.

TABLE 2

Dietary modification of platelet membrane phospholipid fatty acid composition in healthy human volunteers.

| | Platelet phospholipid fatty acids (%) | | | |
	18:2w-6	20:4w-6	20:5w-3	22:6w-3
Control (Western) diet	9	24	0.4	2.3
Cod liver oil				
10 ml, 4 wks	8	23	1.3(A)	3.0
20 ml, 4 wks	7	22(A)	2.5(A)	4.1
40 ml, 4 wks	7	20	4.6	5.5(A)
Mackerel diet				
750 g, 6 days	5(A)	18(A)	5.0(A)	6.0(A)

(A) $p < 0.05$ versus control diet

Omega-3 fatty acids in the diets ranged from 2 to 16 g per day (EPA+DHA). Data adapted from refs. (14,20,23).

Dietary EPA and DHA, partially replace AA in a time-and dose-dependent manner in plasma- and in cellular phospholipids (14, 20,23), (Table 2). Once incorporated into cellular membranes, EPA and DHA are less readily released upon cell stimulation, reducing substrate availability for eicosanoid generation and decreasing potent amplification mechanisms to which AA metabolites contribute, e.g. as a cellular response to injury (4,22).

Differences exist between various cell types (neutrophils/ monocytes, platelets, endothelial cells, parenchymal cells) to incorporate and/or metabolize to eicosanoids w-3

and w-6 fatty acids provided with the diet. Differences
exist also between the w-3 and w-6 fatty acids to be incorporated into specific cellular phospholipid subclasses:
e.g. in vivo the long-chain w-3 fatty acids seem to be
incorporated only to an insignificant extent, e.g. into
phosphatidylinositol (8,22). This might have important
implications for cell function as it relates to the mechanisms of arachidonic acid release and the formation of lipid
mediators, such as eicosanoids or platelet activating factor
(19,27).

DIETARY W-3 FATTY ACIDS AND THE MODIFICATION OF THE
EICOSANOID SYSTEM

In addition to reducing AA levels (23) and most of
its eicosanoid derivatives, the w-3 fatty acid EPA serves as
precursor to a class of eicosanoids with an attenuated
spectrum of biological activity: TXA3 formed in small
quantities in w-3 fatty acid enriched platelets in addition
to a significantly reduced TXA2 formation (5) is almost
inactive as platelet aggregator and vasoconstrictor. PGI3,
which is formed from dietary w-3 fatty acids in amounts of
up to 50 % of an unchanged synthesis of PGI2 (6,20), is
biologically as active as PGI2 to inhibit platelet aggregation and to reduce vascular tone. LTB5 formed from EPA in
neutrophils (13,26), and monocytes/macrophages (13) is one
order of magnitude less active as a chemotactic agent as
compared to LTB4 which, in addition, is reduced after
prolonged dietary w-3 fatty acids (13).

Kinetic and dose response studies indicate that the
changes of plasma fatty acid spectra and in eicosanoid
formation occur rapidly within hours, are demonstrable for
the entire supplementation period and are dose related to
the amounts of w-3 fatty acids provived with the diet
(6,14,20,21,23). This indicates a biologically important
relationship.

Using ethylesters of the highly purified EPA and DHA,
which are the major w-3 fatty acids of marine, origin anti-
platelet effects of both compounds were recently demonstrated (21). In the same study it was found that dietary DHA is
retroconverted to EPA in man (21), and formation of PGI3 from
EPA formed by retroconversion of DHA (7) was discovered.
DHA may, therefore, reduce platelet aggregability both by a
mechanism related to formation of PGI3 and by a direct effect
on platelet function related to alterations in membrane
composition.

FUNCTIONAL EFFECTS OF DIETARY W-3 FATTY ACIDS RELATED TO
MODIFIED EICOSANOID FORMATION

The data indicate that it is possible in man to change
plasma and cell membrane phospholipid fatty acid composition, the arachidonic acid cascade and the spectrum of biologically highly active eicosanoids and other lipid mediators by nutritional means inducing a more antithrombotic and
antiinflammatory state (30,31) (Table 3). Formation of PGI3

in addition to unchanged synthesis of PGI2 (6,7,20) may be a mechanism by which diets enriched with w-3 fatty acids reduce vascular reactivity and blood pressure (14,24), blunt blood cell-vessel wall interactions and increase bleeding time (14). Less reactive platelets (5,14,20,23) may result from reduced formation of TXA2 associated with the synthesis of some less active TXA3 (5). Blunted inflammatory and immunological reactions may be related to the synthesis of

TABLE 3

The biochemical and functional effects of w-3 fatty acids in the diet on factors and mechanisms involved in the development of atherosclerosis.

w-3 fatty acids reduce risk/precipitating factors	w-3 fatty acids increase protective factors
Arachidonic acid	
Platelet aggregation Thromboxane formation Monocyte/macrophage function	Prostacyclin formation PGI2 + PGI3
Leukotriene formation PAF formation Il1 formation TNF formation PDGF like protein	
Intimal hyperplasia	
Blood pressure/BP response	
VLDL, LDL Triglycerides	HDL
Fibrinogen	Fibrinolytic activity
Blood viscosity	Red cell deformability

biologically less active LTB5 (13,26). Other effects of dietary w-3 fatty acids discussed in some detail elsewhere (32) include a reduction of plasma triglycerides (17), a reduced synthesis of LDL and VLDL (10,16), an increase in fibrinolytic activity (1), increased red cell deformability, reduced blood viscosity (2), reduced formation of Il1 and TNF (33), reduced PDGF like protein in endothelial cells (34) and reduced plasma levels of fibrinogen (35).

FUTURE STUDIES

The further evaluation of the relationship of w-6 versus w-3 fatty acids and their oxygenated metabolites to membrane receptor function, the modulation of transmembrane signaling mechanisms, formation of 1,2 DAG or PAF and Ca++ release, gene expression and growth factor synthesis, should contribute to a better understanding of the role of these mediators and their precursor fatty acids in cell function. Some interesting areas for future research can already now be identified. These include the study of the effects of dietary eicosanoid precursors on eicosanoid formation from various long-versus short-chain w-6 and w-3 fatty acids; the reabsorption, organ distribution, cellular and phospholipid subclass incorporation of eicosanoid precursor fatty acids and their metabolism during cell formation and differentiation; and the large scale controlled evaluation of their potential role in the modification of the proliferative, prothrombotic and inflammatory response as it occurs clinically after cell injury and during cell repair, such as after PTCA (36).

ACKNOWLEGEMENTS

The studies of the authors laboratory of the University of Munich were supported by DFG, No. We 681.

References

1.   Barcelli U, Glas-Greenwalt P, Pollak V.E: Enhancing effect of dietary supplementation with n-3 fatty acids on plasma fibrinolysis in normal subjects. Thromb.Res. 39: 307-312, 1985.

2.   Cartwright I.J, Pockley A.G, Galloway J.H, Greaves M, Preston F.E: The effects of dietary n-3 polyunsaturated fatty acids on erythrocyte membrane phospholipids, erythrocyte deformability and blood viscosity in healthy volunteers. Atherosclerosis 55: 267-281, 1985.

3.   Dyerberg J, Bang H.O, Stofferson E, Moncada S, Vane J.R: Eicosapentaenoic acid and prevention of thrombosis and atherosclerosis? Lancet 1: 117-119, 1978.

4.   Fischer S, v.Schacky C, Siess W, Strasser T, Weber P.C: Uptake, release and metabolism of docosahexaenoic acid (DCHA, C22:6n-3) in human platelets and neutrophils. Biochem.Biophys.Res.Commun. 120: 907-819, 1984.

5. Fischer S, Weber, P.C: Thromboxane A3 (TXA3) is formed in human platelets after dietary eicosapentaenoic acid (C20:5n-3). Biochem.Biophys.Res.Commun. 116: 1091-99, 1983.

6. Fischer S, Weber P.C: Prostaglandin I3 is formed in vivo in man after dietary eicosapentaenoic acid. Nature (Lond.) 307: 165-168, 1984.

7. Fischer S, Vischer A, Preac-Mursic V, Weber P.C: Dietary docosahexaenoic acid is retroconverted in man to eicosapentaenoic acid, which can be quickly transformed to prostaglandin I3. Prostaglandins 34: 367-375, 1987.

8. Galloway J.H, Cartwright I.J, Woodcock B.E, Greaves M, Russel R.G.G, Preston F.E: Effects of dietary fish oil supplementation on the fatty acid composition of the human platelet membrane: demonstration of selectivity in the incorporation of eicosapentaenoic acid into membrane phospholipid pools. Clin.Sci. 68: 449-454, 1985.

9. Goto Y: Serum cholesterol and nutrition in Japan. Nutrition and Health 3: 255-257, 1985.

10. Illingworth D.R, Harris W.S, Connor W.E: Inhibition of low density lipoprotein synthesis by dietary omega-3 fatty acids in humans. Arteriosclerosis 4: 270-275, 1984.

12. Kromann N, Green A: Epidemiological studies in the Upernavik district, Greenland. Acta Med.Scand. 208: 401-406, 1980.

13. Lee T H, Hoover R.L, Willians J.D, Sperling, R.I, Revalese J, Spur B.W, Robinson D.R, Corey E.J, Lewis R.A, Austen K.F: Effect of dietary enrichment with eicosapentaenoic and docosahexaenoic acids on in vitro neutrophil and monocyte leukotriene generation and neutrophil function. N.Engl.J.Med. 312: 1217-1224, 1985.

14. Lorenz R, Spengler U, Fischer S, Duhm J, Weber P.C: Platelet function, thromboxane formation and blood pressure control during supplementation of the western diet with cod liver oil. Circulation 67: 504-511, 1983.

15. Nagakava Y, Orimo H, Harasawa M, Morita I, Yashiro K, Murota S: Effect of eicosapentaenoic acid on the platelet aggregation and composition of fatty acids in man. Atherosclerosis 47: 71-75, 1983.

16. Nestel P.J, Connor W.E, Reardon M.R, Connor S., Wong S, Boston R: Suppression by diets rich in fish oil of very low density lipoprotein production in man. J.Clin.Invest. 74: 82-89, 1984.

17. Phillipson B.E:, Rothrock, D.W, Connor W.E, Harris N.S, Illingworth R: Reduction of plasma lipids, lipoproteins and apoproteins by dietary fish oils in patients with hypertriglyceridemia. N.Engl.J. Med. 312: 1210-16, 1985.

18. Prickett J.D, Robinson D.R and Steinberg A.D: Dietary enrichment with the polyunsaturated fatty acid eicosapentaenoic acid prevents proteinuria and prolongs survival in NZBxNZW F1 mice. J.Clin.Invest. 68: 556-559, 1981.

19. Robinson R, Snyder F: Metabolism of platelet activating factor by rat alveolar macrophages: lyso-PAF as an obligatory intermediate in the formation of alkyl-arachidonoyl-glycerophosphocholine species. Biochem.Biophys.Acta 837: 52-56, 1985.

20. v.Schacky C, Fischer S, Weber P.C: Long term effects of dietary marine n-3 fatty acids upon plasma and cellular lipids, platelet function and eicosanoid formation in humans. J.Clin.Invest. 76: 1626-31, 1985.

21. v.Schacky C, Weber P.C: Metabolism and effects on platelet function of the purified eicosapentaenoic and docosahenaenoic acids in humans. J.Clin.Invest. 76: 2446-50, 1985.

22. v.Schacky C, Siess W, Fischer S, Weber, P.C: A comparative study of eicosapentaenoic acid metabolism by human platelets in vivo and in vitro. J.Lipid.Res. 26: 457-464, 1985.

23. Siess W, Roth P, Scherer B, Kurzmann I, Boehlig B, Weber P.C: Platelet membrane fatty acids, platelet aggregation and thromboxane formation during a mackerel diet. Lancet 1: 441-444, 1980.

24. Singer P, Wirth M, Voigt S, Richter-Heinrich E, Gödicke W, Berger I, Naumann E, Listing J, Hathrodt M, Taube, Ch: Blood pressure and lipid-lowering effect of mackerel and herring diet in patients with mild essential hypertension. Atherosclerosis 56: 223-235, 1985.

25. Singer P, Berger I, Wirth M, Gödicke W, Jaeger W, Vogt S: Slow desaturation and elongation of linoleic and α-linolenic acids as a rationale of eicosapentaenoic acid-rich diet to lower blood pressure and serum lipids in normal, hypertensive and hyperlipidemic subjects. Prostagl.Leukotr.Med. 24: 173-193, 1986.

26. Strasser T, Fischer S, Weber P.C: Leukotriene B5 is formed in human neutrophils after dietary eicosapentaenoic acid. Proc.Natl.Acad.Sci, USA 82: 1540-1543, 1985.

27. Sugiura T, Masuzawa Y, Waku K: Transacylation of 1-0-alkyl-sn-glycero-3-phosphocholine (lyso platelet activating factor) and 1-0-alkenyl-sn-glycero-3-phosphoethanolamine with docosahexaenoic acid (C22-6w-3). Biochem.Biophys.Res.Commun. 133: 574-580, 1985.

28. Tinoco J: Dietary requirement and functions of α-linolenic acid in animals. Prog.Lipid.Res. 21: 1-45, 1982.

29. Yamori Y, Nara Y, Iritani N, Workman R.J, Inagami T; Comparison of serum phospholipid fatty acids among fishing and farming Japanese populations and American inlanders. J.Nutr.Vitaminol.(Tokyo), 31: 417-22, 1985.

30. Weber P.C, Fischer S, v.Schacky C, Lorenz R, Strasser T: The conversion of dietary eicosapentaenoic acid to prostanoids and leukotrienes in man. Progr.Lipid.Res. 25; 273-276, 1986.

31. Weber P.C: The dietary modification of the arachidonic acid cascade. In: Biology of eicosanoids and related substances in blood and vascular cells. Colloque INSERM 152: 119-126, 1987.

32. Leaf A, Weber P.C: Cardiovascular effects of n-3 fatty acids. N.Engl.J.Med. 318: 549-557, 1988.

33. Endres S, Kelley V.E, Dinarello C.A: Effects of dietary omega-3 fatty acids on the in vitro production of human interleukin I. J.Leukoc.Biol. 42: 617, 1987.

34. Fox P, Di Corleto E: Fish oils inhibit endothelial cell production of platelet-derived growth factor-like protein. Science 241: 453-456, 1988.

35. Høstmark A, Bjerkedal T, Kierulf P, Flaten H, Ulshagen K: Fish oil and plasma fibrinogen. Br.Med.J. 297: 180-181, 1988.

36. Dehmer G.J, Jeffrey J, Popma J, v.d.Berg E.K, Eich-
    horn E.J, Prewitt J.B, Campbell W.B, Jennings L,
    Willerson J.T, Schmitz J.M: Reduction in the rate of
    early restenosis after coronary angioplasty by a
    diet supplemented with n-3 fatty acids. N.Engl.J.
    Med. 319: 733-740, 1988

N-6 AND N-3 FATTY ACIDS IN PLASMA AND PLATELET LIPIDS, AND GENERATION OF INOSITOL PHOSPHATES BY STIMULATED PLATELETS AFTER DIETARY MANIPULATIONS IN THE RABBIT

Claudio Galli, C. Mosconi, L. Medini, S. Colli, and E. Tremoli

Institute of Pharmacological Sciences, Universita' degli Studi, Via G. Balzaretti 9, I-20123 Milano, Italy

INTRODUCTION

Dietary fatty acids modify the fatty acid composition of plasma and tissue lipids, and these changes appear, in turn, to modulate biochemical and functional parameters in various biological compartments. Modifications of the amounts and proportions of saturated and polyunsaturated fatty acids (PUFA) in the diet, for instance, influence the levels of plasma cholesterol and affect the aggregation of platelets (see Goodnight et al., 1982 for a review), possibly through modifications of the eicosanoid cascade (Galli et al.,1981). More specifically, the administration of polyunsaturated fatty acids of the n-3 series, such as eicosapentaenoic acid (EPA, 20:5 n-3) and docosahexaenoic acid (DHA, 22:6 n-3) results in quantitative and qualitative changes of eicosanoid production (Fischer and Weber, 1983), following the accumulation of this fatty acid in cell lipid pools (Siess et al.,1980). This effect reduces blood platelet-vessel wall interactions and the thrombotic potential.

Dietary induced modifications of the fatty acid profiles of structural lipids in biomembranes, in addition to the influences on eicosanoid production, may affect other parameters, such as membrane fluidity (Schaeffer and Curtis,1977; Hornstra and Rand,1986) or other physicochemical characteristics, which could contribute to the reported functional effects. Changes of the lipid moiety of cell membranes could also affect the formation of lipid-derived mediators, such as the products generated through the enzymatic hydrolysis of phosphoinositides, which are known to play a key role in the early processes of cell activation (Kawehara et al.,1980; Berridge and Irvine,1984).

Polyunsaturated fatty acids associated with structural lipids in cell membranes appear to be mainly produced through desaturation and elongation short chain precursors, provided by the diet, in the liver (Bernert and Sprecher,1977) followed by transport to peripheral tissues and incorporation in lipid pools. Although the factors involved in this overall process are complex, evaluation of the differential distribution of selected PUFAs among various pools after manipulations of the dietary intake, would help in elucidating how endogenously produced PUFAs, e.g. arachidonic acid, and PUFAs which are administered, preformed, through the diet, e.g. EPA, are handled and made available to tissues for further utilization.

Aim of the present study was to measure the absolute levels of the major PUFAs of the n-6 and n-3 series in plasma and platelet lipids, after administration of diets enriched in n-9, n-6 and n-3 unsaturated fatty acids, respectively, in the rabbit. In addition, we have measured the influence of dietary induced manipulations of platelet fatty acids on the generation of inositolphosphates by stimulated platelets obtained from the same animals, in order to relate functional changes to modifications of the fatty acid composition of cell lipids.

MATERIALS and METHODS

Three groups of New Zealand male rabbits (10 animals in each group) were fed ad libitum semisynthetic diets containing 5 per cent by weight (12 per cent of the energy) of either olive oil or corn oil or fish oil (MaxEPA, Seven Seas,UK) for a period of five weeks.
The fatty acid composition of dietary fats is presented in Table 1.

Table 1. Fatty acid percentage composition of oil supplements

Fatty acids	Olive oil	Corn oil	Fish oil
14:0	-	-	8.3
16:0	17.1	13.3	17.6
16:1	-	-	8.7
18:0	-	1.9	2.6
18:1	70.1	26.6	14.4
18:2	12.8	58.1	1.7
20:0	-	-	3.5
20:5	-	-	15.8
22:1	-	-	5.0
22:5	-	-	2.0
22:6	-	-	14.0

At the end of treatments, blood was drawn from the cannulated common carotid arteries and under sodium thiopenthal anaesthesia in plastic tubes using ACD as anticoagulant. Washed platelets were prepared by centrifugation techniques, after preparation of PRP, with care to avoid cell activation by including PGI_2 in the media (Watson et al.,1984), and used for both lipid analysis and for studies on inositol phosphate generation. Production of inositol phosphates (IP_3, IP_2 and IP) by stimulated washed platelets was evaluated by measuring the radioactive water-soluble metabolites generated after stimulation (10 and 90 s) of cells, prelabelled with ^{3}H-inositol for 90 min with 2-U NIH/ml thrombin, after extraction and separation of products on anion exchange chromatography.

Lipids were extracted from plasma and washed platelets and used for the analysis of fatty acid methyl esters by GLC. Quantitation of fatty acids was carried out by the use of internal standards and calibration curves obtained with reference compounds.

Table 2. Levels (nmoles/ml plasma) of major polyunsaturated fatty acids in plasma .

| Fatty acids | Groups | | |
	Olive oil	Corn oil	Fish oil
Linoleic acid	1071 + 170[b]	1241 + 240	599 + 66[b]
Arachidonic acid	82 + 14[a]	112 + 19	148 + 20[a]
Eicosapentaenoic acid	< 1	< 1	339 + 93

Values are the average + SEM of determinations carried out in 10 animals/ group. Corresponding values in each group sharing the same letter are significantly dofferent from each other.

RESULTS

The total levels of LA, AA and EPA (nmoles/ml) in plasma of rabbits fed the three oils are shown in Table 2. Levels of LA were highest in the corn oil group and lowest in the fish oil group, whereas AA levels were identical in olive oil and corn oil group and, un-expectedly, were slightly higher in the fish oil group. EPA reached levels significantly higher than those of AA in the fish oil group.

Levels (nmoles/mg of platelet total lipid) of LA, AA and EPA in total lipids from platelets of rabbit fed different oils are shown in Table 3. LA levels were highest in the corn oil group and lowest in the fish oil group. AA was practically identical in the olive oil and corn oil groups and slightly reduced in the fish oil group and, in this group, EPA was accumulated, above AA levels. In general, the accumulation of EPA in the fish oil group replaced LA rather than AA in platelet phospholipids.

Table 3. Levels (nmoles/mg total lipids) of major polyunsaturated fatty acids in platelet lipids.

Fatty acids	Groups		
	Olive oil	Corn oil	Fish oil
Linoleic acid	$238 \pm 9^{a,b}$	$327 \pm 2^{a,c}$	$113 \pm 4^{b,c}$
Arachidonic acid	$151 \pm 4^{d,e}$	156 ± 4^{e}	$116 \pm 4^{d,e}$
Eicosapentaenoic acid	1	1	141 ± 12

Values are the average $\pm$ SEM of determinations carried out in 10 animals/group. Corresponding values in each group sharing the same letter are statistically different from each other.

In Table 4 the percentage increment of inositolphosphate labelling over basal values at 10 and 90 s after stimulation are represented. At 10 s, the highest increment was observed in IP_2, whereas at 90s IP was the major labelled product. When values among different groups were compared the increment of IP_3 labelling was found significantly lower in the corn oil and fish oil vs the olive oil group, whereas the labelling of IP was highest in the fish oil group and lowest in the olive oil group.

Table 4. Percentage increments of inositolphosphate labelling in washed platelets at 10 and 90 s after stimulation.

Groups	IP_3		IP_2		IP	
	10 s	90 s	10 s	90 s	10 s	90 s
Olive oil	$142 \pm 19^{a,c}$	94 ± 33^{a}	464 ± 132	313 ± 114^{a}	98 ± 14^{b}	563 ± 157
Corn oil	68 ± 29^{a}	57 ± 3^{b}	325 ± 136	85 ± 58	135 ± 38	587 ± 61
Fish oil	56 ± 34^{c}	$27 \pm 10^{a,b}$	324 ± 145	109 ± 78^{a}	153 ± 19^{b}	496 ± 100

Values (dpm after stimulation − dpm basal/dpm basal) X 100, are the average $\pm$ SEM . Corresponding values in each group sharing the same letter are significantly different from each other.

At 90 s, increments of IP_3 and IP_2 values were lowest in the fish oil and highest in the olive oil group. Mean percent increment in the corn oil group was significantly different from the other groups of animals.

DISCUSSION

The administration of diets enriched in fatty acids of the various series resulted, as expected, in significant modifications of the concentrations of individual PUFAs in plasma and platelet lipids. Differences between the olive oil and the corn oil fed animals were small and confined to LA levels. In the fish oil fed animals, instead, significant and complex changes of levels and distribution of all three PUFAs were observed. In fact, the expected accumulation of EPA was associated with a reduction of LA levels, but AA levels were even slightly but significantly enhanced. The lack of reduction of AA in our conditions, which may be dependent upon the duration of the feeding period, suggests that the conversion of LA to AA was not significantly inhibited by n-3 fatty acids supplemented by the diet. The tendency to a rise in plasma AA may have been a consequence of a depletion of tissue AA induced by the administration of n-3 fatty acids. This effect was clearly observed in platelets and to an even greater extent should have occurred in liver.

The generation of labelled inositolphosphates in platelets was also significantly affected by dietary treatments. Stimulation of IP_3 production expressed as increment of labelling over basal values, was highest in samples from the olive oil group and lowest in the fish oil samples, with intermediate values in the corn oil group. The distribution of labelling among the various inositolphosphates was also different in samples from the three animal groups, with a greater proportion of inositolmonophosphate in the fish oil samples. Each dietary treatment resulted, thus, in a distinct pattern of inositolphosphate labelling.

These observations indicate that dietary induced manipulations of platelet fatty acids affect a biochemical process involved in early step of cell activation, i.e. the generation of inositol-phosphates. This effect may be related to changes of various parameters at the cellular levels, such as microviscosity, which are affected by modifications of membrane lipids, in addition to changes of eicosa-noid precursor fatty acids.

REFERENCES

Bernert J.T., H.Sprecher (1977) An analisys of partial reaction in the overall chain elongation of saturated and unsaturated fatty acids in rat liver microsomes. J.Biol.Chem. 252, 6736

Berridge M.J., R.F.Irvine (1984) Inositoltriphosphate, a novel second messenger in cellular signal transduction. Nature 312, 315

Fisher S., P.C.Weber (1983) Thromboxane A_3 (TxA3) is formed in human platelets after dietary eicosapentaenoic acid. Biochem.Biophys. Res.Comm. 116, 1091

Galli C., E.Agradi, A.Petroni, E.Tremoli (1981) Differential effects of dietary fatty acids on the accumulation of arachidonic acid and its metabolic conversion through the cycloxygenase and lipoxygenase in vascular tissues. Lipids 16, 165

Goodnight S.H.Jr., W.S.Harris, W.E.Connor, D.R.Illingworth (1982) Polyunsaturated fatty acids, hyperlipidemia, and thrombosis. Arteriosclerosis 2, 87

Hornstra G., M.L.Rand (1986) Effect of dietary n-6 and n-3 polyunsaturated fatty acids on the fluidity of platelet membranes in rat and man, in "Progress in Lipid Research", vol.25, pp. 637-638, R.T.Holman Ed., Plenum Press

Kawehara Y., Y.Takoi, R.Minakuchi, K.Sono, Y.Nishizuka (1980) Phospholipid turnover as a possible transmembrane signal for protein phosphorylation during human platelet activation by thrombin. Biochem.Biophys.Res.Comm. 97, 309

Schaeffer B.E., A.S.Curtis (1977) Effects of cell adhesion and membrane fluidity of changes in plasmalemma lipids in mouse L929. J.Cell Sci. 26, 47

Siess W., B.Scherer, B.Bohlig, P.Roth, I.Kurzmann, P.C.Weber (1980) Platelet-membrane fatty acids, platelet aggregation and thromboxane formation during a mackerel diet. Lancet 1, 441

Watson S.P., R.T.McConnell, E.G.Lapetina (1984) The rapid formation of inositolphosphates in human platelets by thrombin is inhibited by prostacyclin. J.Biol.Chem. 259, 13199 .

DIETARY LINOLENIC ACID AND TISSUE FUNCTION IN RODENTS

W.M.F. Leat

Trinity College
Cambridge
CB2 1TQ

INTRODUCTION

Since the essential fatty acids (EFA) were first discovered over fifty years ago (Burr and Burr, 1930), most investigations have concentrated on the metabolism and function of linoleic acid (18:2n-6), and research on linolenic acid (18:3n-3) has, until recently, been relatively restricted. An armchair scientist could argue that since linolenic acid possesses the essential Δ^{12} double bond it should show full EFA activity. However, the reverse could be argued in that the presence of the Δ^{15} double bond could modify or nullify the effect of the Δ^{12} double bond by, for example, steric hindrance, resulting in reduced or zero EFA activity. In fact, linolenic acid has partial EFA activity (FAO Report, 1977), and the reasons for this are in itself intriguing.

The pertinent questions to answer are:-
(a) does linolenic acid have any function that cannot be fulfilled by linoleic acid; and its corollary:-
(b) does linoleic acid have any function that cannot be fulfilled by linolenic acid?

When animals are reared on EFA deficient diets certain tissues of the body retain their EFA tenaciously. This is particularly true for the n-3 fatty acids in such organs as the retina and central nervous system where any function for linolenic acid might be expected to be found. We concluded that, in order to deplete these tissues of n-3 fatty acids, it would be necessary to breed animals through a number of generations to achieve this objective.

REPRODUCTION IN THE FEMALE RODENT

Female rats, in groups of three, were therefore reared from weaning on an EFA deficient diet supplemented with either 18:2n-6 or 18:3n-3 in order to deplete the tissues of one or other of the two families of EFA, thus:-

> Group 1 : EFA deficient diet
> Group 2 : EFA deficient diet + 18:2n-6
> Group 3 : EFA deficient diet + 18:3n-3
> Group 4 : Commercial diet.

There was no difference in growth rate between the rats of groups 2, 3 and 4, and that of the rats of group 1 was, as expected, less than in the other groups. At thirteen weeks of age the rats were mated and their progress followed through gestation. There was again no difference in growth rate between the rats of groups 2, 3 and 4. The animals of groups 1, 2 and 4 gave birth without difficulty, although the pups from group 1 animals died within a few days of birth. In the rats of group 3 gestation was prolonged compared to the other groups, and parturition was an extended process (Leat and Northrop, 1981). The blood oozing from the vulva during parturition was very dark in colour and did not appear to clot. At caesarian section it was noted that the uterine wall was thin, flaccid and lacking in tone. If the operation was caried out within four hours of the beginning of parturition live pups could be obtained.

However a less traumatic method of obtaining offspring from Group 3 animals was to replace the daily supplement of 18:3n-3 with 18:2n-6 from day 18 of gestation (full term = 21-22 days). The length of gestation was reduced to normal limits and unimpaired parturition then occurred. By this means it was possible to rear rats through a number of generations on 18:3n-3. Thus, 18:3n-3 has nutritional properties similar to those of 18:2n-6, in allowing fertilization, implantation and growth of foetuses to occur normally. However, 18:3n-3 differs from 18:2n-6 in being unable to allow normal parturition to take place, probably because the prostaglandins derived from n-3 EFA, unlike those from n-6 EFA, are unable to stimulate the contraction of the uterine wall.

Incubation, in vitro, of strips of uterine wall from the 18:3n-3 supplemented rats resulted in the release of large amounts of a prostaglandin-like material which proved impossible to identify.

As the 18:3n-3 fed female rats were bred into successive generations the proportion of female offspring progressively declined and by the sixth generation this line had died out. Whether the ratio of n-6 : n-3 EFA in the diet can affect the ratio of male to female progeny merits further investigation.

The results obtained with the rat cannot be extrapolated to other species without due caution. Preliminary studies using female mice reared on an identical dietary protocol revealed no impairment in parturition into a second generation. Whether subsequent generations would show a defect would again be worth investigation.

TESTICULAR DEVELOPMENT AND MALE REPRODUCTION

When male rats were reared on identical diets and experimental protocols to those described for female rats, the body growth of linolenate-fed animals was again similar to that of rats fed linoleate. However, the testes of the linolenate-fed rats failed to develop, and, on palpation, felt small, hard and atrophic. Histologically there was virtually a complete degeneration of the seminiferous tubules, although the Leydig cells appeared normal. This would explain the observation that the linolenate-fed rats could mate with females fed commercial diets, and left copulatory plugs, but the matings were, of course, sterile (Leat et al., 1983). Chemically, the degenerated testes, unlike the normal testes, contained virtually no triglyceride or sulphogalacto-sylglyceride.

Rats fed the EFA deficient diet alone also showed testicular degeneration, as first reported by Burr and Burr (1930), but the degeneration was not as severe as that found in linolenate-fed animals.

Competition between linolenate and linoleate for the common metabolizing enzymes could account for this finding. The relatively large amounts of linolenate supplement would antagonize the small amounts of linoleate present in the EFA-deficient diet, and hence exacerbating the testicular degeneration. This appears to be a unique situation where a nutrient (in this case linolenate) satisfies the EFA requirement for one function (body growth) but at the same time and in the same animal is antagonistic towards the growth and development of a more specific organ (testis). This interpretation is also applicable to parturition in linolenate-fed female rats. The closest parallel to this situation is between the effect of vitamin A alcohol and acid on growth and vision in rats. Vitamin A alcohol is effective in growth and vision whereas vitamin A acid stimulates growth but is ineffective in vision. However, there is no evidence that vitamin A acid is antagonistic to vitamin A alcohol.

Thus, linoleic acid appears to be essential for testis development in the rat, and cannot be replaced by linolenic acid. Whether all the n-6 EFA are equally important in spermatogenesis or whether certain fatty acids have a more potent effect than others was the subject of a further investigation. Analysis of rat testicular lipids showed that the most abundant fatty acids in testis were 20:4n-6 and 22:5n-6, the latter acid appearing during the development of the spermatids and spermatozoa (Aaes-Jorgensen and Holman, 1958; Davis et al., 1966). Linoleic acid could therefore mediate its stimulatory effect on spermatogenesis via 20:4n-6, perhaps by invoking a pathway involving prostaglandins, or via 22:5n-6, a molecule that would provide the essential configuration of fatty acid required for the formation of the spermatid.

To investigate these alternatives weanling male rats, bred from dams reared on 18:3n-3, were fed EFA deficient diets supplemented with 18:2n-6, 20:4n-6, 22:4n-6, 18:3n-3 and 22:6n-3 (Leat et al., 1984). The acid 22:5n-6 was not available commercially, and so its immediate precursor, 22:4n-6, was used in its place. Compared with rats fed the EFA deficient diet alone maximum stimulation of body growth was found with the n-6 fatty acids, 18:2, 20:4 and 22:4. Docosahexanoic acid (22:6n-3) also promoted body growth but the effect was not as great as that with the n-6 fatty acids or with 18:3n-3. The greatest stimulation of testis weight and development occurred with 20:4n-6 followed by 22:4n-6 and 18:2n-6 (Fig. 1). The feeding of 18:3n-3 and 22:6n-3 resulted in no stimulation of the weight of the testes, which showed extensive histological degeneration. Thus, although 22:6n-3 has all the positional double bonds possessed by 22:4n-6 and 22:5n-6, the extra double bond at the Δ^{15} position completely negates any activity in testis development. There is, therefore, a highly specific requirement for the number and position of the double bonds in a fatty acid molecule to enable it to be active in spermatogenesis.

In the rats fed 22:4n-6 very little of this acid was found in testicular lipids but instead appreciable amounts of 20:4n-6 and 22:5n-6 were detected, these acids arising by desaturation or retroconversion (Verdino et al., 1964). These results suggest that the immediate function of 18:2n-6 in testis development is to provide 20:4n-6 which may function in spermatogenesis via prostaglandin formation. The formation of 22:5n-6 may be required at a later stage in development, namely in the formation of spermatids and spermatozoa.

Again, whether these results obtained in the rat can be extrapolated to other species is debatable. Whereas testicular lipids of the rat, rabbit and dog contain mainly n-6 fatty acids, other species such as the mouse, guinea pig and man also contain appreciable proportions of n-3 fatty acids (Bieri & Prival, 1965). Whether the n-3 fatty acids have a function in these species is unknown. The mouse has similar dietary habits to the rat and might be expected, on theoretical grounds, to have similar requirements for EFA in testis development.

To investigate this point groups of three male weanling mice were reared on:-

(a) EFA deficient diet
(b) EFA deficient diet + 18:2n-6
(c) EFA deficient diet + 18:3n-3
(d) EFA deficient diet + (18:2n-6 + 18:3n-3)

After 40 weeks on the diet there was no difference between groups in the weights of the testes, and spermatozoa were present in the testes and epididymides of all mice. This suggests that in the mouse either 18:2n-6 or 18:3n-3 can satisfy the EFA requirement of the testis for development.

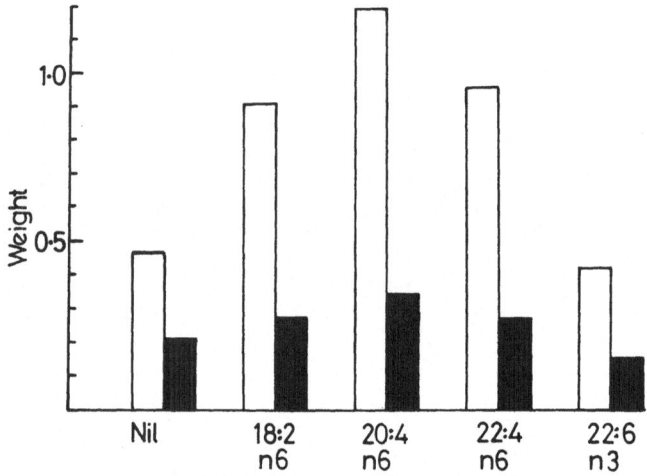

Fig. 1 Weights of testes of rats fed various n-6 and n-3 fatty acids for ten weeks (mean of three observations. ☐ g; ■ g per 100g body weight)

However, although the testes of mice fed 18:2n-6 contained virtually no n-3 metabolites, those fed 18:3n-3 still contained appreciable amounts of 20:4n-6 and 22:5n-6 which may have accounted for the presence of spermatozoa in these mice (Table 1).

RETINAL FUNCTION

During the period when the rats were being reared through a number of generations on either 18:2n-6 or 18:3n-3 their electroretinograms (ERG) were examined, and the fatty acid composition of their retinal lipids determined. The ERGs were very variable and there was no statistical difference between the ERGs of rats fed 18:2n-6 and those fed 18:3n-3. Analysis of retinal fatty acids indicated that even after three generations the retinae of 18:2n-6 fed rats still contained 50% of the 22:6n-3 content of control animals. This level was not reduced further by breeding into subsequent generations (Leat et al., 1986). Even when the ratio of 18:2n-6 : 18:3n-3 in the diet was 1000 : 1 the retinal fatty acids still contained 8-9% 22:6n-3, or approximately 30% of control

Table 1. Major unsaturated fatty acids (% by weight) of testes of rats and mice given similar fatty acid supplements

Fatty acid	Nil		18:2		18:3		18:3 + 18:3	
	Rat	Mouse	Rat	Mouse	Rat	Mouse	Rat	Mouse
18:1n-9	21.0	16.5	14.6	13.0	38.3	19.3	17.4	14.6
18:2n-6	0.6	0.2	1.5	0.5	1.9	0.2	1.8	1.1
18:3n-3	–	–	–	–	–	–	–	–
20:3n-9	14.7	7.6	1.4	0.5	3.4	4.5	0.2	1.1
20:3n-6	0.5	0.5	0.3	0.6	0.3	0.4	1.2	0.7
20:4n-6	6.7	5.6	14.1	11.2	5.0	5.1	12.8	8.5
20:5n-3	–	–	–	–	1.9	0.5	–	
22:4n-6	0.8	2.0	0.8	0.8	0.4	1.1	0.7	0.5
22:5n-6	4.2	9.7	16.0	15.5	1.0	4.0	12.9	8.9
22:5n-3	–	–	–	–	0.6	1.9	0.5	2.0
22:6n-3	–	2.4	–	1.9	0.8	9.6	3.0	6.1

values. It seemed impossible, therefore, to deplete the rat retina of 22:6n-3 by the dietary regime employed, and it proved necessary to seek another experimental animal.

The guinea pig seemed to be the best alternative model. The placenta is freely permeable to fatty acids, and therefore it is feasible to begin depleting the animal of n-3 fatty acids whilst still developing in utero. The maternal guinea pigs were fed a semisynthetic diet containing sunflower oil which had a high content of 18:2n-6 and a very low content of 18:3n-3. On this dietary regime the percentage of 22:6n-3 in retinal lipids fell dramatically by the first generation, and by the third generation no more than a trace (< 0.5%) remained. However, the depleted retinae were still responsive to light, and there were no statistical differences between the ERG responses of retinae containing virtually no 22:6n-3 and those of controls (Leat et al., 1986). Examination of the linolenate-depleted retinae by electron microscopy revealed normal structure of the rods and cones, and of the deeper layer of the retina (R.W. Cox and W.M.F. Leat, unpublished observations). Therefore, in the guinea pig at least, the n-3 EFA are not essential for the detection of light by the retina using the ERG as an index; and 18:2n-6 appears to be able to replace 18:3n-3 in both structure and function. Whether these depleted animals can see and appreciate light is difficult to assess, since objective tests, such as the fixation reflex, are unreliable. However, histological examination of the visual cortex of the brain revealed apparently normal structure in nerve fibres and Nissl substance, with no evidence of demyelination, suggesting that vision was not seriously impaired. Why the retina retains its n-3 EFA so tenaciously during depletion studies is puzzling. The possibility remains that 18:3n-3 might play a more subtle role in vision (e.g. via oxygenated metabolites) which cannot be detected by electroretinography.

BEHAVIOURAL CHANGES

During the course of the depletion studies some subtle behavioural changes were noted in the guinea pigs reared on diets low in linolenic acid. For example, on entering the animal house the noise of the door being opened resulted in the control animals making much noise through vocalising, clattering the food pots and shaking cage doors. In contrast, the depleted animals were very quiet and passive with only an occasional vocalization being heard. However the animals were not mute and vocalized readily when handled.

In an attempt to investigate this difference further the behaviour of the guinea pigs was investigated using an open square procedure. This consisted of an open box 100 cm x 100 cm x 40 cm with the floor marked lightly into sixteen equal squares. The animal under test was placed at the centre point and its behaviour and activity within, and between, squares was recorded over a five-minute period divided into one minute intervals. It was found that the linolenate-deficient animals were less active than controls with less movement between and within squares (Table 2). More notable was the virtual absence of vocalization in the linolenate-deficient animals.

Investigation of the reflexes of the guinea pigs, based on the protocol used in the dog (Palmer, 1976) showed no noticeable differences between the groups except for the startle reflex. Five out of six linolenate-deficient animals showed no response to a loud noise and one gave a much reduced response. All the control animals gave a positive response. This suggested initially that the experimental animals were deaf, and further experiments were designed to investigate the auditory responses.

AUDITORY RESPONSES

Cochlear nerve action potentials were recorded by electrocochleography in control and experimental guinea pigs (by Dr D. Mendel, St Thomas Hospital, London SE1 7EH). Although initially it was thought that there was an increase in latency in the linolenate-depleted animals final analysis showed that there was no difference compared with the control group (Table 3).

Histological preparations of the cochlea of four experimental animals with no startle reflex were compared with similar preparations from four animals in the control group (by Dr M.G. Spencer, Chester Royal Infirmary, Chester CH1 2AZ, in consultation with Professor I. Friedman, Northwick Park Hospital, Harrow). No consistently demonstrable findings clearly distinguished the two groups, but nonetheless the following abnormalities were seemingly more apparent in the cochleae of animals in the linolenate-deficient group.

1. Less fat in the marrow spaces of the temporal bone.
2. Poor condition of the hair cells in the organ of Corti, especially in the basal turn of the cochlea.
3. The presence of protein "lymph" in the cochlear spaces.

Table 2. Behaviour of linolenate-deficient and control guinea pigs in an open field study

(6 x 5 minute observations on five animals/group)

		Control	linolenate deficient
1.	Movement between squares	64	8
2.	Turning in square	24	18
3.	Head lifting	99	144
4.	Sniffing	77	65
5.	Vocalising	11	0

Table 3. Cochlear nerve action potentials in guinea pigs depleted of linolenic acid, and in control animals

Age (months)	Status	Startle Reflex	Average latency (m.sec) to 70 dB stimulus
1.5	depleted	absent	1.27
1.5	depleted	slight	1.33
8	depleted	absent	1.52
9	depleted	absent	1.59
15	depleted	absent	1.80
15	depleted	absent	1.62
6	Control	present	1.22
9	Control	present	1.42
13	Control	present	1.34
18	Control	present	1.06
27	Control	present	1.26

Whether these histological changes have any connection with the absence of the startle reflex is difficult to assess. The presence of apparently normal electrocochleograms in the animals with an absent startle reflex would suggest a defect at some other location, e.g. cochlear nerve or central nervous system. However, histological examination of the auditory cortex did not indicate any obvious abnormality in the nerve cells and axons (R.W. Cox and W.M.F. Leat, unpublished observations.

CONCLUSIONS AND RECOMMENDATIONS

These investigations show that, in the rat at least, there are two functions of linoleic acid that cannot be replaced by linolenic acid. One is at the final stage of female reproduction, i.e. parturition, which was prolonged and impaired in linolenate-fed rats; and this defect is almost certainly the result of a deficiency of prostaglandins of the 2-series. Large amounts of an undefined prostaglandin-like substance (? $PGF_{3\alpha}$) were produced in vitro by the uteri of these linolenate-fed animals, and the identity of this metabolite might be worth determining. Preliminary evidence suggests that mice fed linolenate do not suffer impaired parturition, and further investigations of a comparative nature should be carried out to determine if the rat is unique in this respect.

The second function of linoleic acid not replaced by linolenic acid is in spermatogenesis. This may again involve prostaglandins but it is possible that the number and distribution of double bonds in the fatty acid molecule may be necessary for the production of a specific long chain fatty acid that is structurally essential for spermatogenesis to take place. Again, since linolenate-fed mice do not appear to suffer testicular degeneration, further comparative work is required to determine the relative importance of linoleic and linolenic acid in spermatogenesis in various species.

Both linoleic and linolenic acid stimulate body growth in the rat to a similar extent. However 22:6n-3 has only partial growth promoting properties and the reasons for this difference from the parent acid may be well worth investigating.

There does not appear to be any marked function for linolenic acid that cannot be replaced totally or partially by linoleic acid. If a specific function of linolenic acid is to be found it should be located in the retina which has a high content of metabolites of linolenic acid. However, in guinea pigs where the retina had been virtually completely depleted of linolenic acid the electroretinograms appeared normal. However, it is possible that linolenic acid and its metabolites might be playing a more subtle role in vision that cannot be detected by electroretinography.

The main defect detected in the pilot experiment with guinea pigs was the absence of a startle reflex in linolenate-deficient animals. These preliminary findings should be repeated under more controlled conditions and the locus of defect pinpointed. More controlled studies on other behavioural changes would also seem worthwhile.

ACKNOWLEDGEMENTS

I thank Mrs Stephanie Saunders for skilled secretarial assistance. I also thank the Wellcome Trust and the Medical Research Council for support. I record my dismay at the decision of the Agricultural Research Council to terminate this potentially important line of research.

REFERENCES

Aaes-Jorgensen, E., and Holman, R.T., 1958, Essential fatty acid deficiency: 1. Content of polyenoic acids in testes and heart as an indicator of EFA status, J. Nutr., 65:633.
Bieri, J.G., and Prival, E.L., 1965, Lipid composition of testes from various species, Comp. Biochem. Physiol., 15:275.
Burr, G.O., and Burr, M.M., 1930, On the nature and role of the fatty acids essential in nutrition, J. biol. Chem., 86:587.
Davis, J.T., Bridges, R.B., and Coniglio, J.G., 1966, Changes in lipid composition of the maturing rat testis, Biochem. J., 98:342.
FAO Food and Nutrition Paper No. 3. Rome, 1977.
Leat, W.M.F., Curtis, R., Millichamp, N.J., and Cox, R.W., 1986, Retinal function in rats and guinea-pigs reared on diets low in essential fatty acids and supplemented with linoleic or linolenic acids, Ann. Nutr. Metab., 30:166.
Leat, W.M.F., and Northrop, C.A., 1981, Effect of dietary linoleic and linolenic acids on gestation and parturition in the rat, Q. Jl. exp. Physiol., 66:99.
Leat, W.M.F., Northrop, C.A., Davidson, K., and Harrison, F.A., 1984, The effect of n-6 and n-3 polyunsaturated fatty acids on testicular development in the rat, Proc. Nutr. Soc., 43, 50A.
Leat, W.M.F., Northrop, C.A., Harrison, F.A., and Cox, R.W., 1983, Effect of dietary linoleic and linolenic acids on testicular development in the rat, Q. Jl. exp. Physiol., 68:221.
Palmer, A.C., 1976, "Introduction to Animal Neurology", Blackwell Scientific Publications, Oxford.
Verdino, B., Blank, M.L., Privett, O.S., and Lundberg, W.O., 1964, Metabolism of 4, 7, 10, 13, 16 docosapentaenoic acid in the essential fatty acid-deficient rat, J. Nutr., 83, 234.

THE SUPPLY OF OMEGA-3 POLYUNSATURATED FATTY ACIDS

TO PHOTORECEPTORS AND SYNAPSES

Nicolas G. Bazan*

Louisiana State University Medical Center School of Medicine
LSU Eye Center, and Eye, Ear, Nose and Throat Hospital
New Orleans, Lousiana 70112

INTRODUCTION

Omega-3 (or n-3) fatty acids are accumulated in membrane phospholipids primarily as docosahexaenoate (22:6, omega-3). Brain and retina
contain by far the highest concentrations of this fatty acid of any tissue. The content of docosahexaenoate is third highest in testes, but its
concentration is several fold lower in that tissue than in brain and retina (Salem et al, 1986: Birkle and Bazan, 1986; Bazan and Reddy, 1985).

Omega-3 fatty acids are derived from linolenic acid (18:3, omega-3),
which must be supplied by the diet, as discussed in other contributions
to this book. Dietary deprivation of linolenic acid must be very prolonged (of several months duration, beginning during early development or
gestation) in order to initiate depletion of docosahexaenoic acid in
brain and retina (Sun et al, 1974; Galli et al, 1970, 1971; Bourre et al,
1984; Sanders et al 1974; Tinoco et al, 1977). Although its biochemical
bases are not well understood, this tenacious retention presumably reflects an important functional role. Because brain and retinal tissues
accumulate a polyunsaturated fatty acid from the linoleic-arachidonic
acid family (20:5, omega-6) during dietary deprivation of omega-3 fatty
acids, and due to undefined functional requirements, the essentiality of
the omega-3 fatty acids is often referred to with a question mark (Salem
et al, 1986). This is mainly due to the fact that the appearance of a
fatty acid (20:5, omega 6) during deprivation of omega-3 fatty acids is
seen as the activation of a cellular compensatory mechanism to cope with
the shortage of docosahexaenoic acid by producing the fatty acid that
most closely resembles it, from the point of view of carbon chain length
and the number of double bonds.

Recent studies carried out in rhesus monkeys have shown that diets
deficient in omega-3 fatty acid during gestation result in impaired visual acuity and, as expected, decreased amounts of retinal DHA (Neuringer
et al, 1984, 1985). These findings indicate that DHA depletion alters
photoreceptor function in the central retina, where visual acuity is sustained. Because brain DHA is also decreased under these conditions, it
is possible that there are also effects on the visual cortex or other

---

*Mailing address: LSU Eye Center, 2020 Gravier Street, New Orleans,
LA 70112

electroretinographic signals, and in the regeneration preceding the completion of physiological responses (Neuringer and Connor, 1986; Neuringer et al, 1988).

This chapter deals with the supply of omega-3 fatty acids to the photoreceptor cells and synapses. Omega-3 fatty acids are required by these membranes during:

a) Perinatal development, most specifically during postnatal development in mammals, when a burst requirement of omega-3 polyunsaturated fatty acids takes place during synaptogenesis and photoreceptor membrane biogenesis;

b) Normal functioning of tissues when, due to turnover or use of docosahexaenoic acid for the synthesis of oxygenated metabolites (docosanoids, the functional role of which is not known), small amounts may need to be supplied;

c) Injury to the nervous system (ischemia, convulsions) (Bazan, 1970; Bazan et al, 1983), and during retinal stimulation (Aveldano de Caldironi et al, 1981; Aveldano and Bazan, 1974, 1975; Giusto and Bazan, 1983), both of which trigger the release of docosahexaenoic acid from membrane phospholipids. Some of this docosahexaenoic acid may be peroxidated or lost through washout to the blood, and may need to be replenished.

We summarize here recent studies, mainly from our laboratory, using the retina and photoreceptor cells as models to study the composition, metabolism, supply, alterations in pathological conditions, and possible physiological significance of omega-3 polyunsaturated fatty acids.

Docosahexaenoate and other acyl chains of phospholipids in developing photoreceptor cells

Rod photoreceptor cells of the mouse retina originate from neuroblasts during the first two weeks of postnatal life (Grun, 1982). The development and differentiation of these cells imply the formation of outer segments, axons, and synaptic terminals. By five postnatal days, retinal cell division has concluded, except in the peripheral retina, and rod cells have begun to establish synaptic contact with horizontal cells. About one week later, synaptic contact with bipolar cells has begun (Blank et al, 1974a); at this time a rod outer segment has formed and has begun to elongate. The rod photoreceptors complete their differentiation by 3-4 postnatal weeks.

Photoreceptors were isolated after dissociation from retinas by mechanical agitation after mild protease exposure and characterized by morphology using light- and electron microscopy. The dissociation and characterization of developing rod photoreceptor cells from the mouse retina (Scott et al, 1988) is based on a modification of the procedure of Lolley et al (1986). Figure 1 shows the isolated cells.

The protein/dry weight ratio of dissociated photoreceptors from mice remained relatively unchanged with age; however, the lipid phosphorus/dry weight ratio was enhanced from 2.3 at 5-6 days to 5.1 at adulthood (Scott et al, 1988).

The phospholipid content of photoreceptors increased from 147 nmole/mg protein at 5-6 days to 292 nmole/mg protein by adulthood, and the proportion of individual phospholipids changed during development. Phosphatidylcholine and phosphatidylinositol decreased, whereas phosphatidylethanolamine and phosphatidylserine increased during the same period (Scott et al, 1988).

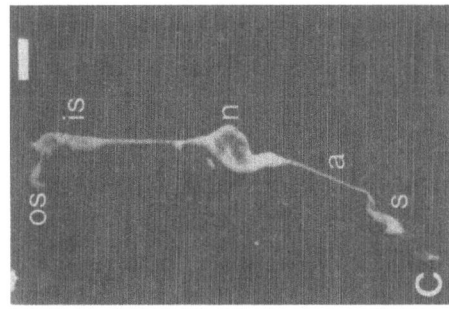
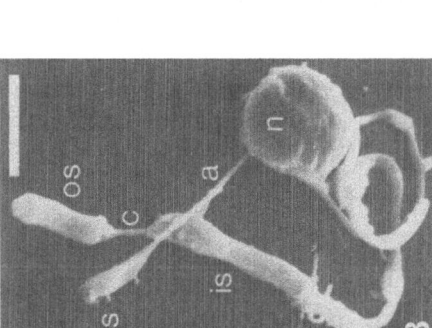
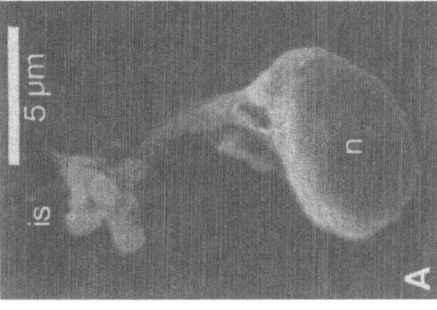

Figure 1. Scanning electron micrographs of dissociated developing photoreceptor cells from five-day-old normal retina. (A) depicts the nucleus (n), short inner segment (is) with ruffled projections, and single cilium; an axonal extension is seen beneath the cell. At 11 postnatal days, photoreceptor cells from normal (B) and rd retinas (C) have an inner (is) and outer segment (os), connecting cilium (c), an axon (a) and synaptic spherule (s); however, the inner and outer segment of the rd visual cell is smaller than normal. Scale bars = 5um. (Reprinted with publisher's permission from Scott et al, 1988.)

229

The fatty acyl chains of phospholipids from rod photoreceptor cells were significantly modified during development (Fig. 2).

The prevalent fatty acids in phosphatidylcholine at 5-6 days of age were palmitate and oleate; in adult cells they changed to palmitate, stearate, and docosahexaenoate. In phosphatidylethanolamine at 5-6 days palmitate, stearate, oleate, arachidonate, and docosahexaenoate were in similar proportions; however, in adult cells the major fatty acids were docosahexaenoate and stearate. In phospatidylserine at 5-6 days, the major fatty acyl chains were stearate and oleate; in adult cells, the stearate content decreased and docosahexaenoate represented more than 50% of the total. In phosphatidylinositol of 5-6-day-old cells, stearate and arachidonate prevailed, whereas in adult cells, the major fatty acids were palmitate, stearate and arachidonate.

Developing photoreceptor cells accumulate all four major phospholipids (phosphatidylcholine, phosphatidylethanolmine, phosphatidylserine, phosphatidylinositol), and the shift in fatty acyl composition becomes more unsaturated. Therefore, photoreceptor cells are mainly responsible for the developing lipid profile of the total retina.

The developmental patterns of the polyunsaturated fatty acids, docosahexaenoic and arachidonic, are very different in photoreceptor cells. Arachidonate was relatively constant or decreased as a function of development, whereas docosahexaenoate increased steadily and significantly in all major membrane phospholipids as a function of age (Figure 2). Therefore the developing photoreceptor cell seems to handle the two major polyunsaturated fatty acids, arachidonate and docosahexaenoate, in different manners. Immature cells are already endowed with adult concentrations of arachidonate; thus, arachidonate does not become further enriched in photoreceptor cells during development.

Arachidonic acid is the precursor of biologically active metabolites such as prostaglandins, leukotrienes, and other oxygenated products (Lands, 1979; Samuelsson, 1983; Corey et al, 1983; Birkle and Bazan, 1986). In developing visual cells, the extent to which arachidonate is released and channeled through the enzyme-mediated oxygenation pathway, the arachidonic acid cascade, is undefined. It is not known whether arachidonic acid, as part of the inositol lipids, bears a functional role during photoreceptor differentiation through cell signaling-linked events.

Arachidonate may be supplied to photoreceptor cells prior to birth and/or during early postnatal life (the first five days). This pattern is remarkable in contrast to that of docosahexaenoate. Developing photoreceptors rapidly and selectively accumulate docosahexaenoate and yield much higher docosahexaenoic acid levels in adult cells than in immature ones.

Docosahexaenoate and other omega-3 fatty acids decrease the synthesis of eicosanoids from arachidonic acid (Lands et al, 1973; Corey et al, 1983); therefore one function of the rapid accumulation of docosahexaenoate during the maturation of visual cells may be to modulate eicosanoid formation in the cells, as well as to play a structural role in photoreceptor membranes. It has been shown that docosahexaenoic acid results in the synthesis of oxygenated compounds, termed docosanoids by analogy to eicosanoids (Bazan et al, 1984b), from arachidonic acid. The biological significance of docosanoids in the visual cell, if any, has not been defined, although these lipids may serve as second messengers or as mediators in the retina (Bazan and Scott, 1987). In other cells, similar metabolites have been described (Aveldano and Sprecher, 1983). The accumulation of docosahexaenoic acid in visual cells could reflect an in-

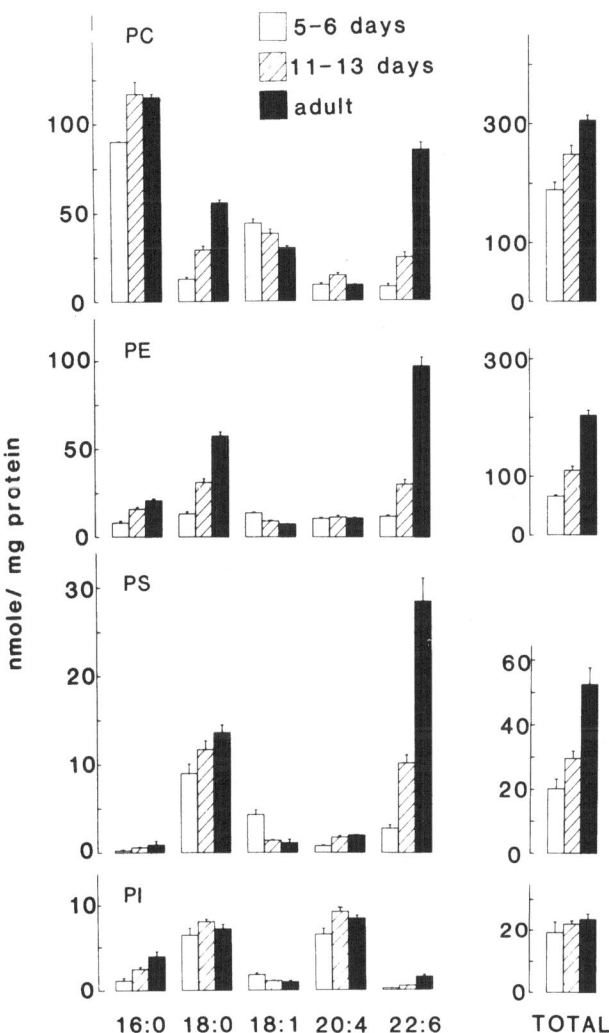

Figure 2. Fatty acyl chain content of phospholipids from photoreceptor cells of mouse retina isolated at 5-6 postnatal days (white bars), 11-13 days (hatched bars) and in adult cells (block bars). Phosphatidylcholine and plasmalogens (PC), phosphatidylethanolamine and plasmalogens (PE), phosphatidylserine (PS), and phosphatidylinositol (PI) were extracted from normal C57BL/6J retinas. Values represent 3-6 independent preparations and are expressed as nmole/ug lipid phosphorus $\pm$ standard deviation. (Reprinted with publisher's permission from Scott et al, 1988.)

creased docosanoid requirement in maturing cells. Docosahexaenoic acid
may also replenish an eicosapentaenoate (EPA 20:5, omega 3) pool, that in
turn synthesizes 3-series of eicosanoids (Weber et al, 1986). Retroconversion of docosahexaenoate to eicosapentaenoate may accomplish this.

Decreased polyunsaturated fatty acid content in photoreceptor cells from mutant mice during early postnatal development

Retinal degenerations comprise a heterogeneous group of diseases
that lead to photoreceptor cell degeneration. Some of them are age-related (e.g. senile macular degeneration); others are inherited, such as
retinitis pigmentosa and Usher's syndrome (deafness from birth and blind-
ness later in life due to retinitis pigmentosa).

Decreased docosahexaenoic and arachidonic acid content in plasma
phospholipids of patients with inherited retinal degenerations has been
observed (Converse et al, 1983; Bazan et al, 1986c; Anderson et al, 1987).
Therefore, it was of interest to investigate the supply, composition and
metabolism of essential fatty acids, particularly of the omega-3 series,
to rod photoreceptor cells from mutant rd mice at the time of development
when most of the omega-3 fatty acids are accumulated.

In the inherited retinal degeneration model, the rd mouse, developing photoreceptor cells fail to differentiate outer segments, and they
degenerate before reaching maturity (Sidman and Green, 1965; Sanyal and
Bal, 1973; LaVail and Sidman, 1974). The abnormalities can be detected
histologically at about postnatal day 12, after a rudimentary outer segment has been formed and synaptic junctions with horizontal cells have
been established (Blank et al, 1974b). The molecular alterations causing
degeneration are unclear; however, a defect in the phototransduction cascade takes place which results in accumulation of cyclic guanosine monophosphate (cGMP) (Lee et al, 1987; Lolley et al, 1980). This accumulation may, in turn, initiate degeneration that interferes with metabolic
events such as lipid or protein synthesis. If lipid metabolism is altered by the rd mutation, photoreceptor membranes may reflect abnormal
lipid composition.

The fatty acyl chains of phospholipids in developing rod photoreceptor cells dissociated from rd mutant retinas has been compared with those
of control mice. The objectives of these studies were to identify alterations in fatty acyl groups in rod photoreceptor cells of the rd mutant
and to ascertain if any changes were detectable prior to the postnatal
time when degeneration can be detected morphologically.

Lipid content and phospholipid classes of 5-6 day-old photoreceptors
from rd retinas were not different from those of controls, although the
concentration of phosphatidylcholine, phosphatidylethanolamine, and phosphatidylserine were somewhat higher in rd photoreceptors. The fatty acyl
composition of individual phospholipids, however, showed several significant alterations at 5-6 days of age. Myristate (14:0) was significantly
increased in phosphatidylcholine and phosphatidylethanolamine of rd cells
as compared with controls; palmitate (16:0) was higher in phosphatidylethanolamine and phosphatidylserine of rd cells. However, stearate
(18:0) levels were similar in rd and control cells at this age. Linoleate (18:2) was significantly increased and 20:3 was lower in phosphatidylethanolamine of rd cells. Phosphatidylinositol from rd photoreceptors
contained much less arachidonate (20:4), whereas phosphatidylcholine from
mutant cells contained more 22:5 6 than controls. Also, the content of
22:5 3 was decreased in phosphatidylcholine and higher in phosphatidylinositol of rd cells, compared with controls. These results depict altered

polyunsaturated fatty acid metabolism associated with the expression of the mutation of retinal degeneration. This view is strengthened because the fatty acid changes are detected early during development, preceding morphological changes by several days.

At 10-12 days of age, photoreceptors are beginning to enter the degenerative phase in rd mice. They have a small inner segment, a disorganized outer segment comprising less than 10% of the size in controls, and an immature synaptic terminal. By this age, a pronounced trend toward greater fatty acid saturation in rd photoreceptors is very clear; saturated and mono-unsaturated fatty acyl chains are higher, and docosahexaenoic acid is lower in rd cells than in controls. Moreover, the changes in omega-6 fatty acids seen in rd cells at 5-6 days (enhanced 18:2 and decreased 20:3 in phosphatidylethanolamine) persist in 10-12 day-old cells. Thus, immature rd photoreceptors may be retarded in the maturation of lipid metabolism, since the fatty acid composition of 10-12 day old rd cells is similar to that predicted for normal visual cells of 6-11 days of age. At 10-12 days of age, rd photoreceptor cells contain only about two-thirds the amount of docosahexaenoic acid found in normal photoreceptors of the same age; however, this decreased level is still about 50% higher than in normal cells at 5-6 days of age. Rd photoreceptor membranes result in abnormally high contents of saturated fatty acids; thus the milieu of membranes is altered and fluidity is decreased in rd cells, as compared with normal ones.

The events triggering photoreceptor degeneration in the rd retina are not understood; they may be linked to the very high levels of cGMP in these cells before they degenerate (Farber and Lolley, 1974). Cyclic GMP may impair ion channels and membrane polarization (Fesenko et al, 1985), may stimulate protein kinase (Lee et al, 1987) and may result in enhanced phosphoproteins. The morphological development of rd photoreceptors is disrupted at about the same time as cGMP accumulates. The establishment of synaptic contacts stops after horizontal dendrites have formed junctions with affected photoreceptors (Blanks et al, 1974b), and inner and outer segment development fails to differentiate beyond that of 8-9 day-old normal visual cells. The inability of rd photoreceptor cells to maintain normal synthesis and utilization of polyunsaturated fatty acids is probably another manifestation of the aborted maturation of the cells. The pleiotropic effects of the rd gene on developing photoreceptors may result in altered metabolism of essential fatty acids, leading to impaired cell function and viability.

Cellular Retention of Omega-3 Polyunsaturated Fatty Acids

Cellular retention of omega-3 polyunsaturated fatty acids is selective and tenacious. The molecular bases for these events are not fully understood, but they are somehow linked to the supply of these fatty acids.

The tissues containing the largest amounts of omega-3 polyunsaturated fatty acid may have receptor-like mechanisms at the microvasculature and/or cells (retinal pigment epithelium, photoreceptor cells, neurons, glia, sperm cells) to selectively take up the fatty acids. This may be why the rest of the tissues, although exposed similarly to blood stream lipids, do not become enriched in docosahexaenoate.

The supply of polyunsaturated fatty acids to developing visual cells may involve: 1) synthesis of polyunsaturated fatty acids directly in photoreceptor cells, from dietary essential fatty acid precursors taken up by the cells; 2) uptake by photoreceptor cells of polyunsaturated fatty acids synthesized elsewhere, such as in the liver, and distributed via the blood; or 3) a combination of these two mechanisms. It is not certain which of these routes are involved in the supply of polyunsaturated fatty acids to the photoreceptors and synapse. However, recent data suggests that during postnatal development, photoreceptor cells are mainly supplied by the liver with docosahexaenoic acid, after this organ has taken dietary linolenic acid through elongation and desaturation (Scott et al, 1986; Scott et al, 1987a,b,c,d; Dobard et al, 1987; Cai et al, 1988; Marcheselli et al, 1988; Scott et al, 1988).

Similar events may also take place in the brain (Scott et al, 1987b). Mice injected after birth with $(1-^{14}C)$ linolenic acid do not take up this fatty acid in brain; rather, ^{14}C-docosahexaenoate appears after the liver has taken up most of the precursor. Docosahexaenoate is increasingly detectable in the blood as a function of development time (Scott et al, 1987b).

Therefore, a combination of both pathways, the synthesis of docosahexaenoic acid and the supply of this fatty acid by the liver, may operate in the retina and brain. During the time of synaptogenesis in mice, the bulk of the supply of docosahexaenoic acid seems to come from the liver, after linolenic acid has been elongated and desaturated. The retina, however, does have enzymes to elongate and desaturate omega-3 fatty acids. After an intravitreal injection of $(1-^{14}C)$eicosapentaenoic acid, the adult rat retina has been shown to elongate and desaturate it to docosahexaenoic acid (Bazan et al, 1982a). Other in vitro studies with entire retinas have shown that both immature and adult retinas incorporate exogenous radiolabeled 22:6 into their lipids (Scott et al, 1987e).

Docosahexaenoyl Coenzyme A Synthetase

Once docosahexaenoic acid enters retinal or brain cells, it is rapidly converted into the thiol ester of coenzyme A by docosahexaenoyl CoA synthetase. This is a low Km enzyme that is very active in brain and retina that has been proposed to be responsible for cellular retention of the fatty acid after arrival in the cell (Reddy et al, 1984; Reddy and Bazan, 1984, 1985a and b). This mechanism may operate by preventing the fatty acid from leaving the cell after it is converted into the coenzyme A derivative. Docosahexaenoyl coenzyme A is not accumulated in cells since the fatty acid is subsequently channeled to the acyl transferases that esterify docosahexaenoic acid into phospholipids (Bazan et al, 1984a; Bazan and Bazan, 1985).

The Polyunsaturated Fatty Acid Binding Properties of Interphotoreceptor Retinoid Binding Protein

The metabolic handling of polyunsaturated fatty acids mainly implies turnover in membrane phospholipids, uptake after arrival in the cells and assurance of retention under conditions in which there is accumulation of free docosahexaenoic acid and other free fatty acids. This occurs during cell stimulation (Bazan, 1970; Aveldano et al, 1981) or injury, such as hypoxia or ischemia (Bazan, 1970; Giusto and Bazan, 1979). Because essential fatty acids are needed for cell function and are not easily replenished, there are fatty acid binding proteins that retain the fatty acids nearby the site of utilization. There is an intracellular fatty acid binding protein (Rhodes, 1982) and an extracellular protein that may also perform such a role. The latter is IRBP (interphotoreceptor retinoid binding protein), which is located in the space between the photoreceptors and retinal pigment epithelial cells, the interphotoreceptor matrix. This protein contains fatty acids that are covalently and noncovalently bound (Bazan et al, 1985). The latter fraction comprises poly-

unsaturated fatty acids, including docosahexaenoic acid. The interphotoreceptor matrix may also contain other extracellular fatty acid binding
proteins with docosahexaenoate (Bazan et al, 1985). IRBP may constitute
a shuttling mechanism to bring back docosahexaenoate from the retinal
pigment epithelium to the photoreceptor cell after the shedding and phag-
ocytosis of outer segments. In mammals, there is a circadian shedding
event through which the discs of photoreceptor tips are shed and phagocytized by the retinal pigment epithelium (Bok, 1985). At the same time,
the length of the outer segment remains constant by the activation of
membrane biogenesis. Because there is an avid retention of docosahexaenoic acid in the photoreceptor cell even after this renewal process, the
IRBP may function as an intercellular carrier of docosahexaenoic acid
(Bazan et al, 1985).

CONCLUDING REMARKS

Studies using isolated photoreceptor cells (visual cells) from the
mouse retina have shown that the accumulation of the two polyunsaturated
fatty acids, arachidonic and docosahexaenoic acid, in phospholipids
during postnatal development and cell differentiation follow different
time courses. Docosahexaenoic acid is accumulated in a profile closely
following photoreceptor membrane (outer segment) biogenesis. The ability
to accumulate polyunsaturated fatty acids is thus achieved during photoreceptor maturation and may represent a specialization of visual cells
which facilitates their function and ensures their viability.

Photoreceptor cells are a useful model to investigate metabolism and
supply of this fatty acid to one specific cellular component of the
retina. This model also allows the study of how a mutation affecting the
differentiation of the rod photoreceptor cell impairs docosahexaenoate
metabolism, and of phospholipid metabolism in general. Since the major
consequence of the rd mutation is a failure to develop the outer segment,
studies on how membranes are assembled and on phospholipid utilization
are also possible. In fact, there is evidence that by day 5 of postnatal
development, cells isolated from the mutant have an abnormal fatty acyl
composition in phospholipid, which is characterized by a relative shortage of docosahexaenoate, and overall decreased unsaturation (Scott et al,
1988). These studies may help in gaining an understanding of the metabolism of polyunsaturated fatty acids in cells of the nervous system, and
the physiological role and pathophysiological significance of their endowment with such large amounts of docosahexaenoic acid.

ACKNOWLEDGMENTS

These studies were supported by NEI grant EY04428 and by the Edward
G. Schlieder Educational Foundation.

REFERENCES

Anderson, R. E., Maude, M. B., Lewis, R. A., Newsome, D. A., and Fishman,
G. A., 1987, Abnormal plasma levels of polyunsaturated fatty acid in
autosomal dominant retinitis pigmentosa. Exp. Eye Res. 44:155-159.
Aveldano, M. I., and Bazan, N. G., 1974, Displacement into incubation
medium by albumin of highly unsaturated retina free fatty acids
arising from membrane lipids. Febs Letters 40:53-56.
Aveldano, M. I., and Bazan, N. G., 1975, Differential lipid deacylation
during brain ischemia in a homeotherm and a poikilotherm. Content

and composition of free fatty acids and triacylglycerols. Brain Res. 100:99-110.

Aveldano, M. I., and Sprecher, H., 1983, Synthesis of hydroxy fatty acids from 4,7,10,13,16,19-[1-^{14}C]docosahexaenoic acid by human platelets. J. Biol. Chem. 258:9339-9343.

Aveldano de Caldironi, M. I., and Bazan, N. G., 1977, Acyl groups, molecular species and labeling by ^{14}C-glycerol and ^{3}H-arachidonic acid of vertebrate retina glycerolipids. Adv. Exp. Med. Biol. 83:397-404.

Aveldano de Caldironi, M. I., Giusto, N. M., and Bazan, N. G., 1981, Polyunsaturated fatty acids of the retina. Progress in Lipid Research 20:49-57.

Bazan, H. E. P., and Bazan, N. G., 1985, Metabolism of docosahexaenoyl groups in phosphatidic acid and in other phospholipids of the retina, in: "Phospholipids in the Nervous System, Vol. 2, Physiological Roles", L. Horrocks, J. Kanfer, G. Porcellati, eds., Raven Press, New York, pp. 209-217.

Bazan, H. E. P., Careaga, M. M., Sprecher, H., and Bazan, N. G., 1982a, Chain elongation and desaturation of eicosapentaenoate to docosahexaenoate and phospholipid labeling in the rat retina in vivo. Biochim. Biophys. Acta 712:123-128.

Bazan, H. E. P., Ridenour, B., Birkle, D. L., and Bazan, N. G., 1986a, Unique metabolic features of docosahexaenoate metabolism related to functional roles in brain and retina, in: "Phospholipid Research and the Nervous System. Biochemical and Molecular Pharmacology", L. Horrocks, L. Freysz and G. Toffano, eds., Liviana Press, pp. 67-78.

Bazan, H. E. P, Sprecher, H., and Bazan, N. G., 1984a, De novo biosynthesis of docosahexaenoyl phosphatidic acid in bovine retinal microsomes. Biochim. Biophys. Acta 796:11-19.

Bazan, N. G., 1970, Effects of ischemia and electroconvulsive shock on free fatty acid pool in the brain. Biochim. Biophys. Acta 218:1-10.

Bazan, N. G., and Birkle, D. L., 1987, Polyunsaturated fatty acids and inositol phospholipids at the synapse in neuronal responsiveness, in: "Molecular Mechanisms of Neuronal Responsiveness", Y. Erlich et al., eds., Plenum Press, pp. 45-68.

Bazan, N. G., Birkle, D. L., and Reddy, T. S., 1984b, Docosahexaenoic acid (22:6, n-3) is metabolized to lipoxygenase reaction products in the retina. Biochem. Biophys. Res. Comm. 125:741-747.

Bazan, N. G., Birkle, D. L., and Reddy, T. S., 1985, Biochemical and nutritional aspects of the metabolism of polyunsaturated fatty acids and phospholipids in experimental models of retinal degeneration, in: "Retinal Degeneration: Contemporary Experimental and Clinical Studies", M. M. LaVail, G. Anderson, J. Hollyfield, eds., Alan R. Liss, Inc., New York, pp. 159-187.

Bazan, N. G., di Fazio de Escalante, M. S., Careaga, M. M., Bazan, H. E. P., and Giusto, N. M., 1982b, High content of 22:6 (docosahexaenoate) and active [2-^{3}H]glycerol metabolism of phosphatidic acid from photoreceptor membranes. Biochim. Biophys. Acta 712:702-706.

Bazan, N. G., and Giusto, N. M., 1980, Docosahexaenoyl chains are introduced in phosphatidic acid during de novo synthesis in retinal microsomes, in: "Control of Membrane Fluidity", M. Kates and A. Kuksis, eds., Humana Press, New Jersey, pp. 223-236.

Bazan, N. G., Morelli de Liberti, S. G., Rodriguez de Turco, E. B. and Pediconi, M. F., 1983, Free arachidonic and docosahexaenoic acid accumulation in the central nervous system during stimulation, in: "Neural Membranes", G. Y. Sun, N. G. Bazan, J. Wu, G. Porcellati and A. Y. Sun, eds., Humana Press, New Jersey, pp. 123-140.

Bazan, N. G., and Reddy, T. S., 1985, Retina, in: "Handbook of Neurochemistry", Vol. 8, A. Lajtha, ed., Plenum Press, New York, pp. 505-575.

Bazan, N. G., Reddy, T. S., Bazan, H. E. P., and Birkle, D. L., 1986b,

Metabolism of arachidonic and docosahexaenoic acids in the retina. Prog. Lipid Res. 25:595-606.

Bazan, N. G., Reddy, T. S., Redmond, T. S., Wiggert, B., and Chader, G. J., 1985, Endogenous fatty acids are covalently and non-covalently bound to interphotoreceptor retinoid-binding protein in the monkey retina. J. Biol. Chem. 260:13677-13680.

Bazan, N. G., and Scott, B. L., 1987, Docosahexaenoic acid metabolism and inherited retinal degenerations, in: "Degenerative Retinal Disorders: Clinical and Laboratory Investigations", J. G. Hollyfield, R. E. Anderson and M. M. Lavail, eds., Alan R. Liss, New York, pp. 103-118.

Bazan, N. G., Scott, B. L., Reddy, T. S., and Pelias, M. Z., 1986c, Decreased content of docosahexaenoate and arachidonate in plasma phospholipids in Usher's Syndrome. Biochem. Biophys. Res. Comm. 141:600-604.

Birkle, D. L., and Bazan, N. G., 1986, The arachidonic acid cascade and phospholipid and docosahexaenoic acid metabolism in the retina, in: "Progress in Retinal Research", Vol 5, N. Osborne, G. Chader, eds., Pergamon Press, London, pp. 309-335.

Blank, J. C., Adinolfi, A. M., and Lolley, R. N., 1974a, Synaptogenesis in the photoreceptor terminal of the mouse retina. J. Comp. Neurol. 156:81-94.

Blank, J. C., Adinolfi, A. M., and Lolley, R. N., 1974b, Photoreceptor degeneration and synaptogenesis in retinal-degenerative (rd) mice. J. Comp. Neurol. 156:95-101.

Bok, D., 1985, Retinal photoreceptor-pigment epithelium interactions. Invest. Ophthalmol. Vis. Sci. 26:1659-1693.

Bourre, J. M., Pascal, G., Duran, G., Masson, M., Dumont, O., Piciotti, M., 1984, Alterations in the fatty acid composition of rat brain cells (neurons, astrocytes, and oligodendrocytes) and of subcellular fractions (myelin and synaptosomes) induced by a diet devoid of n-3 fatty acids. J. Neurochem. 43:342.

Cai, F., Scott, B. L., and Bazan, N. G., 1988, Delivery of omega-3 fatty acids to developing photoreceptor cells. Invest. Ophthalmol. Vis. Sci. [Suppl] 29:245.

Converse, C. A., Hammer, H. M., Packard, C. J., and Shepherd, J., 1983, Plasma lipid abnormalities in retinitis pigmentosa and related conditions. Trans. Ophthalmol. Soc. UK 103:508-512.

Corey, E. J., Shih, C., and Cashman, J. R., 1983, Docosahexaenoic acid is a strong inhibitor of prostaglandin but not leukotriene biosynthesis. Proc. Natl Acad. Sci. USA 80:3581-3584.

Dobard, G., Scott, B. L., Gebhardt, P. C., Reddy, S. T., and Bazan, N. G., 1987, Synthesis of docosahexaenoic acid in the developing mouse retina. Invest. Ophthalmol. Vis. Sci. [Suppl], 28:340.

Farber, D. B., and Lolley, R. N., 1974, Cyclic guanosine monophosphate: Elevation in degenerating photoreceptor cells of the C3H mouse retina. Science 186:449-451.

Fesenko, E. E., Kolesnikov, S. S., and Lyubarsky, A. L., 1985, Induction by cyclic GMP of cationic conductance in plasma membrane of retinal rod outer segment. Nature 313:310-313.

Galli, C., White, H. B. Jr., and Paoletti, R., 1970, Brain lipid modifications induced by essential fatty acid deficiency in growing male and female rats. J. Neurochem. 17:347-355.

Galli, C., Trezciak, H. I., and Paoletti, R., 1971, Effects of dietary fatty acid composition of brain ethanolamine phosphoglyceride: Reciprocal replacement of n-6 and n-3 polyunsaturated fatty acids. Biochim. Biophys. Acta 248:449.

Giusto, N. M., and Bazan, N. G., 1979, Phosphatidic acid of retinal microsomes contains a high proportion of docosahexaenoate. Biochem. Biophys. Res. Comm. 91:791-794.

Giusto, N. M., and Bazan, N. G., 1983, Anoxia-induced production of methylated and free fatty acids in retina, cerebral cortex and white

matter. Comparison with triglycerides and with other tissues. Neurochem. Pathol. 1:17-41.

Grun, G., 1982, The development of the vertebrate retina: A comparative
study. Adv. Anat. Embryol. Cell. Biol. 78:7-85.

Lands, W. E. M., Letellier, P. E., Rome, L. H., and Vanderhoek, J. Y.,
1973, Inhibition of prostaglandin biosynthesis. Adv. Biosci. 9:15-
28.

LaVail, M. M., and Sidman, R. L., 1974, C57BL/6J mice with inherited
retinal degeneration. Arch. Ophthalmol. 91:394-400.

Lee, R. H., Lieberman, B. S., and Lolley, R. N., 1987, A novel complex
from bovine visual cells of a 33,000-dalton phosphoprotein with
beta- and gamma transducin: Purification and subunit structure.
Biochemistry 26:3983-3990.

Lolley, R. N., Lee, R. H., Chase, D. G., and Racz, E., 1986, Rod photoreceptor cells dissociated from mature mice retinas. Invest. Ophthalmol. Vis. Sci. 27:285-295.

Lolley, R. N., Rayborn, M. E., Hollyfield, J. G., and Farber, D. B.,
1980, Cyclic GMP and visual cell degeneration in the inherited
disorder of rd mice: A progress report. Vision Res. 20:1157-1161.

Marcheselli, V. L., Scott, B. L., Racz, E., Lolley, R., and Bazan, N. G.,
1988, Early changes in membrane fatty acids of developing photoreceptor cells of rd mice. Invest. Ophthalmol. Vis. Sci. [Suppl]
29:383.

Neuringer, M., and Connor, W. E., 1986, N-3 fatty acids in the brain and
retina: Evidence for their essentiality. Nutrition Rev. 44:285-294.

Neuringer, M., Connor, W. E., Daigle, D., and Barstad, L., 1988, Electroretinogram abnormalities in young infant rhesus monkeys deprived of
omega-3 fatty acids during gestation and postnatal development or
only postnatally. Invest. Ophthalmol. Vis. Sci. [Suppl] 29:145.

Neuringer, M., Connor, W. E., and Luck, S. J., 1985, Suppression of ERG
amplitude by repetitive stimulation in rhesus monkeys deficient in
retinal docosahexaenoic acid. Invest. Ophthalmol. Vis. Sci. [Suppl]
54:31.

Neuringer, M., Connor, W. E., Van Petten, C., and Barstad, L., 1984, Dietary omega-3 fatty acid deficiency and visual loss in infant rhesus
monkeys. J. Clin. Invest. 73:272-276.

Reddy, T. S., and Bazan, N. G., 1984, Synthesis of arachidonoyl coenzyme
A and docosahexaenoyl coenzyme A in retina. Curr. Eye Res. 3:1225-
1232.

Reddy, T. S., and Bazan, N. G., 1985, Synthesis of arachidonoyl coenzyme
A and docosahexaenoyl coenzyme A in synaptic plasma membranes of
cerebrum, cerebellum and brain stem of rat brain. J. Neurosci. Res.
13: 381-390.

Reddy, T. S., and Bazan, N. G., 1985b, Synthesis of docosahexaenoyl-,
arachidonoyl- and palmitoyl-coenzyme A in ocular tissues. Exp. Eye
Res. 41:87-95.

Reddy, T. S., Sprecher, H., and Bazan, N. G., 1984, Long-chain acyl
coenzyme A synthetase from rat brain microsomes: Kinetic studies
using $[1-^{14}C]$docosahexaenoic acid substrate. Eur. J. Biochem.
145:21-29.

Rhoads, D. E., Kaplan, M. A., Peterson, N. A., and Raghupathy, E., 1982,
Effects of free fatty acids on synaptosomal amino acid uptake systems. J. Neurochem. 38:1255-1260.

Salem, N. Jr., Kim, H.-Y., and Yergey, J. A., 1986, Docosahexaenoic acid:
Membrane function and metabolism, in: "Health Effects of Polyunsaturated Fatty Acids in Seafoods", Vol. 15, Academic Press, Inc., London, England, pp. 263-317.

Samuelsson, B., 1983, Leukotrienes: Mediators of immediate hypersensi-
tivity reactions and inflammation. Science 220:568-575.

Sanders, T. A. B., Mistry, M., and Naismith, D. J., 1974, The influence of a maternal diet rich in linolenic acid on brain and retinal docosahexaenoic acid in the rat. Br. J. Nutr. 51;57.

Sanyal, S., and Bal, A. K., 1973, Comparative light and electron microscope study of retinal histogenesis in normal and rd mutant mice. Z. Anat. Entwickl. Gesch. 142:219-238.

Scott, B. L., and Bazan, N. G., 1988, Developing retinal photoreceptor cells accumulate polyunsaturated fatty acids. Amer. Soc. Neurochem. 79:108.

Scott, B. L., and Bazan, N. G., 1987b, Docosahexaenoate synthesis in the developing mouse brain. J. Neurochem. [Suppl] 48:S80C.

Scott, B. L., and Bazan, N. G., 1987a, Polyunsaturated fatty acids in retinal development, in: "Polyunsaturated Fatty Acids and Eicosanoids. Proceedings of the American Oil Chemists' Society", pp. 534-539.

Scott, B. L., Moises, J., and Bazan, N. G., 1987d, Maternal supply of n-3 essential fatty acids to the developing mouse retina. Soc. Neurosci. 13:239.

Scott, B. L., Moises J, Lolley, R. N., and Bazan, N. G., 1987c, Selective accumulation of docosahexaenoic acid (DHA) in dissociated rod photoreceptor cells during mouse postnatal development. Invest. Ophthalmol. Vis. Sci. [Suppl] 28:340.

Scott, B. L., Racz, E., Lolley, R. N., and Bazan, N. G., 1988, Developing rod photoreceptors from normal and mutant rd mouse retinas: Altered fatty acid composition early in development of the mutant. J. Neurosci. Res. 20:202-211.

Scott, B.L., Reddy, T. S., and Bazan, N. G. 1986, Docosahexaenoate in developing retinas of visual cell mutant mice. Trans. Amer. Soc. Neurochem. 17:305.

Scott, B. L., Reddy, T. S., and Bazan, N. G., 1987e, Docosahexaenoate metabolism and fatty-acid composition in developing retinas of normal and rd mutant mice. Exp. Eye Res. 44:101-113.

Sidman, R. L., and Green, M. C., 1965, Retinal degeneration in the mouse: Location of the rd locus in linkage group XVII. J. Hered. 56:23-29.

Sun, G. Y., and Sun, A.Y., 1974, Synaptosomal plasma membranes: Acyl group composition of phosphoglycerides and $(Na^+ + K^+)$-ATPase activity during fatty acid deficiency. J. Neurochem. 22:15.

Tinoco, J., Mijanich, P., and Medwadowski, B., 1977, Depletion of docosahexaenoic acid in retinal lipids of rats fed a linolenic acid-deficient, linoleic acid-containing diet. Biochim. Biophys. Acta 486:575.

Weber, P. C., Fischer, S., Von Schacky, C., Lorenz, R., and Strasser, T. S., 1986, Dietary omega-3 polyunsaturated fatty acids and eicosanoid formation in man, in: "Health Effects of Polyunsaturated Fatty Acid in Seafoods", Vol. 3, Academic Press, Inc., London, England, pp. 49-60.

OMEGA-3 AND OMEGA-6 FATTY ACIDS IN SERUM LIPIDS AND THEIR RELATIONSHIP TO HUMAN DISEASE

Kristian S. Bjerve[*], Ole-Lars Brekke[*], Kristian J. Fougner[*], and Kristian Midthjell[**]

[*]Department of Clinical Chemistry, Regional Hospital, University of Trondheim, Trondheim, Norway, [**]Norwegian Institute for Public Health, Department for Health Services Research, District Unit, Verdal, Norway

INTRODUCTION

It is now nearly 80 years since omega-6 acids were found to be an essential nutrient in rats[1]. It was later shown that they were essential also in man, and that omega-3 fatty acids were essential nutrients in several animal species[2,3]. Although omega-3 fatty acids have been suggested to protect against cardiovascular disease, rheumatoid arthritis and some types of cancer[4], they have not until recently been considered essential in man. The FAO/WHO recommendations on essential fatty acid requirement in man accordingly do not mention α-linolenic or other omega-3 fatty acids as essential nutrients, and only state the requirement for linoleic and other omega-6 fatty acids[5].

Omega-3 fatty acids gained new and widespread interest when Dyerberg et al.[6] suggested that eicosapentaenoic acid (EPA), an unsaturated fatty acid belonging to the omega-3 family, could prevent thrombosis and atherosclerosis. Later, studies in both man and other species have shown that supplementing with either crude or purified fish oils to an otherwise unchanged western diet increases the content of EPA in tissue lipids, and simultaneously decreases the content of linoleic and arachidonic acids belonging to the omega-6 fatty acid family[7,8]. Simultaneously, serum triglycerides (TG) decreased, bleeding time increased, platelet aggregability was reduced, and leukotriene B₄ mediated functions in neutrophils was inhibited[6,8,9,10].

The possible protective effect of omega-3 fatty acids against thrombosis and atherosclerosis is of special interest in diabetics, who have an increased incidence of cardiovascular diseases. The platelet content of arachidonic acid is increased in diabetics[11], and some reports have indicated that omega-3 fatty acids are more efficient than omega-6 acids in preventing development of late complications in diabetes[12,13].

Both omega-3 and omega-6 fatty acids serve as precursors for similar, very biologically active prostanoids and leukotrienes, and omega-3 acids seem to prevent cardiovascular disease, rheumatoid arthritis and some types of cancer[4]. It was therefore puzzling that

Table 1. Criteria used to Diagnose Omega-3 Fatty Acid Deficiency

---

A: Fed only by gastric tube for 2 to 12 years.
B: Low omega-3 intake (28mg to 120mg daily) confirmed by fatty acid analysis of plasma and cellular lipids.
C: Disappearance of clinical symptoms within short time (10 days) after supplementing with omega-3 fatty acids.
D: Supplementation should increase omega-3 fatty acids in plasma and cellular lipids.

---

omega-6 fatty acids should be essential in both animals and man, while omega-3 acids were essential only in animals. In 1982, Holman et al.[14] described α-linolenic acid deficiency in a 6-year old girl, and we have lately reported omega-3 fatty acid deficiency in a total of 8 adults and 1 child[15-18].

In the following will be presented some observations made in our patients treated for an iatrogenic omega-3 fatty acid deficiency, followed by some initial results obtained from a prospective study on omega-3 and omega-6 fatty acids in total serum lipids. The study includes 325 newly diagnosed Type 2 (Insulin independent) diabetics and 200 non-diabetic controls. The purpose of the study is to investigate whether there is any correlation between the concentration of omega-3 and omega-6 acids in serum phospholipids, and the risk of developing cardiovascular disease and late diabetic complications in Type 2 diabetes.

OMEGA-3 FATTY ACID DEFICIENCY

A major problem in searching for patients with possible omega-3 fatty acid deficiency, is that there exist no clinical or laboratory findings that will show whether there is a state of deficiency or not, in man. Fatty acid analysis of serum, blood cells or tissues will certainly reflect the dietary intake of omega-3 fatty acids, but can not be used to diagnose whether an individual actually is in a state of dietary deficiency or not. We have therefore used the criteria shown in Table 1 to establish whether overt omega-3 deficiency was present. Factor C in Table 1 was the most important of these in establishing whether the patient was in a state of nutritional deficiency, while paragraphs A, B and D served to verify that there was indeed a low intake of omega-3 fatty acids, and that blood and cellular lipids started to normalize their content of omega-3 acids upon supplementation.

The typical fatty acid changes in omega-3 fatty acid deficiency are low concentrations of all omega-3 acids in lipid fractions in plasma and cellular lipids, increased concentrations of 22:5n-6 and 22:4n-6, while 20:3n-9 is normal or only slightly increased[14-19]. Table 2 shows the concentration of fatty acids in total erythrocyte lipids in one woman aged 28 years who had been fed by gastric tube for 8 years due to an extensive brain damage. She had received 300 g of a powdered diet mixed with 2.4 l of water for at least 72 months. This supplied 115 mg 18:3n-3 and 6 mg long-chain n-3 fatty acids daily (0.09% and 0.01% of total calories, respectively). At the same time, she

Table 2. Effect of Supplementing with Pure Alpha-Linolenic Acid and Purified Fish Oil on Fatty Acid Concentration in Erythrocyte Total Lipids

		μg Fatty Acid /10^{10} Erythrocytes					
		Linolenate[a]		EPA-oil[a]			
	Before	0.1[b]	0.5[b]	0.5[b]	2.5[b]	Controls(n=9)[c]	
16:0	380	321	324	323	237	242	– 314
18:0	239	155	213	228	224	203	– 235
18:1	354	286	307	301	372	181	– 241
20:3n9	21.5	8.1	17.1	16.8	27.1	1	– 2
18:2n6	68	40.7	59.6	63.4	105	140 ·	– 220
20:3n6	41.3	17.1	34.4	43.6	84.2	14	– 30
20:4n6	287	84	238	327	643	178	– 268
Totn6	568	191	483	640	1170	405	– 567
18:3n3	0.2	0	7.1	1.5	0	1.4	– 2.8
20:5n3	1.4	0	1.6	3.7	37	11	– 23
22:5n3	15.3	4.0	12.6	26.4	91	41	– 57
22:6n3	26.0	5.6	21.9	38.8	183	90	– 136
Totn3	43.9	9.5	43.2	70.4	311	148	– 214

[a]The patient was supplemented with 0.1 ml/24h with pure alpha-linolenic acid ethyl ester for 2 weeks, followed by 0.5 ml/24h for another 2 weeks. The supplement was then changed to a purified fish oil (EPA-oil) containing 49.4% of fatty acids as omega-3 (21.4% 20:5n-3, 17.2% 22:6n-3), starting with 0.5 ml/24h for 2 weeks, and thereafter increasing to 2.5 ml/24h. [b]ml/24 h. [c]Range given as mean ± 2SD of nine healthy controls.

received 2.8 g 18:2n-6 and 16 mg long-chain n-6 acids daily (2.8% and 0.01% of total calories, respectively).

She was supplemented for two weeks with 0.1 ml daily of 99% pure alpha-linolenic acid ethyl ester, followed for another two weeks with 0.5 ml daily. The supplement was thereafter changed to a purified fish oil (EPA-oil), starting with 0.5 ml daily for 2 weeks followed by another 2 weeks with 2.5 ml daily.

Supplementing with 0.1 ml alpha-linolenic acid daily resulted in a pronounced decrease in the concentration of all unsaturated fatty acids in total erythrocyte lipids, as shown in Table 2. This was also observed in two other n-3 deficient patients[16]. Healthy controls analyzed simultaneously showed normal fatty acids, indicating that the decrease was not an analytical artifact. We have previously reported that 14 days supplement with 0.1 ml pure alpha-linolenic acid reduced the concentration of n-6 acids in total plasma lipids in a 90-yr old omega-3 deficient woman[16]. Similar observations have previously also been reported by others[20]. One possible explanation is that supplying small amounts of 18:3n-3 to n-3 deficient patients induces redistribution of unsaturated fatty acids from erythrocytes to other tissues with a higher demand.

This 28-yr old patient had a pressure ulcer on one buttock, initially measuring 73 x 48 mm. It had been refractory to treatment for the last 1.5 years. One week after starting supplementing with 0.1 ml daily of alpha-linolenic acid, the nurses reported that the ulcer had started to heal. After 4 weeks, its size was reduced to 47 x 32 mm, decreasing further to 34 x 20 mm after 4 weeks of EPA-oil supplementation. The ulcer had healed completely when she was reexamined 6 months later. A slight, dandruff like, scaly dermatitis on arms and dorsal aspects of the feet also disappeared upon supplementing with alpha-linolenic acid ethyl ester.

Combined Omega-3 and Omega-6 Fatty Acid Deficiency

A 27 year old man who had received gastric tube feeding for 13 years because of extensive brain damage, showed biochemical findings and clinical symptoms of a combined omega-3 and omega-6 fatty acid deficiency. He had received a low fat diet the first 11.5 years, and later changed to a powdered diet mixed with water and skimmed milk giving a daily supply of 1.2 g 18:2n-6 (1.3% of total calories) and 46 mg 18:3n-3 (0.05% of total calories). He had an extensive, partly hemorrhagic, maculo-papulous dermatitis of the face, especially localized to the forehead and cheeks. His scalp was heavily seborrhoeic and showed an extensive and widespread hemorrhagic folliculitis. On both legs, there was a thick, crusty and scaly dermatitis typical of omega-6 fatty acid deficiency, and very different from the light, dandruff-like scaly dermatitis previously reported in omega-3 deficient patients[15,16].

He was now supplemented daily with 0.1 ml pure alpha-linolenic acid ethyl ester for 2 weeks followed by 0.5 ml for another 2 weeks. The supplement was then changed to 0.5 ml EPA-oil for 2 weeks and finally 2.5 ml EPA-oil for 2 weeks. The hemorrhagic folliculitis of the scalp started to heal rapidly, and had completely disappeared after 4 weeks of alpha-linolenic acid supplement. The facial dermatitis improved dramatically, and disappeared after another 4 weeks of EPA-oil supplement.

The coarse, crusty dermatitis on his legs had, however, not started to heal at all. The supplement was therefore changed to 10 ml of soya oil daily. This resulted in healing and normalization of the skin changes on his legs as well.

As seen in Table 3, four weeks supplement with pure alpha-linolenic acid increased the total plasma lipid concentration of all omega-3 fatty acids. Simultaneously, the concentration of omega-6 fatty acids also increased, arachidonic acid by as much as 60%. This indicates that omega-6 fatty acid accumulation in lipids might be impaired in omega-3 deficiency[15-16]. The clinical symptoms started to disappear 2 to 3 weeks before alpha-linolenic acid supplement had increased omega-3 concentration in plasma lipids. This suggest that there is an initial redistribution of essential fatty acids to tissues with the highest demand for them. This might explain the initial, extensive decline in erythrocyte unsaturated fatty acids seen after 2 weeks of low-dose alpha-linolenic acid supplement (Table 2).

The trienoic acid 20:3n-9 typical of omega-6 fatty acid deficiency, was highly increased in this 27-yr old man. Its concentration in total plasma lipids was similar to arachidonic acid before supplementation started (Table 3). Supplementing with alpha-linolenic acid and EPA-oil increased its concentration further by approximately 45%, although the extensive skin changes in the face and the scalp disap-

Table 3. Effect of Supplementing with Pure Alpha-Linolenic Acid, Purified Fish Oil, and Soya Oil on Fatty Acid Concentration in Plasma Total Lipids

mg Fatty Acid /L Plasma

	Before	Linolenate[a] 0.1[b]	0.5[b]	EPA-oil[a] 0.5	2.5[b]	Soya oil 10[b]	Controls(n=9)[c]
16:0	299	276	447	347	243	266	261 - 463
18:0	100	105	148	138	119	141	109 - 231
18:1	618	585	931	886	695	436	207 - 409
20:3n9	56	51	68	87	77	16	1.7 - 3.5
18:2n6	86	78	114	136	119	247	304 - 754
20:3n6	24	22	29	40	44	41	19 - 41
20:4n6	66	64	106	162	171	89	74 - 166
Totn6	189	176	268	367	362	393	444 - 954
18:3n3	1.4	1.7	2.3	3.5	2.9	5.4	5.0 - 14
20:5n3	6.6	5.9	11	34	66	43	8.0 - 30
22:5n3	6.3	5.8	13	25	30	17	9.1 - 19
22:6n3	21	20	34	77	111	69	44 - 80
Totn3	38	34	65	145	222	141	76 - 140

[a]The patient was supplemented with 0.1 ml/24h with pure alpha-linolenic acid ethyl ester for 2 weeks, followed by 0.5 ml/24h for another 2 weeks. The supplement was then changed to a purified fish oil (EPA-oil) containing 49.4% of fatty acids as omega-3 (21.4% 20:5n-3, 17.2% 22:6n-3), starting with 0.5 ml/24h for 2 weeks, and thereafter increasing to 2.5 ml/24h.
[b]ml oil/ 24 h.
[c]Range given as mean ± 2SD for nine healthy controls.

peared completely. 20:3n-9 started to normalize only after supplementing with approximately 5 g daily of 18:2n-6 in the form of soya oil, and this coincided with the disappearance of the typical omega-6 deficiency skin symptoms on his legs.

The observations made in this patient strongly indicates that the clinical symptoms of omega-3 and omega-6 fatty acid deficiency are not the same. They also suggest that omega-3 fatty acid deficiency impairs cellular and tissue functions in a different way than omega-6 deficiency.

The skin changes seen in two of the omega-3 deficient patients suggested that they might have an increased immunological reactivity. This hypothesis was investigated in three patients by measuring the effect of mitogen stimulation in isolated lymphocytes (Table 4). The response to stimulation with Con A (Concanavalin A) was normal, although high, in the three patients. Supplementing with alpha-linolenic acid for 4 weeks decreased the Con A stimulated incorporation of ³H-Thymidine to 15% of initial, and was attenuated further to 8% after 4 weeks supplement with EPA-oil. These results are consistent with

Table 4. Mitogen Stimulation of Isolated Lymphocytes.

	Cpm (Mean ± SEM, n=3)	
	No Addition	Con A[a]
Before	202 ± 9	58 178 ± 3 808
4 weeks 18:3n-3	462 ± 122	8 576 ± 1 854*
4 weeks EPA-oil	–	4 433 ± 1 811**

[a]Con A = Concanavalin A. *P<0.003, **P<0.0002. Mitogen stimulation was measured as the effect of Con A upon incorporation of ^{3}H-Thymidine into isolated lymphocytes before and after supplementation for 4 weeks with alpha-linolenic acid followed by 4 weeks with EPA-oil.

the clinical observations that the skin symptoms in omega-3 deficiency might partly be due to a changed responsiveness of immunological competent cells.

Function of N-3 Essential Fatty Acids in Man

The information on n-3 fatty acid deficiency in man is still sparse. It seems clear, however, that they are different from those seen in n-6 deficiency. Supplying n-3 fatty acids induced within 10 days healing of a pressure ulcer previously resistant to treatment, and induced rapid weight gain and growth in a 7-year old girl[17]. This indicates that n-3 fatty acids are essential for normal cell growth in man as they are in other species[3]. There also seems to be some sort of defect regulation of immune competent cells in these patients, which is not surprising from our present knowledge on leukotrienes formed from n-6 compared to those formed from n-3 fatty acids[4,10].

The observed changes in plasma and erythrocyte lipids in n-3 deficient patients indicate that the incorporation of n-6 fatty acids into lipids is also impaired (Tables 2 and 3, Refs. 15-18,20). Although the supporting data are few, results obtained in cultured human monocytes indicate that supplementing with pure n-3 acids increases the ability of the cells to accumulate n-6 fatty acids from the medium (O.-L. Brekke and K. S. Bjerve, unpublished results).

Dietary Requirements of Omega-3 Fatty Acids in Man

Information on the dietary requirement of omega-3 fatty acids in man is very scanty, and based on case studies. The results available to date have been summarized in Table 5. In one 6 year girl, Holman et al.[14] estimated the minimal dietary requirement of alpha-linolenic acid to be 0.54% of total calories, or 44 mg/kg body weight. This is in the same range as the values estimated for minimal and optimal dietary intake of omega-3 acids observed from a series of 9 patients[15-16].

Table 5. Optimal and Minimal Dietary Requirements Reported for Omega-3 Fatty Acids in Man.

Type of N-3 Acid	Optimal Requirement	Minimal Requirement	Reference
Linolenic acid			
mg/24 h	860 – 990	290 – 390	15-18
mg/kg Body Wt.	–	44	14
% of Calories	1.0 – 1.2	0.2 – 0.3	15-18
	–	0.54	14
Long-Chain N-3 Acids			
mg/24 h	350 – 400	100 – 200	15-18
% of Calories	0.4	0.1 – 0.2	15-18

The minimal dietary requirement has been estimated mainly from patients who had received a low, but not extremely low, intake of n-3 acids[15], but showed no clinical symptoms that could be related to n-3 deficiency. The calculation of an optimal requirement assumes that the concentration of omega-3 fatty acids in plasma and erythrocyte lipids increases linearly with increasing supplement of omega-3 acids, and that at each time point studied, steady state had been achieved.

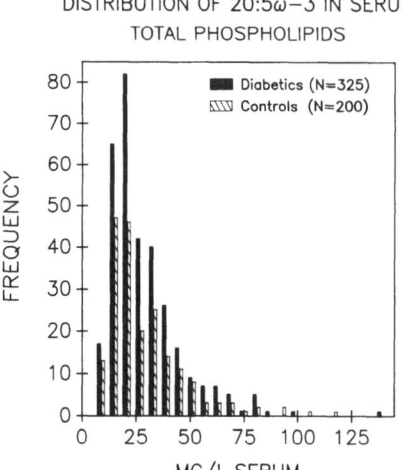

DISTRIBUTION OF 20:5ω−3 IN SERUM
TOTAL PHOSPHOLIPIDS

Fig. 1. Concentration distribution of eicosapentaenoic acid in serum total phospholipids in 325 Type 2 diabetics and 200 none-diabetic controls.

The amount of dietary n-3 acids required to obtain a mid-normal concentration of that fatty acid was calculated using these assumptions. The figures given in Table 5 are thus based on rather few observations, and are calculated indirectly assuming linearity and steady state. No solid data exist to prove that these assumptions are correct. However, they probably reflect the true situation rather close, because the same range of results was obtained after both 2 and 4 weeks supplementation, and when calculated from total N-3 acids as well as from 18:3n-3, 20:5n-3, 22:5n-3, or 22:6n-3[16].

OMEGA-3 FATTY ACIDS IN NEWLY DIAGNOSED TYPE 2 DIABETES

The effects on thrombosis, atherosclerosis, platelet aggregability and serum lipids, make omega-3 fatty acids particularly interesting in patients with Type 2 (Non insulin dependent) diabetes, because of the high incidence of cardiovascular disease in these patients. It is also possible that omega-3 acids might decrease the risk of developing late diabetic complications like nephropathy, neuropathy, and retinopathy[12].

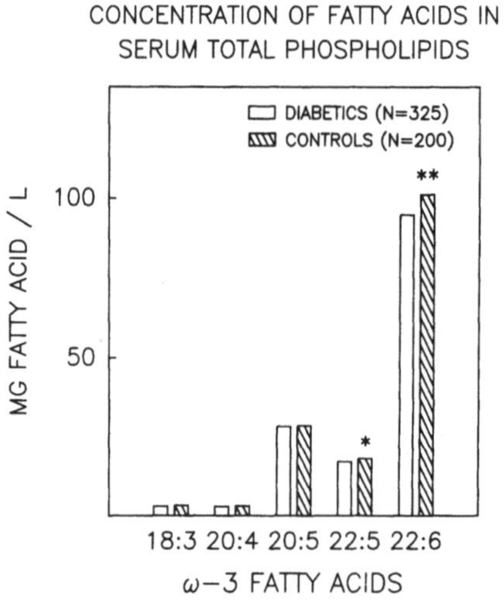

CONCENTRATION OF FATTY ACIDS IN
SERUM TOTAL PHOSPHOLIPIDS

Fig. 2. Concentration of 18:3n-3, 20:4n-3, 20:5n-3, 22:5n-3, and 22:6n-3 in serum total phospholipids. Comparison between Type 2 diabetics and none-diabetic, controls. *P<0.01, **P<0.02

We have tried to address this question in a prospective study including 325 newly diagnosed Type 2 diabetics and 200 controls. Several studies indicate that the fatty acid composition of blood lipids reflects the dietary intake of omega-3 acids[21,22]. We have therefore measured the concentration of omega-3 fatty acids in serum total phospholipids, using a column extraction technique followed by quantitative capillary GLC analysis.

The Nord-Trøndelag Omega-3 Study includes 325 of a total of 423 newly diagnosed Type 2 diabetics who were found during the screening period (1984 to 1986) of The Nord-Trøndelag Diabetes Study. All had been conventionally treated for their diabetes for 3 to 18 months before reexamination and blood sampling, which took place in 1985 to 1987. In addition, 200 non-diabetic controls matched for age, sex, and living area were included. The concentration distributions of eicosa-pentaenoic acid in total serum phospholipids in Type 2 diabetics and none-diabetic controls are shown in Fig. 1. The mean concentration of serum phospholipid eicosapentaenoic acid was 29.0 mg/l (SD=18.1) in controls, and 28.8 mg/l (SD=17.1) in diabetics. Both controls and diabetics showed a skewed distribution (Fig. 1), with the 80-percentile at 40.6 mg/l in controls, and at 38.8 mg/l in the controls. The only omega-3 acids showing statistically significant different concentrations between diabetics and controls, were 22:5n-3 and 22:6n-3 (Fig. 2).

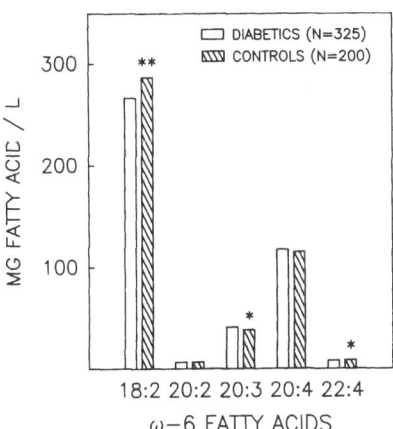

CONCENTRATION OF FATTY ACIDS IN
SERUM TOTAL PHOSPHOLIPIDS

Fig. 3. Concentration of 18:2n-6, 20:2n-6, 20:3n-6, 20:4n-6, and 22:4n-6 in serum total phospholipids. Comparison between Type 2 diabetics and none-diabetic, controls. *P<0.05, **P<0.001

The concentrations of different omega-6 acids in diabetics and controls, have been compared in Fig. 3. The Type 2 diabetics showed a lower concentration of 18:2n-6 and 22:4n-6, while the concentration of 20:3n-6 was higher than in the controls (Fig. 3). There was no statistically significant differences in the concentration of arachidonic acid.

Fatty acid analysis of platelet lipids has indicated a relative inability to convert linoleic acid to arachidonic acid in diabetics[23]. It has also been reported a decreased conversion of alpha-linolenic to eicosapentaenoic acid in diabetics[24]. It has therefore been suggested that diabetics have a reduced activity of the delta-6 desaturase. Our results do not show any difference in the serum phospholipid con-centration of arachidonic and eicosapentaenoic acid in Type 2 diabetics and controls (Figs. 1 - 3). The 18:2n-6/20:3n-6 ratio was 6.99 in diabetics and 8.21 in controls (P<0.01), and the 18:2n-6/18:3n-6 ratio was 367 in diabetics compared to 449 in the controls (P<0.01). The fatty acid concentrations in total serum phospholipids thus do not indicate a general reduction of the delta-6 desaturase activity in patients with Type 2 diabetes.

REFERENCES

1. G. O. Burr and M. M. Burr, On the nature and role of the fatty acids essential in nutrition. J. Biol. Chem., 86:587 (1930).
2. W. K. Yamanaka, G. W. Clemans, and M. L. Hutchinson, Essential fatty acids deficiency in humans. Prog. Lipid Res., 19:187 (1981).
3. J. Tinoco, Dietary requirements and functions of alpha-linolenic acid in animals. Prog. Lipid Res., 21:1 (1981).
4. Editorial: Eskimo diets and diseases. Lancet, i,1139 (1983).
5. FAO/WHO. Dietary fats and oils in human nutrition. Paper 3. Geneva, Switzerland: Food and Agriculture Organization, World Health Organization, 1977.
6. J. Dyerberg, H. O. Bang, E. Stoffersen, S. Moncada, and J. R. Vane, Eicosapentaenoic acid and prevention of thrombosis and atherosclerosis? Lancet, ii:117 (1978).
7. H. M. Sinclair, The relative importance of essential fatty acids of the linoleic and the linolenic families: studies with an Eskimo diet. Progr. Lipid Res., 20:897 (1981).
8. C. von Schacky, S. Fischer, and P. C. Weber, Long-term effects of dietary marine omega-3 fatty acids upon plasma and cellular lipids, platelet function, and eicosanoid formation in humans. J. Clin. Invest., 76:1626 (1985).
9. T. A. B. Sanders, M. Vickers, and A. P. Haines, Effect on blood lipids and haemostasis of a supplement of codliver-oil, rich in eicosapentaenoic and docosahexaenoic acid, in healthy young men. Clin. Sci., 61:317 (1981).
10. T. H. Lee, R. L. Hoover, J. D. Williams, R. I. Sperling, J. Ravalesee III, B. W. Spur, D. R. Robinson, E. J. Corey, R. A. Lewis, and F. Austen, Effect of dietary enrichment with eicosapentaenoic and docosahexaenoic acids on in vitro neutrophil and monocyte leukotriene generation and neutrophil function. New Engl. J. Med., 312:1217 (1985).
11. I. Morita, R. Takahashi, H. Ito, H. Orimo, and S. Murota, Increased arachidonic acid content in platelet phospholipids from diabetic patients. Prostaglandines Leukotrienes Med. 11:33 (1983).
12. Editorial Retrospective: Platelets and diabetes mellitus. New Engl. J. Med., 311:665 (1984).
13. A. J. Houtsmuller, J. van Hal-Ferwerda, K. J. Zahn, and H. E. Henkes, Favorable influences of linoleic acid on the progression of diabetic micro- and macro-angiopathy in adult onset diabetes mellitus. Prog. Lip. Res., 20:377 (1981).
14. R. T. Holman, S. B. Johnson, and T. F. Hatch, A case of human linolenic acid deficiency involving neurological abnormalities. Am. J. Clin. Nutr., 35:617 (1982).
15. K. S. Bjerve, I. Løvold Mostad, and L. Thoresen, Alpha-linolenic acid deficiency in patients on long-term gastric-tube feeding: estimation of linolenic acid and long-chain unsaturated n-3 fatty acid requirement in man. Am. J. Clin. Nutr., 45:66 (1987).
16. K. S. Bjerve, S. Fischer, and K. Alme, Alpha-linolenic acid deficiency in man: effect of ethyl linolenate on plasma and erythrocyte fatty acid composition and biosynthesis of prostanoids. Am. J. Clin. Nutr., 46:570 (1987).
17. K. S. Bjerve, L. Thoresen, and S. Børsting, Linseed and cod liver oil induce rapid growth in a 7-year old girl with n-3 fatty acid deficiency. J. Ent. Parent. Nutr., In press.
18. K. S. Bjerve, S. Fischer, F. Wammer, and T. Egeland, Alpha-

linolenic acid and long-chain n-3 fatty acid supplementation in three patients with n-3 fatty acid deficiency. Am. J. Clin. Nutr., In press.

19. M. Neuringer, W. E. Connor, D. S. Lin, L. Barstad, and S. Luck, Biochemical and functional effects of prenatal and postnatal omega-3 fatty acid deficiency on retina and brain in rhesus monkeys. Proc. Natl. Acad. Sci. USA, 83:4021 (1986).

20. J. A. O'Neill, M. D. Caldwell, and H. C. Meng, Essential fatty acid deficiency in surgical patients, Ann. Surg., 185:535 (1977).

21. T. Moilanen, Short-term biological reproducibility of serum fatty acid composition in children. Lipids, 22:250 (1987).

22. M. Laserre, F. Mendy, D. Spielmann, and B. Jacotot, Effects of different dietary intake of essential fatty acids on $C20:3\Omega-6$ and $C20:4\Omega-6$ serum levels in human adults. Lipids, 20:227 (1985).

23. D. B. Jones, R. D. Carter, B. Haitas, and J. I. Mann, Low phospholipid arachidonic acid values in diabetic platelets. Brit. Med. J.,286:173 (1983)

24. H. J. Mest, J. Beitz, I. Heinroth, H. U. Block, and W. Förster, The influence of linseed oil diet on fatty acid pattern in phospholipids and thromboxane formation in platelets in man. Klin. Wochenschr., 61:187 (1983).

OMEGA-3 FATTY ACIDS AND LIPOPROTEINS

T.A.B. Sanders

Department of Food and Nutritional Sciences

King's College, University of London

Campden Hill Road, London W8 7AH

INTRODUCTION

There are five major classes of lipids in blood: free fatty acids, phospholipids, triglycerides, cholesterol esters and free cholesterol. Lipids are generally insoluble in aqueous solution and need to be bound to proteins to facilitate their transport. Free fatty acids are transported in the blood bound to albumen, the other lipids are carried by the plasma lipoproteins. The plasma lipoproteins are lipid - protein complexes of density $d < 1.21 g/ml$. They are classified in terms of their hydrated density, as determined in the preparative ultracentrifuge. Four classes of lipoprotein particles are present in the plasma of normal fasting human subjects: very low density lipoprotein (VLDL) $d < 1.006 g/ml$, intermediate density lipoprotein (IDL) d 1.006-$1.019 g/ml$, low density lipoprotein (LDL) d 1.019-1.063 g/ml and high density lipoprotein (HDL) $d > 1.063$ g/ml. Particles of a fifth class, the chylomicrons (d $< 1.006 g/ml$), appear in the plasma a few hours after a fatty meal and are gradually cleared from the circulation.

VLDL are secreted mainly by the liver and are responsible for the endogenous transport of triglycerides to peripheral tissues. Hepatic VLDL particles contain apoB100 and a triglyceride rich core. As soon as nascent VLDL enter the plasma it acquires apoC and cholesterol ester, probably from HDL. In the capillaries of adipose tissue and muscle, the VLDL triglycerides are hydrolyzed by endothelial lipase which is activated by apoC, resulting in 80-90% loss of triglyceride and the loss of apoC. The product of this conversion, IDL is relatively enriched in cholesteryl ester and contains apoB100 and apoE. IDL particles are rapidly cleared from the plasma by hepatic uptake facilitated by receptors to apoE. Large VLDL particles are more effectively cleared from the circulation than small particles. IDL particles that escape hepatic uptake are further delipidated and lose apoE to form LDL.

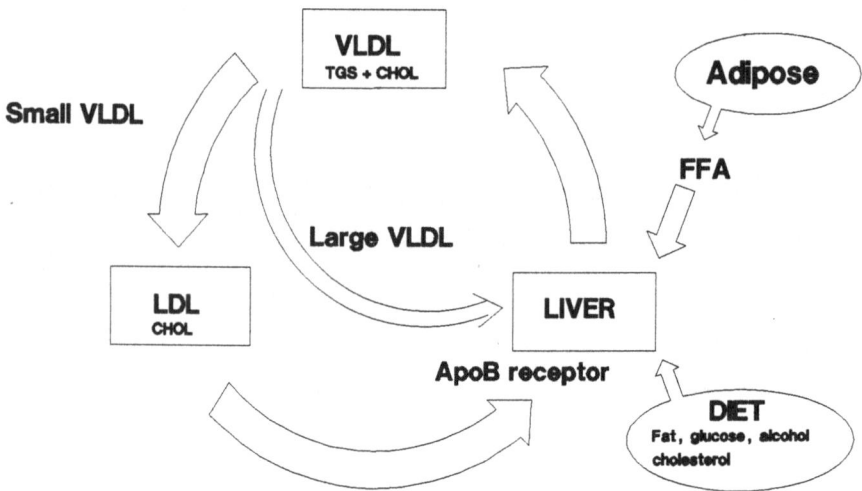

Figure 1. Metabolism of lipoproteins.

The LDL particle is enriched in cholesterol ester and contains apoB100 (Figure 1). About two thirds of LDL particles are metabolized after binding to specific receptors to apoB100 that are located on the surface of liver cells and other body cells. Binding leads to cellular uptake and lysosomal degradation of the LDL by receptor-mediated endocytosis.

The transport of exogenous lipid differs from the endogenous pathway. Dietary fatty acids (>C10) together with dietary cholesterol are incorporated as triglycerides into chylomicrons in the small intestine. Chylomicrons are similar to VLDL except that they contain apo B48 instead of apo B100. The triglycerides are hydrolyzed by endothelial lipoprotein lipase leaving a chylomicron remnant. The chylomicron remnant is rapidly cleared by uptake into the liver. Chylomicron remnant particles are not converted to LDL.

The concentration of a lipoprotein in plasma is dependent on its rate of formation as well as its rate of removal. The biological function of plasma lipoproteins is to enable the transport of lipid soluble nutrients to tissues where they are required. They also play a major role in the transport of lipid soluble foreign compounds such as drugs and pesticides. Plasma lipoprotein concentrations are major risk factors for cardiovascular disease[1] : high concentrations of cholesterol associated with VLDL, IDL and LDL but not with HDL are associated with increased risk. Most investigators believe that chylomicrons are not atherogenic. Raised concentrations of triglycerides when associated with increased VLDL or IDL are also good predictors of risk. A high concentrations of LDL apoB is associated with coronary atherosclerosis in patients with hypertriglyceridaemia.

The plasma lipoproteins of Eskimos consuming their traditional seafood diet differ markedly from those of Eskimos consuming a Western diet, especially with regard to the fatty acid

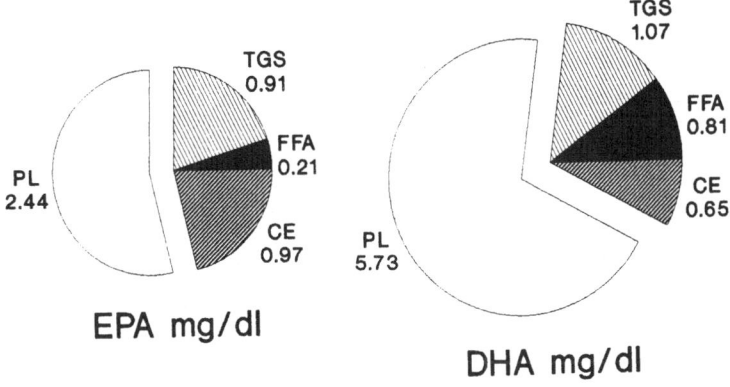

Figure 2. Distribution of EPA and DHA in plasma lipids.

Table 1

VLDL arising from a lower rate of VLDL Table 1 Changes in plasma phospholipid fatty acid composition with different intakes of w3 fatty acid in healthy volunteers (mean values ± SEM).

TREATMENT	N	EPA (wt%)		DHA (wt%)	
		PRE	POST	PRE	POST
6g linolenic/d	5	1.3 ± 0.37	2.7* ± 0.50	4.3 ± 0.33	3.7 ± 1.0
2.5g EPA 2.5g DHA/d	4	1.5 ± 0.27	6.7** ± 0.96	3.9 ± 0.45	4.5 ± 0.34
1.8g EPA 1.2g DHA/d	12	1.0 ± 0.12	4.9** ± 0.31	2.1 ± 0.27	4.4* ±+ 0.21
0.8g EPA 2.2g DHA/d	9	1.1 ±. 0.60	4.3** ± 0.25	2.23 ± 0.2	5.4** ± 10.3
1.8g EPA 0.4g DHA/d	8	1.1 ± 0.10	5.1** ± 0.17	3.3 ± 0.45	4.4* ± 0.25

*P<0.05, **P<01.

composition of the cholesterol esters and phospholipids [2]. The most pronounced change was the replacement of polyunsaturated fatty acids of the w-6 series by those of the w-3 series, notably eicosapentaenoic acid (20:5w3; EPA) and docosahexaenoic acid (22:6w3; DHA), which were provided by the Eskimo diet. Their daily diet provided about 5g EPA and 7g DHA. The concentrations of plasma VLDL and triglycerides were also markedly lower and that of HDL was higher. However, their diet differed in several other respects: it was high in protein and monounsaturated fatty acids and low in carbohydrates and saturated fatty acids. This paper considers the effects on plasma lipoproteins of supplementing the diet with different doses and types of w3 fatty acids.

METHODS

The influence of w3 fatty acid supplements on plasma lipoproteins has been studied in healthy volunteers , insulin dependent diabetics and hyperlipidaemic patients. The majority of studies have used fish oil triglyceride concentrates usually MaxEPA (Seven Seas Health Care Ltd, Hull, U.K.), which has a standard composition of 18% EPA and 12% DHA. In most studies the effects of fish oil have been compared with either treatment with vegetable oil or no treatment. In double-blind placebo controlled studies, the oil was provided in opaque, soft gelatin capsules each containing 1g of oil. A peppermint flavour was added to both types of capsule to disguise the taste of the fish oil. Compliance to supplements has usually been monitored by measuring the incorporation of EPA and DHA into erythrocyte or plasma phospholipids. Lipoprotein turnover studies have also been carried out to elucidate the mechanisms involved. These studies have been complemented where appropriate by animal feeding studies.

RESULTS AND DISCUSSION

Concentrations of w3 fatty acids in different lipid fractions

Linolenic acid is a minor constituent of cholesterol esters and phospholipids even when present in the diet in considerable amounts. It is found in its highest concentration in triglyceride fraction. EPA and DHA are found in the highest proportions in plasma phospholipids (Figure 2) . In subjects consuming a typical western diets, DHA is the major w3 fatty acids found in plasma lipids. The consumption of fish oil greatly increases the proportion of EPA in plasma phospholipids mainly at the expense of linoleic acid and other w6 fatty acids (Table 1). This fraction is probably the best marker of recent intake of EPA. The consumption of a similar amount of DHA leads to a less marked increase. There appears to be a relatively higher rate of incorporation of EPA into cholesterol esters and plasma phospholipids than DHA. Whereas the proportion of DHA in plasma triglycerides is greater than that of EPA (Table 2).

Plasma triglycerides and VLDL

Fish oil supplements containing EPA and DHA result in a consistent dose-dependent plasma triglyceride lowering effect both in normal volunteers and in patients with hypertriglyceridaemia (Figure 3). This decrease is due to lower concentrations of VLDL arising from a lower rate of VLDL

Table 2. Proportions of EPA and DHA in VLDL triglycerides and phospholipids in 5 hypertriglyceridaemic patients before and after treatment with 15g/d MaxEPA (mean value ± SEM).

		Triglycerides	Phospholipids
EPA	PRE	1.0 ± 0.12	1.7 ± 0.27
	POST	2.1 ± 0.28*	4.2 ± 0.88*
DHA	PRE	2.2 ± 0.15	4.4 ± 0.61
	POST	4.2 ± 0.39*	4.9 ± 0.38

* P.05. Data from reference 3.

triglyceride synthesis (Table 3). Both EPA and DHA show this triglyceride lowering effect in man. Linolenic acid in comparable doses does not share this effect either in man [4] or animals [5].

HDL Cholesterol

Moderate intakes of MaxEPA (about 10-15g/day) lead to a slight increase in HDL cholesterol concentration. This increase appears to be in the HDL_2 fraction. Our studies using concentrates with different ratio of EPA/DHA found that only the oils with a low ratio produced this effect (Table 4). This suggest that DHA is responsible for the increase in HDL cholesterol concentration. High intakes of EPA and DHA may, however, lower HDL cholesterol similar to very high intakes of linoleic acid.

Plasma cholesterol and LDL

Total plasma and LDL cholesterol concentrations are unaffected by moderate intakes of linolenic acid, EPA and DHA [6,7]. Total plasma cholesterol concentrations in some instances are reduced by moderate fish oil consumption (2-5g EPA and DHA/day) but most of the reduction occurs in the

Table 3. Kinetics of VLDL triglycerides before and after treatement with a 15g daily supplement of fish oil (MaxEPA) for 4 weeks in 5 hypertriglyceridaemic patients (mean values ± SD)

	Triglyceride pool g	Synthetic rat e mg/kg/hr	Fractional catabolic hr^{-1}
PRE	25 ± 10.8	48 ± 19.0	0.137 ± 0.012
POST	14 ± 6.1*	29 ± 12.0*	0.148 ± 0.024

* P.05. Data from reference 3.

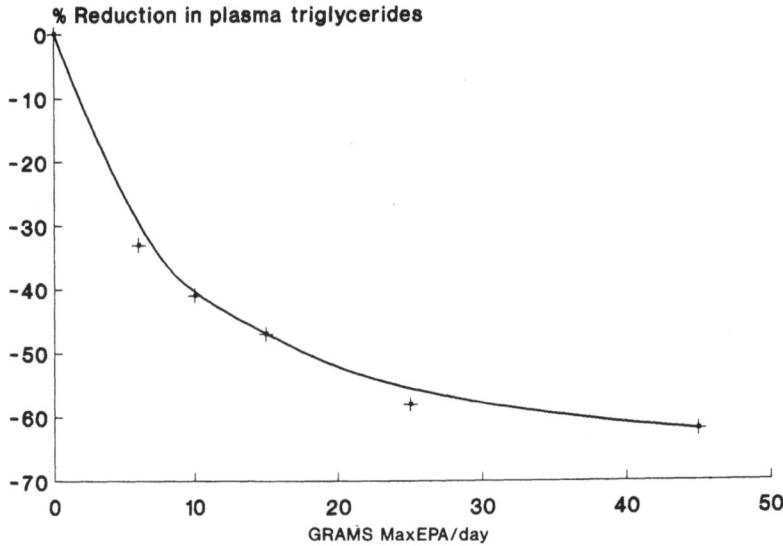

Figure 3. Plasma triglyceride lowering effect of fish oil.

VLDL or IDL fractions. However, at very high intakes (about 23g EPA and DHA/day) total and LDL cholesterol concentrations are lowered. Turnover studies show that this to be the result of a decreased rate of LDL synthesis [8]. Such high intakes also prevent the rise in plasma cholesterol obtained with dietary cholesterol [9]. In some subjects fish oil supplements lead to increases in the concentrations of LDL cholesterol and LDL apoB [10,11] particularly in hypertriglyceridaemic patients. The increase in LDL

Table 4. Influence of different intakes of w3 fatty acids on plasma HDL concentration in male volunteers (mean values ± SEM)

		HDL CHOLESTEROL mg/dl	
TREATMENT	N	PRE	POST
1.8g EPA 2.2g DHA/d	12	52 ± 2.5	57 ± 2.3**
2.7g EPA 2.2g DHA/d	8	52 ± 3.5	56 ± 5.0*
0.8g EPA 2.2g DHA/d	9	55 ± 4.2	61 ± 4.4**
1.8g EPA 0.4g DHA/d	8	46 ± 4.2	43 ± 2.4

*P.05,**P.01.

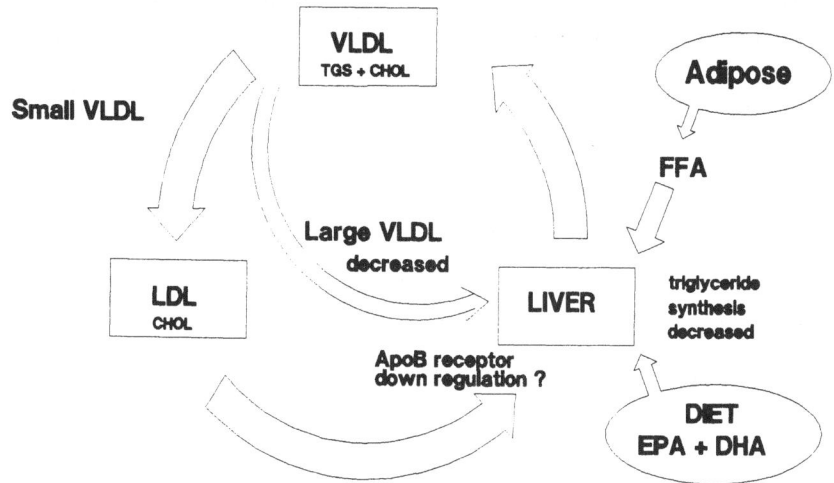

Figure 4.Influence of EPA and DHA on lipoprotein metabolism.

apoB concentration may occur without an increase in LDL cholesterol. We were the first to report an increase in LDL cholesterol in insulin dependent diabetics given 15g MaxEPA [12] and this has subsequently been confirmed by others[13]. This paradoxical increase in LDL cholesterol concentrations with fish oil treatment has been attributed by some workers to the cholesterol in the fish oil. It is extremely unlikely that the increase in LDL could be attributed to the effect of cholesterol in the fish oil as the amounts present are so trivial: 10g of fish oil contains about 60mg cholesterol an amount that would be predicted to increase plasma cholesterol concentrations by between 2-5 mg/100ml.

Table 5. Paradoxical increase in LDL cholesterol in patients with Type V
hyperlipoproteinaemia treated with large amounts of fish oil (mean value $\pm$ SEM).

	LDL CHOLESTEROL mg/dl	
TREATMENT	TYPE IIB	TYPE V
	N = 4	N = 8
CONTROL	173 $\pm$ 25	77 $\pm$ 55
FISH OIL	156 $\pm$ 19	110 $\pm$ 34*

*P < .05. Data from reference 15.

Table 6. Changes in density distribution and diameter of VLDL after ingestion of fish oil (MaxEPA 15g/d) for 2 weeks in 3 healthy volunteers (mean values ± SD or ranges)

	Sƒ 100-400	Sƒ 60-100	Sƒ 20-60	Diameter Å
PRE	39 (30-50)	33 (28-41)	28 (20-35)	1124 ± 267
POST	33 (26-39)	29 (25-32)	38 (37-41)	881 ± 181

Data from reference 10.

EPA and DHA could effect LDL clearance possibly by altering liver membrane composition (Figure 4). The LDL turnover studies[8] show that the reduction in LDL pool size is smaller than would be predicted from the decrease in synthetic rate. Moreover, despite the reduced LDL pool size there was no increase in fractional catabolic rate of LDL. This might imply down regulation of the LDL receptors by fish oils. Indeed Wong & Nestel[14] have reported decreased binding of LDL to Hep G2 cells treated with EPA.

LDL cholesterol levels rise even at high intakes of fish oil in patients with Type V hyperlipoproteinaemia [15] but not in those with Type IIB (Table 5). This appears to be a general phenomenon seen with most forms of triglyceride lowering therapy including caloric restriction[16]. The most likely explanation is that fish oil decreases hepatic triglyceride synthesis so that smaller than normal VLDL particles are secreted. Our studies[10] show that 15g MaxEPA/day reduces VLDL size and increases the proportion of VLDL 20-60 Sƒ (Table 6). LDL is derived mainly from these small VLDL particles with little being derived from the large triglyceride rich particles of Sƒ 100-400 . An increase in the proportion of smaller denser VLDL particles, as occurred after ingestion of fish oil, would be expected to increase the proportion of VLDL converted to LDL. Further LDL turnover studies in hypertriglyceridaemic subjects are needed to confirm this hypothesis.

It is uncertain whether the changes in plasma lipid concentrations brought about by moderate intakes of fish oil will reduce risk of coronary atherosclerosis. While the reduction in plasma triglyceride may be regarded as beneficial it must be offset by the potentially adverse effect of an increase in LDL concentration. It is of course possible that fish oil protects against against coronary heart disease by mechanisms independent of plasma lipoprotein concentrations.

In conclusion, moderate intakes of fish oil concentrates are not useful for the treatment of hypercholesterolaemia. However, fish oil offer a safe and effective means of treatment of hypertriglyceridaemia resulting from excessive VLDL synthesis especially in patients with Type III and Type V hyperliproteinaemias

REFERENCES

1. European Atherosclerosis Society, Strategies for the prevention of coronary heart disease: a policy statement by the European Atherosclerosis Society, *Eur. Hrt. J.* 8: 77-88 (1987).

2. H.O. Bang and J. Dyerberg, Lipid metabolism in Greenland Eskimos, *Adv. Nutr. Res.* 3:1-40 (1980).

3. T.A.B. Sanders and D.R. Sullivan, J. Reeve, and G.R. Thompson, Triglyceride-lowering effect of marine polyunsaturates in patients with hypertriglyceridemia, *Arteriosclerosis* 5: 459-465 (1985).

4. T.A.B. Sanders and F. Roshanai, The influence of different types of w3 fatty polyunsaturated fatty acids on blood lipids and platelet function in healthy volunteers, *Clin. Sci.* 64:91-99 (1983).

5. F. Roshanai and T.A.B. Sanders, Influence of different supplements of n-3 polyunsaturated fatty acids on blood and tissue lipids in rats receiving high intakes of linoleic acid, *Ann. Nutr. Metab.* 29; 189-196 (1985).

6. T.A.B. Sanders, Fish and coronary artery disease, *Br. Hrt. J.* 57:214-219 (1987).

7. S. Rogers, J.S. James, B.K. Butland, M.D. Etherington, J.R. O'Brien and J.G. Jones, Effects of fish oil supplement on serum lipid, blood pressure, bleeding time and rheological variables. A double blind randomised controlled trial in healthy volunteers, *Atherosclerosis* 63 , 137-143 (1987).

8. D.R. Illingworth, W.R. Harris, and W.E. Connor, Inhibition of low density lipoprotein synthesis by dietary omega-3 fatty acids in humans, *Arteriosclerosis* 4: 270-275 (1984).

9. P.J. Nestel, Fish oil attenuates the cholesterol induced rise in lipoprotein cholesterol, *Am. J. clin. Nutr.* 43: 752-757 (1986).

10. D.R. Sullivan, T.A.B. Sanders, I.M. Trayner,I.M. and G.R. Thompson, Paradoxical elevation of LDL apoprotein B levels in hypertriglyceridaemic patients and normal subjects ingesting fish oil, *Atherosclerosis* 61:129-134 (1986).

11. D.M. Demke, G.R. Peter, O.I. Linet, C.M. Metzler, and K.A. Klott, Effects of a fish oil concentrate in patients with hypercholesterolaemia, *Atherosclerosis* 70, 73-80 (1988).

12. A.P. Haines,A.P., T.A.B. Sanders, J.D. Imerson, et al., Effects of a fish oil supplement on platelet function, haemostatic variables and albuminuria in insulin-dependent diabetics, *Thromb. Res.* 43 :643-655 (1986).

13. R. Vandongen, T.A. Mori, J.P. Codde, K.G. Stanton,K.G. and J.R.L. Maserei, Hypercholesterolaemic effect of fish oil in insulin dependent diabetic patients, *Med. J. Aus.* 148: 141-143 (1988).

14. S. Wong,S., and P.J. Nestel, Eicosapentaenoic acid inhibits the secretion of triacylglycerol and of apoprotein B and the binding of LDL in Hep G2 cells, Atherosclerosis 64: 139-146 (1987).

15. B.E. Phillipson, D.W. Rothrock, W.E. Connor, W.S.Harris, and D.R. Illingworth, Reduction of plasma lipids, lipoproteins, and apoproteins by dietary fish oils in patients with hypertriglyceridemia, *N. Eng. J. Med.* 19:1210-1216 (1985).

16. Y.A. Kesaniemi, W.F. Beltz, and S.M. Grundy, Comparison of clofibrate and caloric restriction on kinetics of very low density lipoprotein triglycerides, *Arteriosclerosis* 5:153-161 (1985).

ALPHA-LINOLENIC ACID, PLATELET LIPIDS AND FUNCTION

Serge Renaud

INSERM, Unit 63
22 av. Doyen Lépine
69500 Bron, France

INTRODUCTION

In recent years, it has been clearly demonstrated that platelets play in coronary heart disease (CHD) a more prominent role than formerly suspected. Decreasing their activity by aspirin administration in unstable angina has resulted in a decrease of coronary events by as much as 50%, and total mortality by 71% (1), a success apparently not duplicated so far by any other substance including hypolipemic drugs. This result can be explained by the fact that unstable angina seems to be an intermittent thrombotic event as shown recently by angioscopy (2). This last technique has also demonstrated that stenotic lesions usually attributed to atherosclerotic plaque at angiography were frequently due to a fresh thrombus (3). Finally, in acute myocardial infarction, more than 90% of coronary arteries become patent after thrombolysis (4) clearly demonstrating that myocardial infarction is really due to thrombosis. Since in addition, platelet reactivity has been shown to be significantly increased in prevalent cases of myocardial infarction (5) confirmed in 2500 subjects from the Caerphilly study (to be published elsewhere), all those results suggest a primary role for platelet reactivity in CHD.

DIETARY FATS AND PLATELET FUNCTIONS

In addition to increasing serum cholesterol and inducing atherosclerotic lesions, saturated fats are associated with an increased reactivity of platelets as shown in animal (6) as well as in human studies (7-10). On an individual basis, the results obtained in 250 male farmers from France and Great Britain studied over a period of two years are reported in Table 1. As shown in that Table, the intake of saturated fatty acids was highly significantly positively associated with the clotting activity of platelets (F_3-CT) and their response to thrombin induced aggregation, while serum cholesterol as in many other studies (12) was not significantly related to the intake of saturated fat. By contrast, cholesterol was significantly inversely related to 18:2 (linoleic acid), and positively associated with the platelet function tests determined in confirmation of previous studies (13). Finally, the two main platelet function tests F_3-CT and thrombin aggregation were significantly inversely related to 18:3(n-3) (αlinolenic acid).

Table 1. Nutrients and blood parameters.
Stepwise multivariate regression analysis
in 250 French and British farmers (10).

	Clotting F_3-CT	Aggregation THR	ADP	Serum Cholesterol
Sat	****.54	****.47	***-.20	—
18:2	—	—	***.21	****-.32
18:3	****-.29	****-.44	—	—
Alcohol	*-.16	**-.19	****-.27	**.21

*p<.05   **p<.01   ***p<.001   ****p<.0001
Standard partial regression coefficients.
F_3-CT = clotting activity of platelets.
THR = thrombin
Sat = saturated fatty acids (14:0 + 16:0 + 18:0)
The nutrients, determined by chemical analysis of
the food taken over a 24 hour period were expressed
as % Kcalories.

COMPARISON BETWEEN THE EFFECTS OF LINOLEIC AND α-LINOLENIC ACIDS

The results mentioned above indicated that the eventual beneficial
effects of the two main polyunsaturated fatty acids could be directed

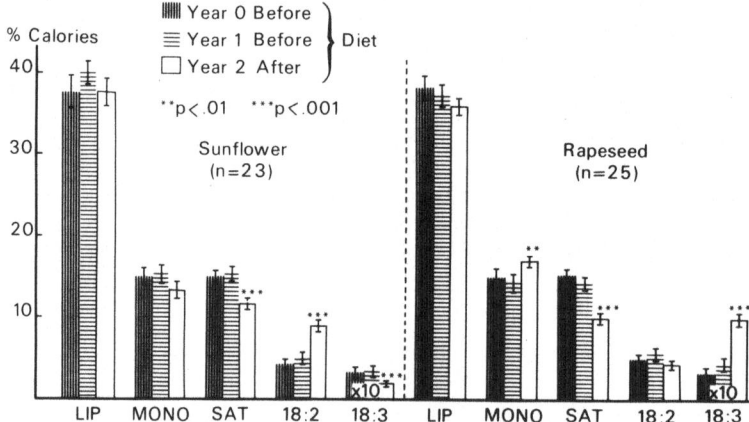

Fig. 1. Diet lipid composition in two groups of Moselle male farmers.
Lip = total lipids   Mono = monounsaturated fatty acids
Sat = saturated fatty acids. Adapted from Renaud et al (14).
Courtesy of Amer. J. Clin. Nutr.

towards different blood parameters, 18:2 against cholesterol and 18:3
(n-3) against certain platelet function tests. To verify that
hypothesis, a one year intervention trial was settled in two groups of
Moselle male farmers as indicated in figure 1, with 23 subjects in one
group (sunflower) and 25 in the second group (rapeseed). For the first
two years, diet was not changed. The third year, a decrease in the
intake of saturated fat (mostly dairy fat such as butter, cream, whole

milk) was recommanded and replaced on bread by margarine (containing only 3 to 4% of trans 18:1, and no trans 18:2) from sunflower seed in the first group, and new rapeseed (without erucic acid) in the second group. For cooking, oil and margarine were used. The results shown in figure 1 indicate that the intake of saturated fat was decreased similarly in the two groups. In the first group, only the intake of 18:2 was increased. In the second group, both the intake of 18:1 (mono-unsaturated) and 18:3 (n-3) were significantly increased, owing to the content of rapeseed oil (and margarine) in 18:1 (59%) and 18:3(n-3) (8%). In both groups, one year after diet modification, all the platelet function tests examined (F_3-CT, aggregation to thrombin, ADP, collagen) were considerably decreased (Fig. 2).

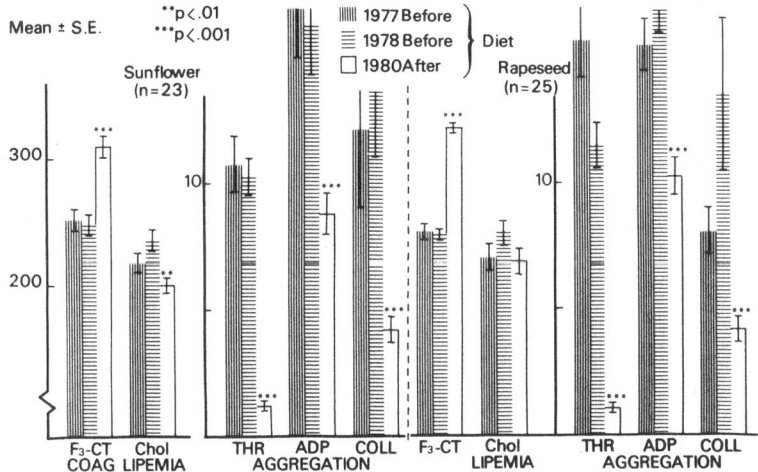

Fig. 2. Platelet function tests and serum cholesterol in two groups of Moselle farmers, two years before, and one year after diet modification as shown in Fig. 1. Aggregation to thrombin (THR), ADP, collagen (COLL). Adapted from Renaud S. et al. (14) courtesy of Amer. J. Clin. Nutr.

Nevertheless, in the group with rapeseed fat, the clotting activity of platelets (F_3-CT) was more prolonged and the aggregation to thrombin more reduced than in the group with sunflower oil and margarine. By contrast, serum cholesterol was significantly decreased only in the sunflower group. These results confirm our previous findings in 250 farmers that linoleic acid lowers mostly serum cholesterol while α-linolenic acid decreases mainly certain platelet function tests. Therefore further studies seem to be required to evaluate until what extent an increased level of α-linolenic acid in the diet could effectively prevent CHD since decreasing platelet reactivity by aspirin (1) has been shown to significantly reduce coronary events in pending myocardial infarction.

DIETARY FATTY ACIDS, PLATELETS AND CHOLESTEROL IN 18 GROUPS OF FARMERS

The study over a period of 4 years under similar conditions of 18 groups of male farmers from France, Great Britain and Belgium, comprising a total number of 460 subjects, gives the opportunity to examine in detail the relationship between nutrients, platelet function and composition, and plasma lipids. In figure 3 are shown the 18 groups classified by increasing intake of saturated fatty acids. Also shown are the intake of 18:2, 18:3(n-3) and the long chain (n-3) fatty acids 20:5 + 22:5 + 22:6. The first 4 groups, before analysis, had been submitted to a one year diet modification whith a decrease intake of saturated fat, and an increase in the intake of both 18:2 and 18:3(n-3). For the other groups, the values obtained are representative of their usual fatty acid consumption. To be noted is that the level of 18:3(n-3) in these samples of population was much higher in all the groups than that of the sum of the long chain (n-3) fatty acids.

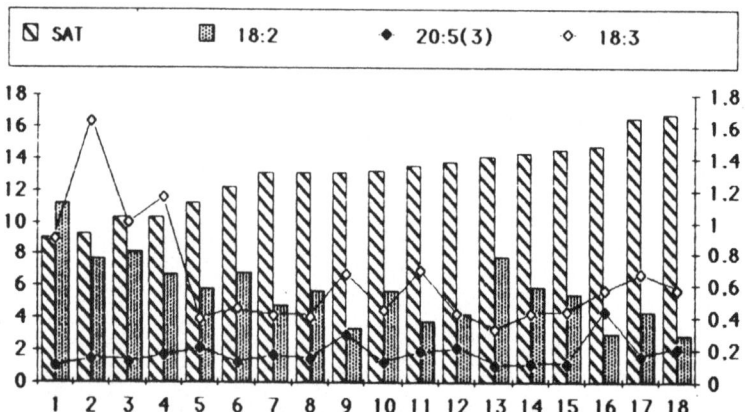

Fig. 3. The intake of dietary fatty acids in 18 groups of French-British-Belgian farmers. The level of these fatty acids was determined by chemical analysis of a duplicate sample of food collected over a 24 hour period (10). The results are expressed in percentage of K-calories. Sat = saturated fatty acids 14:0 + 16:0 + 18:0.
20:5(n-3) is in fact the sum of 20:5 + 22:5 + 22:6 (n-3).

In Table 2 are shown some of the univariate correlation coefficients between dietary fatty acids and blood parameters. Usually 14:0, 16:0 and 18:0 are so closely related that it is impossible to obtain for each of them separately, reliable relationship with blood

Table 2. Univariate correlation coefficients between dietary fatty acids and blood parameters in 18 groups of French-British-Belgian   farmers

	14:0	16:0	18:0	18:1	18:2	18:3	20:5
20:3(n-9	.56**	.75***	.85***	-.16	-.53*	.25	.21
Chol	.61**	.54*	.38	-.43	-.80***	.34	.42
F_3-CT	.55*	.68***	.65**	.04	-.36	-.70***	-.03
THR	.33	.54*	.58**	.24	-.17	-.58**	-.12

*p<.05 **p<.01 ***p<.001
Chol = serum cholesterol
Other abbreviations as in Fig. 2.

parameters, even on an individual basis. In the present study by contrast, the diet composition in the 18 areas of the 3 countries is sufficiently different for being able to somewhat separate the independent effects of 14, 16 and 18:0.

The first blood parameter examined is platelet phospholipid 20:3 (n-9), the polyunsaturated fatty acid derived from 18:0 and also to some extent from 16:0 by elongation and desaturation to constitute the (n-9) family as shown below (15).

$$16:0$$
$$\downarrow$$
$$18:0 \longrightarrow 18:1 \longrightarrow 18:2 \longrightarrow 20:2 \longrightarrow 20:3\,(n-9)$$

It can be seen in Table 2 that 20:3(n-9) is most closely associated with 18:0 and 16:0, its precursors. By contrast, serum cholesterol is more closely associated with 14:0 than 18:0, a result which confirms on a large number of subjects recent results indicating that 18:0 is not hypercholesterolemic (16). By contrast, cholesterol is strongly inversely related to dietary 18:2.

Concerning the 2 platelet function test F_3-CT and thrombin aggregation, they are mostly positively related to the 2 long chain saturated fatty acids 16:0 and 18:0 as observed in animals years ago (17) and inversely related to 18:3 as in our previous studies. As to 20:5(n-3), it was not significantly related to any of the blood parameters shown here.

Finally, none of the other platelet function tests examined (aggregation to ADP, collagen and epinephrine) were significantly related to any of the diet fatty acids, at least in univariate analysis.

As a confirmation of the inhibitory role of (n-3) fatty acids on platelet tests, in Table 3 is shown the univariate analysis between those tests and the level of the 3 main (n-3) fatty acids in plasma total lipids. Among all the tets, F_3-CT and thrombin aggregation were significantly inversely correlated with 18:3, while epinephrine aggregation was inversely related to 22:6(n-3).

Table 3. Platelet function tests and plasma (n-3) fatty acids in 18 groups of farmers. Univariate correlation coefficients.

	18:3(n-3)	20:5(n-3)	22:6(n-3)
F_3-CT	-.61**	-.08	-.04
THR	-.59**	-.08	-.02
EPI	-.34	-.24	-.52*

*p<.05   **p<.01
EPI = aggregation to epinephrine

PLATELET FATTY ACIDS AND FUNCTIONS IN RELATION TO NUTRIENTS

The relationship in the 18 groups of farmers of the platelet function tests with the platelet composition is shown in Table 4. Univariate and multivariate stepwise regression analysis were performed. For F_3-CT and thrombin aggregation, they were positively related to 20:3(n-9) and inversely to platelet cholesterol as in previous studies

(10,14). It has to be emphasized that 20:3(n-9) is the precursor of a 12-OH derivative, a substance potentiating markedly platelet aggregation to thrombin, at physiological concentration (18). That explains at least part of the mechanism involved in the effect of saturated fats to increase platelet reactivity to thrombin. In multivariate analysis only aggregation to collagen and epinephrine were significantly related to the fatty acid composition of platelets as shown in Table 4. In these studies, the level of (n-3) fatty acids, either 22:5 or 22:6(n-3) were inversely related to all the platelet aggregation tests. However this inverse relationship was significant only for thrombin and epinephrine.

Table 4. The main platelet fatty acids (and cholesterol) related to platelet function tests. Multivariate analysis in 18 groups of farmers.

	20:3-9	20:3-6	20:4-6	22:3-9	22-3	P-chol
F_3-CT	.43 **	—	—	(.46) *	—	-.56 ***
THR	.66 ***	—	—	(.69) ***	-.57 ** †	-.57 ***
ADP	—	(-.31)	(-.39)	—	(-.32) ††	—
Coll	.64 **	-.69 **	—	—	(-.30) ††	—
EPI	(-.25)	(-.26)	-.83 ****	—	-.64 ** ††	-.35 *

† 22:5 †† 22:6(n-3) $*p<.05$, $**p<.01$, $***p<.001$, $****p<.0001$
Partial regression coefficients. In parentheses, univariate correlation coefficients. Coll = aggregation to collagen. Other abbreviations as in previous tables.

Table 5. Relationship between nutrients and platelet fatty acids (and cholesterol). Multivariate analysis in 18 groups of farmers.

Nutrients	20:3-9	22:3-9	20:3-6	20:4-6	20:5-3	22:5-3	22:6-3	P-chol
18:0	.62 *	.64 ****	.46 *	(.33)	(.53) *	(.64) **	(.48) *	-.79 ***
18:1	(-.16)	(-.28)	-.33 *	—	(-.58) **	(-.34)	-.32 *	—
18:2	(-.53) *	-.28 *	-.68 ***	—	-.29 *	(-.68) ***	-.64 ***	(.46) *
18:3	(-.25)	(-.36)	(-.26)	—	—	—	.35 *	(.40)
20:5	(.20)	(.37)	(.68) **	-.36 *	.28 *	(.34)	(.61) **	—
Ca++	(.59) *	(.54) *	(.34)	.72 ***	.27 **	.47 **	.41 **	.91 ****
Alco	(-.30)	-.22 *	(-.60) **	—	-.50 ***	-.58 ***	(-.48) *	-.33 *

$*p<.05$, $**p<.01$, $***p<.001$, $****p<.0001$. Partial regression coefficients
In parentheses, univariate correlation coefficients.
Dietary 20:5 = in fact, 20:5 + 22:5 + 22:6(n-3). Alco = Alcohol.

The relationship between nutrients and platelet fatty acids is shown in Table 5. In multivariate analysis 20:3(n-9) is significantly related only to 18:0 although in univariate analysis it was also inversely related to 18:2. 22:3(n-9) is similarly associated with diet nutrients but more significantly than 20:3(n-9), these 2 fatty acids being strongly associated (r=.77) in the platelet phospholipids.

It is surprising to note that 20:5(n-3) in the platelet phospholipids of the present study was more strongly inversely associated with the intake of alcohol than it was positively related to the intake of 20:5(n-3) in multivariate analysis. The other determinants of platelets 20:5(n-3) were 18:2 (negatively) and calcium.

In multivariate analysis, 22:6(n-3) was significantly related to 18:3(n-3), but to 20:5(n-3) only in univariate association. However, the level of 22:6(n-3) was strongly determined by other nutrients such as 18:2, 18:1 and calcium. Concerning 22:5(n-3), the only significant determinants were calcium and alcohol although several dietary fatty acids were related to its level.

Platelet cholesterol was mostly inversely related to 18:0 and positively to calcium in multivariate analysis. It is known that calcium impedes the absorption of 18:0 (19). Therefore the main determinant of platelet cholesterol can be the level of 18:0 really absorbed. As to the mechanism through which 18:0 could regulate the level of platelet cholesterol, it appears to be totally unknown.

Finally of additional interest is that the intake of alcohol was inversely related either in univariate or multivariate analysis, to most of the platelet polyunsaturated fatty acids examined and platelet cholesterol.

CONCLUSIONS

The recent findings on the role of thrombosis and platelet activity in myocardial infarction suggest that more attention should be paid to the effects of dietary factors on platelet behavior and mechanisms involved. The intake of saturated fat, the environmental factor the most closely associated with CHD, appears to be more closely related to certain platelet functions such as their clotting activity, aggregation to thrombin than to serum cholesterol. In addition, it is apparently not the same fatty acids which increase serum cholesterol (14:0, 16:0) and platelet activity (18:0, 16:0), as well as those decreasing (18:2 vs 18:3) these same blood parameters.

The dietary fatty acid 18:3(n-3) in one year intervention trial as well as in normal populations studied in France, Great-Britain and Belgium, appears to have a specific inhibitory effect on certain platelet function tests such as clotting activity and aggregation to thrombin. This is also confirmed by the inverse relationship found between these tests and the level of 18:3(n-3) in plasma lipids.

The intake of α-linolenic acid, in the populations studied, represent a substantial source of n-3 fatty acids since its intake is 2 to 3 fold that of the long chain n-3 fatty acids (20:5 + 22:5 + 22:6).

As to the possible mechanism involved in the protective effect of 18:3(n-3) on platelet functions it is not obvious from the present studies. Apparently 18:3(n-3) is positively associated only with 22:6 (n-3) in the platelet phospholipids, which is significantly inversely related only to epinephrine induced aggregation. A surprising finding is the close relationship between the level of all the long-chain polyunsaturated n-3 fatty acids in the platelet phospholipids and the intake of calcium and alcohol. Their association positive with 18:0 and negative with 18:2 can be more easily interpreted, 18:2(n-6) having, for desaturases, a strong competitive effect with both the n-9 and the n-3 family fatty acids. The present results suggest that for a same dietary supply of n-3 fatty acids, the level of long-chain n-3 in platelets will be more elevated in a saturated fat diet than in diet rich in 18:2(n-6). These results could enligten some of the discrepancies observed in fish eating populations, protected or not for CHD as shown in this Symposium by Dr. Kromhout, depending on the intake of other nutrients.

ACKNOWLEDGMENTS

These studies were supported in part by CETIOM Organization and La Fondation pour la Recherche Médicale.

REFERENCES

1. J.A. Cairns, M. Gent, J. Singer et al, Aspirin, sulfinpyrazone, or both in unstable angina. Results of a Canadian multicenter trial, N. Engl. J. Med. 313:1369 (1985).
2. C.T. Sherman, F. Litvack, W. Grundfest et al, Coronary angioscopy in patients with unstable angina pectoris, N. Engl. J. Med. 315:913 (1986).
3. J.S. Forrester, F. Litvak, W. Grundfest, P. Hickey, A perspective of coronary disease seen through the arteries of living man, Circulation 75:505 (1987).
4. A.D. Timmis, B. Griffin, J.C.P. Crick, E. Sowton, APSAC in acute myocardial infarction: a placebo-controlled arteriographic coronary recanalization study, J.A.C.C. 10:205 (1987).
5. J.W.G. Yarnell, P.C. Elwood, Renaud S., Platelet function and ischaemic heart disease in the Caerphilly study, in: "Emerging problems in human nutrition", J.C. Somogyi, S. Renaud, M. Astier-Dumas, ed., Biblthca. Nutr. Dieta, Basel, 40:19 (1987).
6. L. McGregor, R. Morazain, and S. Renaud, A comparison of the effects of dietary short and long chain saturated fatty acids on platelet functions, platelet phospholipids, and blood coagulation in rats, Lab. Invest., 43:438 (1980).
7. S. Renaud, E. Dumont, F. Godsey, A. Suplisson, and C. Thevenon, Platelet functions in relation to dietary fats in farmers from two regions of France, Thromb. Haemost. 40:518 (1979).
8. S. Renaud, R. Morazain, F.Godsey, E. Dumont, I.S. Symington, E.M. Gillanders, and J.R. O'Brien, Platelet functions in relation to diet and serum lipids in British farmers, Br. Heart J. 46:562 (1981).
9. S. Renaud, E. Dumont, F. Baudier, and I.S. Symington, Effect of smoking and dietary saturated fats on platelet functions in Scottish farmers, Cardiovasc. Res. 19:155 (1985).
10. S. Renaud, R. Morazain, F. Godsey, et al, Nutrients, platelet function and composition in nine groups of French and British farmers, Atherosclerosis 60:37 (1986).
11. S. Renaud, Dietary fatty acids and platelet composition and function in human studies, in: "Fat production and consumption", C. Galli and E. Fedeli, ed., Plenum Publishing Co., 83 (1987).
12. K. Liu, J. Stamler, A. Dyer, J. McKeever, P. McKeever, Statistical methods to assess and minimize the role of intra-individual variability in obscuring the relationship between dietary lipids and serum cholesterol, J. Chron. Dis. 31:399 (1978).
13. S. Renaud, L. McGregor, J.L. Martin, Influence of alcohol on platelet functions in relation to atherosclerosis, in: "Diet, Diabetes and Atherosclerosis", Raven Press, New York, N.Y., 177 (1984).

14. S. Renaud, F. Godsey, E. Dumont, C. Thevenon, E. Ortchanian, and J.L. Martin, Influence of long-term diet modification on platelet function and composition in Moselle farmers, <u>Amer. J. Clin. Nutr.</u> 43:136 (1986).

15. H. Sprecher, The mechanisms of fatty acid elongation and desaturation in animals, <u>in</u>: 'High and Low Erucic Acid Rapeseed Oils, Academic Press, Canada, 385 (1983).

16. A. Bonamone, S.M. Grundy. Effect of dietary stearic acid on plasma cholesterol and lipoprotein level, <u>N. Engl. J. Med.</u> 318: 1244 (1988).

17. P. Gautheron, and S. Renaud, Hyperlipemia induced hypercoagulable state in rat. Role of an increased activity of platelet phosphatidyl-serine in response to certain dietary fatty acids, <u>Thromb. Res.</u> 1:353 (1972).

18. M. Lagarde, M. Burtin, M. Rigaud, H. Sprecher, M. Dechavanne, and S. Renaud, Prostaglandin E_2-like activity of 20:3 n-9 platelet lipoxygenase end-product, <u>FEBS Letters</u> 181:53.(1985).

19. S. Renaud, M. Ciavatti, C. Thevenon, and J. P. Ripoll, Protective effects of dietary calcium and magnesium on platelet function and atherosclerosis in rabbits fed saturated fat, <u>Atherosclerosis</u> 47: 187 (1983).

FISH (OIL) CONSUMPTION AND CORONARY HEART DISEASE

Daan Kromhout

Department of Epidemiology, National Institute of Public Health and Environmental Protection, P.O. Box 1, 3720, Bilthoven, The Netherlands

INTRODUCTION

The classic epidemiologic studies by Bang and Dyerberg carried out among Greenland Eskimos suggested that a high intake of N-3 polyunsaturated fatty acids through a high intake of marine foods is associated with a low mortality from coronary heart disease (1,2). Also the low mortality from coronary heart disease and the high fish intake in Japan is taken as evidence for the inverse relation between N-3 polyunsaturated fatty acids and coronary heart disease (3). It can, however, not be ruled out that confounding factors may explain the relation between fish intake and coronary heart disease. Therefore epidemiologic studies, both cross-cultural and within populations, are needed in which confounding factors are taken into account. In this paper the relation between fish consumption and coronary heart disease in between and within population studies will be reviewed. Finally recommendations for future research will be given.

CROSS-CULTURAL STUDIES IN GREENLAND AND JAPAN

The studies carried out by Bang and Dyerberg in the 1970's showed that the consumption of marine foods was extremely high among Greenland Eskimos and amounted to about 400 g/day (4). They calculated that a diet of 3,000 Kcal per day contained on average about 14 g of N-3 polyunsaturated fatty acids per day (5). The intake of N-6 polyunsaturated fatty acids was relatively low and amounted to 5 g per day. The total polyunsaturated fatty acid intake amounted to 7 per cent of total energy. The diet of the Eskimos can be characterized as a high fat diet (39 per cent of energy from fat) with a low amount of saturated fat (9 per cent of total energy) and a high P/S ratio (Polyunsaturated/Saturated fat ratio = 0.84). The intake of dietary cholesterol was also high and amounted to 260 mg/1,000 Kcal per day.

Kromann and Green showed that the age-adjusted incidence for coronary heart disease among Eskimos was about 10 per cent of that among Danes (6). The diet of the Danes is characterized by a high amount of saturated fat (22 per cent of energy) and a low P/S ratio of 0.24 (5). These results are interpreted as evidence for the inverse relation between N 3 polyunsaturated fatty acids and coronary heart disease (5).

The per capita fish consumption in Japan is high, about 100 g per day. Mortality from coronary heart disease is low in Japan compared to Western countries (7). Comparison of people in a fishing village with an average fish consumption of 250 g per day with people in a farming village with an average fish consumption of 100 g per day showed that mortality from coronary heart disease was lower in the fishing area than in the farming area (Hirai A. et al., unpublished results). These results also suggest that an inverse relation may exist between fish consumption and mortality from coronary heart disease.

There are several problems with respect to the comparative studies carried out between Greenland Eskimos and Danes and between persons in different areas within Japan. These comparisons do not answer the question whether on the population level a dose response relation exists between fish consumption and coronary heart disease. It may also be possible that the inverse relation between fish consumption and coronary heart disease could be explained by confounding factors not taken into account in the reported comparison between populations. Therefore results of cross-cultural studies including as many populations as possible, are needed. Also analytic epidemiologic studies are needed to investigate whether the inverse relation between fish consumption and coronary heart disease is also present at the individual level.

The relation between fish consumption and coronary heart disease was investigated using data from 21 countries (7). In this study per capita fish consumption data and mortality data from coronary heart disease were used. A moderate negative association was found which appeared stable over different periods. This correlation was very much dependent on the inclusion of the Japanese data. Exclusion of Japan reduced the correlation between fish consumption and coronary heart disease mortality. Multiple regression analyses including data of all 21 countries showed no relation between fish consumption and coronary heart disease mortality after inclusion of milk products and meat in the multivariate model.

A major disadvantage of this study is that per capita fish consumption data are used. These are not real intake data because no allowance was made for offal and waste. The per capita fish consumption can also not be broken down by age and sex. Average per capita fish consumption data were related to male age-adjusted coronary heart disease mortality data. The drawback of this analysis is that per capita fish consumption data not adjusted for age and sex are related to age-adjusted mortality data for males only. Therefore results are needed of a cross-cultural study using high quality dietary and mortality data.

THE SEVEN COUNTRIES STUDY

At the end of the 1950's the Seven Countries Study was designed to investigate relations betwween diet and cardiovascular diseases (8-10).
Sixteen cohorts were selected in seven countries: Finland, Greece, Italy, Japan, The Netherlands, United States and Yugoslavia. Information about the food consumption pattern was collected by the dietary record method in subsamples of the study populations. Mortality data of the studied populations are complete now for the subsequent 15 years (11). This provides the possibility to relate the fish consumption data collected at baseline to 15 years coronary heart disease mortality data.

Large differences were observed in fish consumption patterns between the 16 cohorts (12). Less than 10 grams of fish per day was consumed in Velika Krsna (Yugoslavia), U.S. Railroad, Western Finland and Zrenjanin (Yugoslavia). Between 10 and 35 grams of fish per day was consumed in Zutphen (The Netherlands), Crete (Greece), Belgrade and Slavonia (Yugoslavia) and Crevalcore, Montegiorgio and Rome (Italy). About 60 grams per day was consumed in Corfu (Greece) and Eastern Finland. About 95 grams per day in Dalmatia (Yugoslavia) and Tanushimaru (Japan). The highest fish intake of about 200 grams per day was observed in the fishermen's village of Ushibuka in Japan.

A weak non-significant correlation coefficient of -0.28 was found between fish consumption at baseline and 15-year mortality from coronary heart disease (Figure 1). Several remarkable features can be deduced from this figure. The six cohorts with a very low mortality from coronary heart disease had very different fish consumption patterns. No fish was consumed in Velika Krsna (Yugoslavia). The average fish intake in Crete (Greece) amounted to about 20 grams per day. In Corfu (Greece), Tanushimaru (Japan) and Dalmatia (Yugoslavia) average fish intake varied between 60 and 100 grams per day. In Ushibuka (Japan) an average fish intake of 200 grams per day was observed. These results suggest that low levels of coronary heart disease are not primarily dependent on the level of fish consumption.

In Eastern Finland the average fish intake was about 60 grams per day. In spite of this relatively high fish intake 15-year mortality from coronary heart disease was highest among this cohort of the Seven Countries Study. Earlier reported results of this study have shown that on the population level saturated fat is strongly related to 15-year mortality from coronary heart disease (11). These results suggest that a relatively high level of fish consumption does not protect from coronary heart disease in cultures with a very high saturated fat intake. This conclusion is supported by the results of dietary intervention studies showing a serum total cholesterol

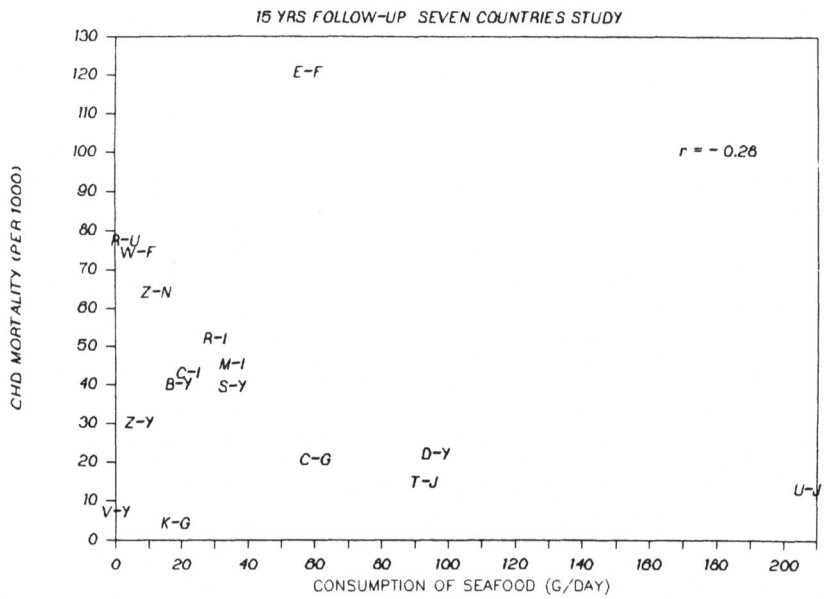

Figure 1. Consumption of seafood and CHD.
E-F East Finland; W-F West Finland; C-G Corfu; K-G Crete;
C-I Crevalcore, Italy; M-I Montegiorgio, Italy; R-I Rome, Italy;
T-J Tanushimaru, Japan; U-J Ushibuka, Japan; Z-N Zutphen,
Netherlands; R-U Railroad, USA; B-Y Belgrade, Yugoslavia; D-Y
Dalmatia, Yugoslavia; S-Y Slavonia, Yugoslavia; V-Y Velika Krsna,
Yugoslavia

elevating effect of saturated fat (13). Serum cholesterol is one of the major risk factors for coronary heart disease (14). Animal experiments have shown a direct relation between dietary saturated fat and experimental thrombosis (15).

COHORT STUDIES ON THE RELATION BETWEEN FISH CONSUMPTION AND CORONARY HEART DISEASE

Analytic epidemiologic studies are needed to answer the question whether fish consumption is related to coronary heart disease mortality on the individual level. We have examined this question in the Zutphen Study, the Dutch contribution to the Seven Countries Study (16). In 1960 information about the food consumption pattern of all men, who participated in the Zutphen Study was collected by the cross-check dietary history method. This method provides information about the usual food consumption pattern. The vital status of all men was verified 20 year after the baseline survey. Of the 852 men free of coronary heart disease at entry, 78 died from this disease during 20 years of follow-up. The average fish consumption of the Zutphen men amounted to 20 grams per day in 1960. About 20 per cent of the men did not eat fish. About two thirds of the fish consisted of lean fish and one third of fatty fish.

Fish consumption in 1960 was inversely related to mortality from coronary heart disease in the period 1960-1980. Fish consumers can differ in many respects from non-users. The intake of monounsaturated fat, polyunsaturated fat, dietary cholesterol, animal protein and alcohol was significantly positively and the intake of polysaccharides was significantly inversely related to fish consumption. No association was found between fish consumption and age, serum total cholesterol, systolic blood pressure, cigarette smoking, subscapular skinfold, physical activity, occupation, energy intake and prescribed diet. The inverse relation between fish consumption and 20-year mortality from coronary heart disease persisted after multivariate logistic regression analyses. The risk ratios decreased with increasing fish consumption. Mortality from coronary heart disease was about two times lower among men who eat at least 30 grams of fish per day compared with non-eaters. A dose-response relation was observed between 0 and 30 grams of fish per day.

In the Western Electric Study, a cohort study carried out among 2,000 men in Chicago aged 40-55 in 1957, also an inverse relation was found between fish consumption in 1957 and mortality from coronary heart disease during 25 years of follow-up (17). The Western Electric Study is in many respects comparable to the Zutphen Study. A similar dietary survey method was used and the average fish consumption level of the men in Chicago was comparable to that of the men in Zutphen. In a 14-year follow-up study on 10,966 subjects carried out in Sweden an inverse relation was found between fish consumption and mortality from myocardial infarction (18). A risk ratio of 0.70 was found for subjects with a high level of fish consumption compared with those of a low level of fish consumption.

In two cohort studies carried out in Hawaii and Norway, no relation was found between fish consumption and coronary heart disease (19,20). In these studies information about fish consumption was collected by food frequency methods. Among these populations, fish consumption was higher compared with that among men in Zutphen and Chicago. Only a very few men did not eat fish in Hawaii and Norway. In a recently reported study from Norway two communities with a different level of fish intake were compared (21). The average fish intake amounted to 55 and 134 grams of fish per day respectively. The authors claimed that coronary heart disease mortality was highest in the community with the lowest fish intake. However, after adjustment for differences in the age distribution between the two communities no difference in coronary heart disease mortality was found.

The results of the cohort and cross-cultural epidemiologic studies published so far may be summarized as follows. A low level of fish consumption e.g. 30 grams of fish per day decreases mortality from coronary heart disease by about 50% compared with eating no fish at all. Mortality from coronary heart disease does not seem to vary with a fish intake in the range between 30 and 150 grams per day. At very high levels of fish intake a further reduction in coronary heart disease mortality seems to take place. The relation between fish consumption and the risk ratio of coronary heart disease is summarized in Figure 2.

MECHANISMS

Greenland Eskimos have significantly lower levels of serum total cholesterol and triglycerides and significantly higher High Density Lipoprotein (HDL) levels than Danish controls (22). Also bleeding time, platelet aggregation and the fatty acid composition of platelets differed between Eskimos and Danes (23). These results suggest that the low mortality rate from coronary heart disease among Eskimos may be due to a more favourable lipid and haemostatic profile compared with Danes. It is, however, not known whether these mechanisms can also be responsible for the inverse relation between a low level of fish consumption and coronary heart disease mortality. Therefore the relations between a low level of habitual fish consumption, serum lipids, serum fatty acids and platelet function were investigated within the Zutphen Study (24,25).

For the present study 40 healthy elderly men, whose fish consumption pattern over the previous 26-year period was known, were selected. The 15 men in the control group consumed on average about 2 grams of fish per day, and the 25 habitual fish consumers used, on average, about 33 grams of fish per day during that period. The cholesterol content of the different lipoprotein fractions did not differ between the two groups. The total triglyceride levels and the triglyceride levels of the atherogenic Intermediate Density Lipoprotein (IDL) fraction were significantly lower among habitual fish consumers compared with controls. No difference between the two groups was observed

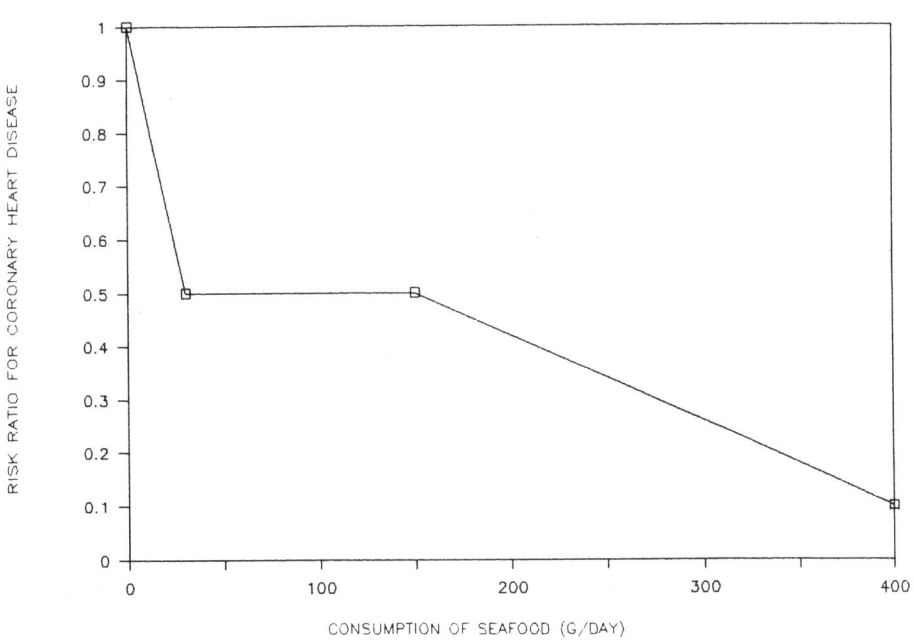

Figure 2. Consumption of seafood and CHD.

in bleeding time and different platelet function tests. The N-3 poly-unsaturated fatty acid content of the phospholipids was significantly higher among the habitual fish consumers compared with the controls. These results suggest that a habitual fish consumption of about 30 grams per day influences serum triglyceride metabolism and the fatty acid content of the phospholipids significantly.

Currently an intervention study is carried out to test the hypothesis that a small amount of N-3 polyunsaturated fatty acids influences serum triglycerides and the fatty acid composition of the phospholipids significantly. For this trial about 100 healthy men aged 20-60 who consumed usually no fish or a very small amount of fish, were selected. These men are asked not to eat fish at all during the intervention period of five weeks and are randomized in two groups. One group adds every day 3 capsules of fish oil containing 1 gram of N-3 polyunsaturated fatty acids to their diet. The control group adds 3 capsules with corn oil and olive oil containing 1 gram of N-6 polyunsaturated fatty acids to the diet. The effect on total triglycerides, total and HDL cholesterol and the fatty acid content of phospholipids will be investigated.

CONCLUSIONS

The results of epidemiologic studies carried out so far suggest that a small amount of fish e.g. less than 30 grams per day is inversely related to coronary heart disease mortality. Very high amounts of fish consumption e.g. about 200-250 grams per day among Japanese fishermen and about 400 grams of marine foods among Greenland Eskimos seem to be associated with very low rates of coronary heart disease mortality. These low mortality rates for coronary heart disease may also be due to a low intake of saturated fat because the diets of the Japanese and the Greenland Eskimos are characterized by both a low saturated fat and a high N-3 polyunsaturated fat content.
Future research activities should therefore focus on the interrelation-ships between saturated and· N-3 polyunsaturated fatty acids on blood lipids, lipoproteins, platelet function, prostaglandin metabolism and their effects on coronary heart disease incidence and mortality. Special attention should be paid to the amount of fish and fish oils needed for prevention and treatment of coronary heart disease.

REFERENCES

1. Bang H.O., J.Dyerberg, N.Hjorne. The composition of food consumed by Greenland Eskimos. Acta Med.Scand. (1976) 200, 69
2. Bang H.O., J.Dyerberg, H.M.Sinclair. The composition of the Eskimo food in north western Greenland. Am.J.Clin.Nutr. (1980) 33, 2657
3. Hirai A., T.Hamazaki, T.Terano, T.Nishikawa, Y. Tamura, A.Kuma-gai. Eicosapentaenoic acid and platelet function in Japanese. Lancet (1980) 2, 1132
4. Bang H.O., J.Dyerberg. Plasma lipids and lipoprotein in Green-landic west coast Eskimos. Acta Med.Scand. (1972) 192, 85

5. Dyerberg J..Linolenate-derived polyunsaturated fatty acids and prevention of atherosclerosis. Nutr.Rev. (1986) 44, 125

6. Kromann N., A.Green. Epidemiological studies in the Upernavik district, Greenland. Acta Med.Scand. (1980) 208, 401

7. Crombie I.K., P.Mcloone, W.C.S.Smith, M.Thomson, H.Tunstall Pedoe. International differences in coronary heart disease mortality and consumption of fish and other foodstuffs. Eur.Heart J. (1987) 8, 560

8. Keys A., C.Aravanis, H.W.Blackburn, F.S.P.Van Buchem, R.Buzina, B.S. Djordevic, A.S.Dontas, F.Fidanza, M.J.Karvonen, N.Kimura, D.Lekos, M.Monti, V.Puddu, H.L.Taylor. Epidemiological studies related to coronary heart disease: characteristics of men aged 40-59 in seven countries. Acta Med.Scand. (1967) 460, Suppl.1-392

9. Keys A.. Coronary heart disease in seven countries. Circulation (1970) 41, Suppl.I-1-211

10. Keys A. . Seven countries: A multivariate analysis of death and coronary heart disease. Harvard University Press, Cambridge, Massachusetts (1980)

11. Keys A., A.Menotti, M.J.Karvonen, C.Aravanis, H.Blackburn, R.Buzina, B.S.Djordevic, A.S.Dontas, F.Fidanza, M.H.Keys, D.Kromhout, S. Nedeljkovic, S.Punsar, F.Seccareccia, H.Toshima. The diet and 15-year death rate in the Seven Countries Study. Am.J.Epidemiol. (1986) 124, 903

12. Kromhout D., A.Keys, C.Aravanis, R.Buzina, F.Fidanza, A.Jansen, A.Menotti, S.Nedeljkovic, M.Pekkarinen, B.S. Simic, H.Toshima. Food consumption patterns in the nineteen sixties in seven countries. Am.J.Clin.Nutr. Accepted for publication

13. Keys A., J.T.Anderson, F.Grande. Serum cholesterol response to changes in the diet. IV.Particular saturated fatty acids in the diet. Metabolism (1965) 14, 776

14. Stamler J., D.Wentworth, J.D.Neaton. Is the relationship between serum cholesterol and risk of premature death from coronary heart disease continous and graded ?. J.Am.Med.Ass. (1986) 256, 2832

15. Hornstra G.. Dietary lipids, platelet function and arterial thrombosis in animals and men. Proc.Nutr.Soc. (1985) 44, 371

16. Kromhout D., E.B.Bosschieter, C.De Lezenne Coulander. The inverse relation between fish consumption and 20-year mortality from coronary heart disease. N.Eng.J.Med. (1985) 312, 1205

17. Shekelle R., L.Missell, O.Paul, A.MacMillan-Schryock, J.Stamler. Fish consumption and mortality from coronary heart disease (Letter) N.Eng.J.Med. (1985) 313, 820

18. Norell S.E., A.Ahlbom, M.Feychting, N.L.Pedersen. Fish consumption and mortality from heart disease. Brit.Med.J. (1986) 293, 426

19. Curb J.D., D.M.Reed. Fish consumption and mortality from coronary heart disease (Letter) N.Eng.J.Med. (1985) 313, 821

20. Vollset S.E., I.Heuch, E.Bjelke. Fish consumption and mortality from coronary heart disease. N.Eng.J.Med. (1985) 313, 820

21. Simonsen T., A.Vartun, V.Lyngmo, A.Nordoy. Coronary heart disease, serum lipids, platelets and dietary fish in two communities in Northern Norway. Acta Med.Scand. (1987) 222, 237

22. Bang H.O., J.Dyerberg, A.Brondum-Nielsen. Plasma lipid and lipo-protein pattern in Greenlandic west-coast Eskimos. Lancet (1971) 1, 1143

23. Dyerberg J., H.O.Bang. Haemostatic function and platelet poly-unsaturated fatty acids in Eskimos. Lancet (1979) 2, 433

24. Kromhout D., G.L.Obermann-de Boer, M.B.Katan, L.Havekes, A.Groener, G.Hornstra, E.B.Bosschieter, C.De Lezenne Coulander. The effects of 26 years of habitual fish consumption on serum lipid and lipo-protein levels (The Zutphen Study). Submitted.

25. Van Houwelingen R., G.Hornstra, D.Kromhout, C.De Lezenne Coulander. Habitual fish consumption, fatty acids of serum phospholipids and platelet function. Submitted.

DIETARY POLYUNSATURATES, VASCULAR FUNCTION AND PROSTAGLANDINS

Howard Knapp, Deborah Gregory and Sharina Nolan

Division of Clinical Pharmacology
Vanderbilt University Nashville, TN 37232

The high prevalence of certain cancers, atherosclerosis and hypertension in Europe and North America has been closely linked to the high-fat diet consumed by these populations. The strongest dietary correlate with vascular disease and hypercholesterolemia is the intake of saturate fat (1), and current recommendations include an increased proportion of polyunsaturated fatty acids in a diet with an overall reduction in total fat content of one-fourth or more. In most industrialized countries, the readily available polyunsaturates are of the omega-6 class found in vegetable oils. Increased consumption of such fats in the place of saturated fats in dairy products has been accompanied by a continuous decline in cardiovascular disease rates in the United States (2), although decreased cigarette consumption and better treatment of hypertension have probably also contributed to this.

During the last decade, there has been a great deal of interest in the potential health benefits of omega-3 fatty acids in marine oils due to the initial observations that Greenland Eskimos appeared to have a low prevalence of vascular disease despite a dietary fat content as high as that of Europeans (3). We are now becoming aware that omega-3 and omega-6 polyunsaturates compete with each other at many metabolic steps in the human body, in addition to those involving oxygenation products, and much work will be needed to define the optimal balance between these fatty acid families in our diets (4). For current clinical applications, the majority of studies are directed more at the pharmacological aspects of polyunsaturate supplementation rather than studying this in the context of our overall nutrition, but the broadest health benefits will result from an eventual understanding of findings obtained from both approaches.

Among the most frequently discussed effects of dietary enrichment with polyunsaturated fats are the reduction of blood pressure and vascular reactivity (4,5). Studies demonstrating such effects have been conducted in both animals and in man and although the emphasis has been on studying omega-3 fatty acids most recently, similar effects on these parameters have been reported in the past for omega-6 supplementation as well (Table 1). Since prostaglandins are made from polyunsaturated fatty acids and have effects on many processes related to blood pressure control (6), the hypotensive effects of dietary polyunsaturates are usually ascribed to alterations in the in vivo synthesis of these autocoids. However, in vitro studies have suggested that omega-6 supplements increase the release of arachidonate-derived mediators (7) while omega-3 supplements decrease their release (8) and such findings appear to be inconsistent with the hypothesis that prostaglandins are involved in these physiological effects. Proposed explanations for this disparity include differences between species or between in vivo effects of polyunsaturates. Nevertheless, few studies published so far have examined endogenous prostaglandin synthesis during the lowering of blood pressure or vascular reactivity with polyunsaturate supplements. Prior to discussing our current findings in this area, we will discuss the epidemiologic and experimental background to these studies, focusing upon human studies and omega-3 fatty acid supplementation.

Table 1   Reported Effects of PUFA on Vascular Parameters

	W -3	W -6
Decrease vascular reactivity in humans	+ or -	+
in animals	+	+
Effect blood pressure in humans	↓	↓
in SHR	↓	↓
in other models	↑ or ↓	↓
Prolong bleeding time	++	+ ?

Studies in Populations with a High Intake of Fish

Although epidemiologic investigations can provide information about correlations of parameters, mechanistic interpretations of such data must be made with caution. Eskimos, for example, have consumed large amounts of omega-3 fatty acids for hundreds of generations, so it would not be surprising if they were found to have genetic differences in their metabolism of polyunsaturates from Western populations. Also, blood pressure is influenced by a number of dietary components, particularly sodium, which can confound diet-blood pressure comparisons between populations. Despite these problems, there are three groups which have a high dietary omega-3 intake whose blood pressures have been compared to control groups of some sort, and some of this data is summarized in Table 2.

Table 2 Fish Consumption and Blood Pressure

	Omega-3			
	Intake	Plasma	Blood Pressure	Ref
Eskimos (vs Danes)	↑ 20 x	↑	Same	10-12
Fisherman vs Farmers				
Japan	Same	↑ or Same	↓ or Same	14,15
Norwegian	↑ 4 x	Same	Same	18

The suggestion that hypertension is rare among Eskimos appears to be based upon anecdotal evidence (9) but is frequently mentioned in reviews (4,5). A study of over a thousand Greenland Eskimos in the late 1940's found that their systolic pressures were actually higher than that of a population in a similar climate in Finland (10). The less frequent occurrence of diastolic hypertension and the absence of malignant hypertension among the Eskimos was notable. The author attributed the higher systolic pressures to the premature aging of the Eskimos, with a concurrent loss of arterial compliance. She found a lower incidence of atherosclerosis among the Eskimos (based upon physical examination and x-rays) but felt that most of the difference was due to the fact that only 10% of the Eskimos reached the age of 50, compared with over one-third of the group in Finland. Earlier authors make mention of the short lifespan and poorly balanced diet of most Greenlanders, due to the climate, primitive lifestyle and shortage of fresh fruits and vegetables. A second study on Eskimo blood pressure was carried out 30 years later in nearly 500 occupants of relatively isolated coastal settlements in Greenland (11). These workers found the same distribution of blood pressure at any age as had been found in Copenhagen. Finally, a survey of over 800 Alaskan Eskimo men consuming a high-fish diet found no difference in their blood pressures from those of Caucasian Americans (12).

The Japanese have a higher consumption of fish than is ever likely to be achieved on a large scale in the West but have one of the highest rates of stroke and prevalence of hypertension in the world (13). A major factor determining rates of hypertension in Japan is sodium intake, and studies showing lower blood pressures in fishermen vs farmers indicate a lower sodium intake in the former group (14). The findings that fishermen do not always have more omega-3 fatty acids in their cell lipids (15) and do not make as great an amount of trienoic PGI$_3$ as farmers do (16) is additional evidence against a causative role for omega-3 fatty acid consumption in the lower blood pressures of Japanese fishermen. In other studies of Japanese islanders with a high consumption of fish, dietary omega-3 fatty acids do not seem to counteract the effects of other dietary components on blood pressure; Okinawans moving to the US have higher serum cholesterol but lower blood pressure than those in Okinawa (17), as well as a lowered intake of omega-3 fatty acids. Likewise, Norwegian fishermen who have an intake of fish four times that of their inland countrymen do not have lower blood pressures but do have higher cardiovascular death rates (18).

Patients with essential hypertension have exaggerated reactivity to pressor agents, and enhanced vascular prostacyclin synthesis is often suggested as the mechanism of blood pressure lowering by dietary polyunsaturates. As noted in Table 1, both omega-3 and omega-6 fatty acids have been reported to lower vascular reactivity, so it is not possible at this time to consider this to be a particular property of omega-3 fatty acids. Animal studies have produced quite variable results, with some finding decreased (19) and others increased (20) reactivity to vasopressors after feeding omega-3 supplements. Several reports of diminished pressor response after omega-6 supplements have been published (21,22) and the effects of both omega-3 (19) and omega-6 (22) fatty acids to reduce reactivity of vascular strips ex vivo have been shown to be blocked by indomethacin, implying a role for vasodilator prostaglandins in the altered responsiveness.

Two investigations of vascular responsiveness during omega-3 fatty acid ingestion by volunteers have been published. The first of these found a decrease in upright systolic, but not diastolic or supine pressures, at baseline and a decrease in systolic pressure response during an infusion of norepinephrine into normotensive male subjects while taking fish oil (23). No significant change in the response to angiotensin was found, in contrast to the decrease in this parameter found in a study of pregnant women taking omega-6 fatty acid supplements (24). The other study of omega-3 effects on vascular reactivity was performed in males with mild essential hypertension whose blood pressure appeared to be reduced while consuming a mackerel diet (25). Their pressor response to a standard psychophysiological stress test was the same before and during the study period, indicating unchanged responsiveness to endogenous catecholamines.

In addressing the hypothesis that omega-3 fatty acid supplements alter vascular reactivity via changes in vascular prostacyclin release, we have performed infusion studies in volunteers with essential hypertension taking 50 ml fish oil per day (15 g omega-3 fatty acids) with phenylephrine, an alpha-adrenergic agonist with practically no direct cardiac effects in man. Graded infusion were started at low infusion rates and increased in a stepwise fashion every 10 minutes until there was either an increase in diastolic pressure of 20 mm Hg or a bradycardia below 46 beats per minute. Urine was collected just prior to and 15 minutes after the completion of the infusion for the measurement of 2,3 dinor-6-keto-$PGF_{1\alpha}$ and it's delta-17 analog, major urinary metabolites of PGI_2 and PGI_3, respectively. The details of the methods have been published but, in brief, quantitation was accomplished by capillary gas-chromatography/ negative chemical ionization-mass spectrometry in the selected-ion-monitoring mode, utilizing isotope dilution of a tetradeuterated internal standard.

The blood pressure results of the paired infusion studies are presented in Figure 1. At the low and medium infusion rates, there was no difference in the pressor response during infusions done on or off the fish oil. The blood pressure response after 10 minutes of the 0.5 μg/kg/min infusion was about the same as seen after a 15 minute infusion of 5.0 μg/min norepinephrine by Lorenz and colleagues, but they reported a lower response during the infusion performed while their normotensive subjects were taking fish oil (23). At the higher infusion rate of 0.75 μg/kg/min, our subjects appeared to have more of a tendency to develop bradycardia, resulting in the termination of the infusion, during the study done while they were taking fish oil. This would seem to suggest that the effect of fish oil supplements was not a direct one on arterial responsiveness to adrenergic agonists, but rather was indirect, perhaps involving an alteration of baroreceptor sensitivity. Resetting of baroreceptors does occur after several days of adjustment to new levels of blood pressure, and we will discuss below the lowering of these subjects blood pressures during fish oil ingestion, as presented previously in abstract form (26).

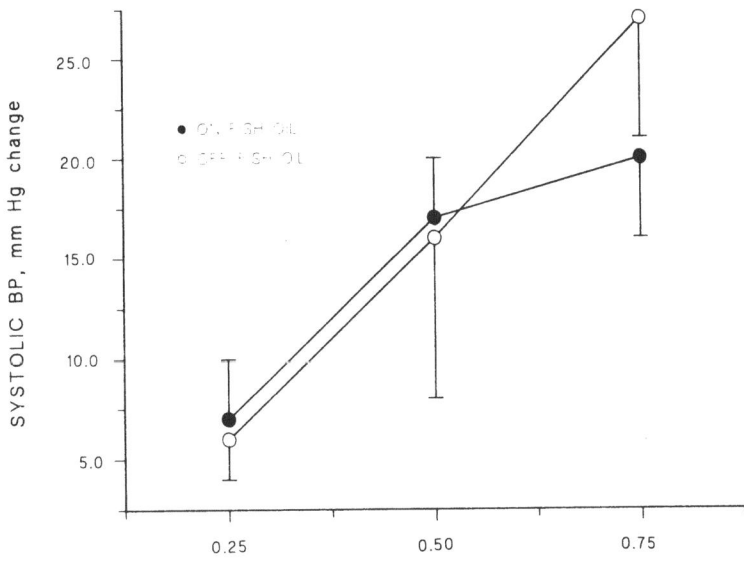

PHENYLEPHRINE, ug/kg/min.

Figure 1. Shows the effect of fish oil supplementation upon the pressor response to graded infusions of the alpha-adrenergic agonist phenylephrine. During the infusion performed while taking fish oil, the subjects developed a more pronounced bradycardia and achieved less of an increase in blood pressure at the highest infusion rate. Lower rates of infusion gave identical responses on the two occasions.

It is often suggested that polyunsaturate supplements lower vascular reactivity by increasing the ability of the blood vessels to produce vasodilator prostacyclin in response to pressor agents. The results of our urinary prostacyclin metabolite measurements during phenylephrine infusion are shown in Figure 2. The values represent total prostacyclin released (PGI_2-M+PGI_3-M) since these two have been reported as having approximately equal vasodilator potency. Baseline values on the two infusion days were the same. During the infusion conducted while on fish oil, no significant change in excretion rate took place, while a small but statistically significant increase was seen during the infusion taking place while off the oil. It seems likely that this difference was due to the higher pressures achieved during the control infusion, and indicated that the reduced response while on fish oil cannot be accounted for by increased prostacyclin formation.

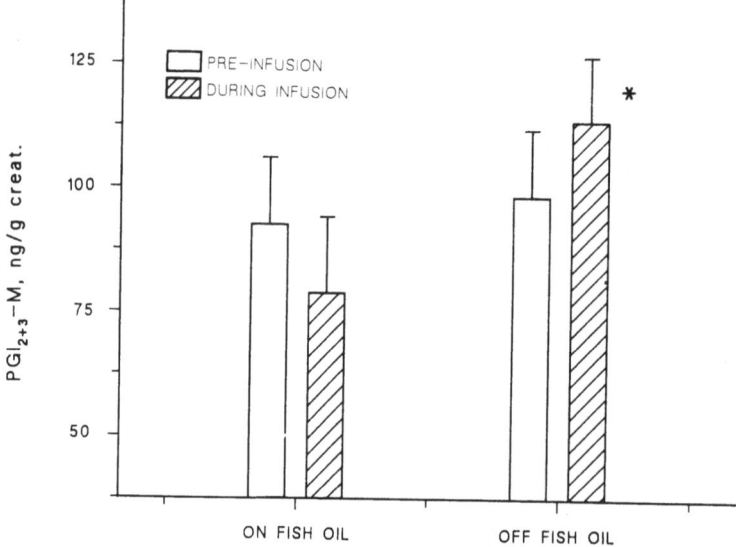

Figure 2. Shows the excretion of 2,3 dinor-6-keto-PGF$_{1\alpha}$ before and during phenylephrine infusion studies performed off and on fish oil, 50 ml per day. Paired analysis of pre and post-infusion values revealed a significant, small increase during the infusion off fish oil, when higher blood pressures were achieved (* p,0.05 by the Mann-Whitney test).

Dietary Polyunsaturate Enrichment and Blood Pressure

Enrichment of the diet with polyunsaturated fatty acids of either the omega-3 or omega-6 classes have been reported to lower blood pressure in both normotensive and hypertensive subjects. Unfortunately, most of these studies contain design flaws which confound their interpretation and prior to a discussion of the actual results it would be useful to mention the problems involved in studying outpatient blood pressure. Some appreciation of these issues will explain the high degree of controversy surrounding the entire area of dietary fat and blood pressure.

Blood pressure has a high degree of intrinsic variability in both normal and hypertensive individuals, and studies on changes in blood pressure are confronted with a number of problems. Observer bias and measurement errors can create or obscure small changes in the mean pressures of small study groups. Since the great variability of blood pressure is well known, for example, reports showing exactly the same mean pressures in small numbers of subjects on numerous occasions over a period of months are sometimes received with skepticism. The effect of habituation to study conditions means that adequate run-in periods must be employed to obtain accurate baseline pressures. Indeed, it has been shown that three weeks of placebo treatment resulted in 10 mm HG drops in blood pressure, as well as a statistically significant decline in catecholamine excretion(27). This drop in blood pressure is larger than that usually reported in studies on the effects of dietary fat. Besides habituation to the study and placebo effects per se, regression upon the mean with repeated measurements results in significant reductions in the mean blood pressures of any study group during a period which can last up to six weeks.

In addition to lacking adequate run-in periods, few studies on dietary fat and blood pressure include recovery periods to see whether the subjects' pressure returns to the presumed baseline after a return to baseline conditions. Another source of variability in blood pressure data is the anxiety of some individuals to blood pressure measurement. It has been found that 30-50% of patients considered to have mild hypertension on the basis of clinic measurements are normotensive when evaluated with ambulatory blood pressure monitors worn by the patients during their routine activities away from the doctor's office (28). Random-zero sphygmomanometers eliminate digit preference and treatment group bias on the part of the investigator, but even these are seldom used in studies on the effect of diet upon blood pressure.

A number of studies on the effects of omega-6 fatty acid, usually linoleate, on blood pressure have been published (29-37) but all failed to take account of the many problems listed above. A series of studies in several population groups seems to be convincing when taken in aggregate but usually more than one dietary component was altered in the studies, making the lowering of blood pressure difficult to attribute to the effected omega-6 supplementation alone (32). Several recent works which did address the above issues, as well as try to have more stringent dietary control, have failed to confirm a hypotensive effect of omega-6 fatty acids in man (33,34).

There have been at least twelve studies published (23,25, 35-44) on the effects of omega-3 fatty acids on blood pressure in volunteers, and some of the characteristics of these are listed in Table 3. It can be seen that only one study used a random-zero sphygmomanometer, and this negative report was from one of only two groups incorporating recovery periods into their study designs. An adequate run-in period seems to have been used in only two studies (36,41), and parallel controls were used in four studies, two of which were negative (39,41). Statistical inadequacies are also frequently encountered, with multiple paired t-tests being performed in several cases without correction for multiple comparison or small sample size, failure to exclude a single extreme outlier upon whom statistical significance depends (36), and the odd use of non-parametric tests on changes in the median (38) to obtain a "p" value.

The results of the twelve studies done thus far on the hypotensive effects of fish oil are summarized in Table 4. There does not appear to be a consistent pattern of change in blood pressure, and the study with the greatest drop (35) had no return to baseline pressures upon withdrawal of the oil. This was interpreted as a negative study by the author. Although the protocols are quite variable, no obvious dose-response relationship is present. Sometimes declines are seen in systolic pressure, sometimes diastolic. One group reported divergent results using the same protocol on two occasions (25,43). Overall, the available literature presents a confusing picture, and it does not appear that a firm conclusion can be reached at this point.

Dietary Polyunsaturates and In Vivo Eicosanoids

Prostaglandins are involved in many aspects of blood pressure regulation (6) and are derived from polyunsaturated fatty acids, so alterations in the production of prostaglandins in vivo is a hypothesis often advanced to explain hypotensive effects of polyunsaturates.

Table 3. Design Characteristics of Omega-3 BP Studies

Ref.	Subjects	Run-in	Recovery	Random-Zero SPHYG	Controls
35	12	1 week	Yes	Yes	None
16	16	8 weeks	No	No	Placebo Crossover
37	28	No	No	No	None
38	20	No	No	No	Placebo Crossover
23	8	No	No	No	None
39	13	No	No	No	Parallel
40	12	No	No	No	No
41	84	2 weeks	No	No	Parallel
42	60	No	No	No	Parallel
43	14	No	Yes	No	Parallel
25	14	No	Yes	No	Crossover
44	12	No	Yes	No	Parallel

There have been few studies, however, in which the endogenous production of prostaglandins was studied during the lowering of blood pressure with dietary polyunsaturate enrichment. Although indirect assessments of this (e.g. serum thromboxane) can provide useful information during pharmacological studies, the most direct and meaningful means of assessing in vivo prostaglandin synthesis is by measurement of urinary metabolites. Since there appears to be some differences in the effects of dietary polyunsaturate supplements upon the different metabolites, however, no clear picture has emerged. Volunteers receiving ethyl arachidonate excreted more PGE metabolite (45) but other workers did not find such changes with linoleic acid (46). Other, less specific metabolite pools do seem to increase with linoleate supplements (47), however, as does excretion of PGE metabolite by EFA-deficient neonates given linoleate (48).

Table 4   Results of Omega-3 BP Studies

Ref.	Dose (EPA/AA)	Duration	Systolic	Diastolic	Comment
35	1.8/2.2	6 wks	↓ 10%	↓ 15%	No ↑ Recovery
36	3.2/2.0	6 wks	↓ 5.9%	No ▲	Outlier
37	3.6/2.4	8 wks	↓ 15%	↓ 9%	Dialysis Pt.
38	2.0/1.3	4 wks	(median) ↓ 3.5%	No ▲	Odd Stats.
23	4.4/6.4	25 days	↓ 8% upright No ▲ supine	No ▲	Controls Obtained "Randomly"
39	1.0/0.6	4 wks	No ▲	No ▲	
40	1.6/1.0	12 wks	No ▲	↓ 9%	Dialysis Pt.
41	1.7/3.0	6 wks	No ▲	No ▲	
42	2.9/1.9	6 wks	No ▲	↓ 7%	
43	2.2/2.8	2 wks	↓ 8%	No ▲	No ↑ Recovery
25	2.2/2.8	2 wks	↓ 9%	↓ 11%	
44	2.2/2.8	Various	↓ 9%	↓ 11%	

In regards to fish oil supplements, it has been shown by two groups that ingestion of eicosapentaenoic acid does not suppress prostacyclin formation and does result in substantial formation of PGI_3 (49,50). No reports on the effects of fish oil supplements upon the systemic production of PGE_2 have yet been reported and, in particular, there have not been any comparisons of the effects of omega-3 and omega-6 fat supplements upon the in vivo synthesis of prostacyclins and prostaglandins of the E-type. Interestingly, the excretion of a pooled prostaglandin metabolite is not different between Greenland Eskimos and Danes (51) and is not decreased during cod-liver oil ingestion by volunteers, so that it does not appear that omega-3 fatty acids are acting as generalized cyclooxygenase inhibitors in man, as was suggested by a large number of animal experiments.

A Study of Polyunsaturates, Eicosanoids and Blood Pressure

In an attempt to address the questions of whether and how polyunsaturates alter blood pressure, we have designed a study utilizing ambulatory blood pressure monitors instead of cuff pressures, and some of our data has been presented in abstract form (26). These monitors eliminate observer bias and reduce the within-individual variability in blood pressure measurements (52). They also provide data which predicts end organ damage from hypertension much more accurately than cuff pressures (53). Our study has a two-week screening period during which the volunteers diastolic pressure must stay above 92 mm Hg. The subjects then enter a three month study with four week run-in, supplementation and recovery periods. We carry out 12-hour ambulatory monitoring weekly, as well as collect urines for measurement of metabolites of prostacyclin and prostaglandin E to determine whether any change in blood pressure could be related to changes in the endogenous production of vasodilator prostaglandins. As reported in our abstract, preliminary data suggests that omega-3 fatty acids do have hypotensive properties not shared by omega-6 fatty acids. The dose-response and time course of the effect will be important to determine so that further studies on the mechanism can have an optimal design. It appears that changes in blood pressure are more closely related to alterations in the synthesis of prostacyclin than that of PGE_2. As described in the section on vascular reactivity, it does not seem likely that dietary fish oil alters the response to endogenous catecholamines. Further work is warranted to determine whether dietary omega-3 fatty acid supplements could have potential as adjunctive treatment of human hypertension.

Acknowledgement

This work was supported by Grant HL-35380 from the National Institutes of Health.

References

1. A. Keys, Seven countries, a multivariate analysis of death and coronary heart disease. Cambridge, MA, and London A Commonwealth Fund Book, Harvard University Press (1980).
2. C. Lenfant and R.J. Hegyeli, Nonpharmacologic prevention of cardiovascular diseases in the United States, in: Nutritional Prevention of Cardiovascular Disease, W. Lovenberg, Y. Yamori eds., Academic Press, Inc., Orlando, FL (1984).
3. N. Kromann and A. Greeland, Epidemiological Studies in the Upernavik District, Greenland, Acta Med Scan 208:401 (1980).
4. A. Leaf and P.C. Weber, Cardiovascular effects of n-3 fatty acids, New Eng J Med, 318:549 (1988).
5. C. Von Schacky, Prophylaxis of atherosclerosis with marine omega-3 fatty acids: a comprehensive strategy. Ann Intern Med, 107:890 (1987).
6. M.C. Smith, and M.J. Dunn, The role of prostaglandins in human hypertension, Hypertension and the Kidney: Proceedings of a Symposium, A32 (1985).
7. S.H. Goodnight, Jr., W.S. Harris, W.E. Connor, and D.R. Illingworth, Polyunsaturated fatty acids, hyperlipidemia, and thrombosis. Arteriosclerosis 2:87 (1982).

8. G. Hornstra, E. Christ-Hazelhof, E. Haddeman, F. ten Hoor and D.H. Nugteren, Fish oil feeding lowers thromboxane- and prostacyclin production by rat platelets and aorta and does not result in the formation of prostaglandin I_3. Prostaglandins 21:727 (1981).

9. L.K. Dahl, Salt and hypertension. Amer J Clin Nutr 25:231 (1972).

10. I. Ehrstrom. Medical studies in North Greenland 1948-1949: VI. Blood pressure, hypertension and atherosclerosis in relation to food and mode of living. Acta Med Scand., 140:416 (1951).

11. P. Bjerager, N. Kromann, K. Thygesen, and B. Harvald. Blood pressure in Greenland Eskimos. Ugeskr Laeger, 142:2278 (1980).

12. E.M. Scott, I.V. Griffith, D.D. Hoskins, R.D. and R.D. Whaley, Serum-cholesterol levels and blood-pressure of Alaskan Eskimo men. Lancet i:667 (1958).

13. Y. Goto and Y. Homma, Recent trends of coronary heart disease in Japan in relation to dietary alteration. In: Nutritional Prevention of Cardiovascular Disease, W. Lovenberg and Y. Yamori eds., Academic Press, Orlando, FL (1984).

14. Y. Kagawa, M. Nishizawa, M. Suzuki, T. Midatake, T. Hamamoto, K Foto, E. Motonaga, H. Izumikawa, H. Hirata and A. Ebihara, Eicosapolyenoic acids of serum lipids of Japanese islanders with low incidence of cardiovascular diseases. J Nutr Sci Vitaminol 28:441 (1982).

15. T. Taminato, T. Matsumoto, K Suzuki, K. Sakamoto, N. Yasuda, S. Tsujimoto, T. Nakamura, and Y. Sasai, Mass spectrometric analysis of eicosapentaenoic acids in the serum of Japanese living on isolated islands and farming villages. In: Nutritional Prevention of Cardiovascular Disease, W. Lovenberg and Y. Yamori eds., Academic Press, Orlando, FL (1984).

16. T. Hamazaki, S. Fischer, M. Urakaze, S. Sawazaki and S. Yano. Comparison of the urinary metabolites of prostacyclin and thromboxane of the 2- and 3- series in a Japanese fishing and a Japanese farming village. Prostaglandins 32:644 (1986).

17. I.A.M. Prior. Migration, hypertension, and pacific perspectives for prevention. In: Nutritional Prevention of Cardiovascular Disease, W. Lovenberg and Y. Yamori eds., Academic Press, Orlando, FL (1984).

18. T. Simonsen, A. Vartun, V. Lyngmo and A. Nordoy, Coronary heart disease, serum lipids, platelets and dietary fish in two communities in Northern Norway. Acta Med Scand 222:237 (1987).

19. W. E. Lockette, R. C. Webb, B. R. Culp and B. Pitt. Vascular reactivity and high dietary eicosapentaenoic acid. Prostaglandins 24:631 (1982).

20. D. E. Mills and R. P. Ward. Effects of eicosapentaenoic acid (20:5w3) on stress reactivity in rats. Proceedings of the Society for Experimental Biology and Medicine 182:127 (1986).

21. B. A. Scholkens, D. Gehring, V. Schlotte and U. Weithmann, Evening primose oil, a dietary prostaglandin precursor, diminishes vascular reactivity to renin and angiotensin II in rats. Prostaglandins, Leukotrines and Medicine 8:273 (1982).

22. L. Somova, P. Hoffman and W. Forster. The reactivity of isolated blood vessels of salt-loaded rats fed low or high linoleic acid diets. European J of Pharmacology 64:79 (1980).

23. R. Lorenz, U. Spengler, S. Fischer, J. Duhm and P.C. Weber. Platelet function, thromboxane formation and blood pressure control during supplementation of the Western diet with cod liver oil, Circulation 67:504 (1983).

24. F.B. Pipkin, R.A. Morrison and P. M. S. O'Brien. The effect of dietary supplementation with linoleic and γ-linolenic acids on the pressor and biochemical response to exogenous angiotensin II in human pregnancy. PRog Lipid Res 25:425 (1986).

25. P. Singer, M. Wirth, S. Voigt, E. Richter-Heinrich, W. Godicke, I. Berger, E. Naumann, J. Listing, W. Hartrodt and C. Taube. Blood pressure- and lipid-lowering effect of mackerel and herring diet in patients with mild essential hypertension. Atherosclerosis 56:223 (1985).

26. H.R. Knapp and G.A. FitzGerald. Dietary polyunsaturates, blood pressure and vasodilator prostanoids. Clin Res 36:357A (1988).

27. V. Hossmann, G. A. FitzGerald, and C. T. Dollery. Influence of hospitalization and placebo therapy on blood pressure and sympathetic function in essential hypertension. Hypertension 3:113 (1981).

28. J. I. M. Drayer, M. A. Weber, and D. K. Nakamura. Automated ambulatory blood pressure monitoring: A study in age-matched normotensive and hypertensive men. Am Heart J 109:1334 (1985).

29. A. I. Fleischman, M. L. Bierenbaum, A. Stier, H. Somol, P. Watson, and A. M. Naso. Hypotensive effect of increased dietary linoleic acid in mildly hypertensive humans. J of the Med Soc of NJ, 76:181 (1979).

30. H. U. Comberg, S. Heyden and C. G. Hames. Hypotensive effect of dietary prostaglandin precursor in hypertensive man. Prostaglandins 15:193 (1978).

31. A. M. Heagerty, J. D. Ollerenshaw, D. I. Robertson, R. F. Bing, and J. D. Swales. Influence of dietary linoleic acid on leucocyte sodium transport and blood pressure. British Med J 293:295 (1986).

32. J. M. Iacono, R. M. Dougherty and P. Puska. Reduction of blood pressure associated with dietary polyunsaturated fat. Hypertension 4 (supp III): III-34 (1982).

33. B. M. Margetts, L. J. Beilin, B. K. Armstrong, I. L. Rouse, R. Vandogen, K. D. Croft and E. J. McMurchie. Blood pressure and dietary polyunsaturated and saturated fats: a controlled trial. Clin Sci 69:165 (1985).

34. F. M. Sacks, M. J. Stampfer, A. Munoz, K. McManus, M. Canessa and E. H. Kass. Effect of linoleic and oleic acids on blood pressure, blood viscosity, and erythrocyte cation transport. Journal of the American College of Nutrition, 6:179 (1987).

35. T. A. B. Sanders, M. Vickers and A. P. Haines. Effect on blood lipids and haemostasis of a supplement of cod-liver oil, rich in eicosapentaenoic and docosahexaenoic acids, in healthy young men. Clin. Science 61:371 (1981).

36. P. G. Norris, C. J. H. Jones, and M. J. Weston. Effect of dietary supplementation with fish oil on systolic blood pressure in mild essential hypertension. Brit Med J, 293:104 (1986).

37. P. B. Rylance, M. P. Gordge, R. Saynor, V. Parsons, and M. J. Weston. Fish oil modifies lipids and reduces platelet aggregability in haemodialysis patients. Nephron 43: 196 (1986).

38. J. Z. Mortensen, E. B. Schmidt, A. H. Nielsen and J. Dyerberg. The effect of N-6 and N-3 polyunsaturated fatty acids on hemostasis, blood lipids and blood pressure. Thromb Haemostas 50:(2) 543 (1983).

39. D. M. Demke, G. R. Peters, O. I. Linet, C. M. Metzler and K. A. Klott. Effects of a fish oil concentrate in patients with hypercholesterolemia. Atherosclerosis 70:73 (1988).

40. T. Hamazaki, R. Nakazawa, S. Tateno, H. Shishido, K. Isoda, Y. Hattori, T. Yoshida, T. Fujita, S. Yano, and A. Kumagai. Effects of fish oil rich in eicosapentaenoic acid on serum lipid in hyperlipidemic hemodialysis patients. Kidney International, 26:81 (1984).

41. R. V. Houwelingen, A. Nordoy, E. van der Beek, U. Houtsmuller, M. de Metz and G. Hornstra. Effect of a moderate fish intake on blood pressure bleeding time, hematology, and clinical chemistry in healthy males. Am J Clin Nutr 46:424 (1987).

42. S. Rogers, K. S. James, B. K. Butland, M. D. Etherington, J. R. O'Brien and J. G. Jones. Effects of a fish oil supplement on serum lipids, blood pressure, bleeding time, haemostatic and rheological variables. Atherosclerosis 63:137 (1987).

43. P. Singer, M. Wirth, S. Voigt, S. Zimontkowski, W. Godicke and H. Heine. Clinical studies on lipid and blood pressure lowering effect of eicosapentaenoic acid-rich diet. Biomed Biochim Acta 43:8/9, S421 (1984).

44. P. Singer, I. Berger, K. Luck, C. Taube, E. Naumann and W. Godicke. Long-term effect of mackerel diet on blood pressure, serum lipids and thromboxane formation in patients with mild essential hypertension. Atherosclerosis 62:259 (1986).

45. W. H. Seyberth, O. Oelz, and T. Kennedy, et al. Increased arachidonate in lipids after administration to man: Effects on prostaglandin biosynthesis. Clin Pharmacol Ther 18:521 (1975).

46. A. Ferretti, J. T. Judd, M. W. Marshall, V. P. Flanagan, J. M. Roman and E. J. Matusik, Jr. Moderate changes in linoleate intake do not influence the systemic production of E prostaglandins. Lipids 20:(5)268 (1985).

47. O. Adam, G. Wolfram, and N. Zollner. Prostaglandin formation in man during intake of different amounts of linoleic acid in formula diets. Ann Nutr Metab 26:315 (1982).

48. Z. Friedman, H. W. Seyberth, E. Lamberth, and J. A. Oates. Decreased prostaglandin E turnover in infants with essential fatty acid deficiency. Pediatr Res 12:711 (1978).

49. H. R. Knapp, I. A. G. Reilly, P. Alessandrini, and G. A. FitzGerald. In vivo idexes of platelet and vascular function during fish-oil administration in patients with atherosclerosis. N Eng J Med 314:937 (1986).

50. S. F Fischer and P. C. Weber. Prostaglandin I_3 is formed in vivo in man after dietary eicosapentaenoic acid. Nature 307:165 (1984).

51. E. Zuccato, G. Hornstra and J. Dyerberg. Long term marine diet in Eskimos is not associated with altered urinary excretion of total tetranor prostagalndin metabolites. Prostaglandins 30(3):465(1985).

52. E. O'Brien, D. Fitzgerald, and K. O'Malley. Blood pressure measurement: current practice and future trends. Brit Med J 290:729 (1985).

53. C. Alicandri, R. Fariello, E. Boni, M. L. Muiesan, A. Zaninelli, E. Agabiti-Rosei, and G. Muiesan. Left ventricular hypertrophy and ambulatory monitoring of blood pressure. J Clin Hypertension, 3:197 (1987).

POLYUNSATURATED OILS OF MARINE AND PLANT ORIGINS AND THEIR USES IN

CLINICAL MEDICINE

David F. Horrobin

Efamol Research Institute
PO Box 818, Kentville
Nova Scotia, Canada B4N 4H8

INTRODUCTION

The two main dietary essential fatty acids (EFAs) are linoleic acid and alpha-linolenic acid (ALA). These two parent fatty acids are metabolized within the body to whar are sometimes called the derived EFAs. The derived EFAs include gamma-linolenic acid (GLA), dihomogammalinolenic acid (DGLA) and arachidonic acid of the n-6 series from linoleic acid, and eicosapentaenoic acid (EPA) and docosahexaenoic acid (DHA) of the n-3 series from ALA. The traditional teaching is that there is no dietary requirement for the derived EFAs. It is assumed that all the amounts of the derived EFAs that are needed can be formed within the body from linoleic acid and ALA.

Linoleic acid is abundant in the diet. It is found particularly in seed oils, including oils derived from corn, sunflower, rape and soy (1). Safflower oil is the richest source, with some strains containing over 80% linoleic acid. Studies of nutritional intake have not always distinguished biologically active cis, cis, 9,12-linoleic acid from its positional and trans isomers which have no EFA activity. Such isomers are often more stable than the EFA-active natural isomer, and so may be deliberately manufactured during processing to increase the shelf lives of processed foods. This is not a trivial issue since the usual intake of these isomers and related compounds in North Americans may be as high as 5 to 10g per day (2).

Alpha-linolenic acid is present in vegetables, especially the dark green varieties, and also in some seed oils. notably soy and rape. Linseed oil is the most abundant source, often containing over 50% ALA. Because of its three double bonds and relatively easy peroxidation, manufacturing processes frequently deliberately destroy ALA in order to improve product stability. Concerns have therefore been expressed that in people who consume large amounts of convenience food and relatively little fresh food, intake of ALA may be inadequate.

Brenner and other investigators (3,4,5) have provided evidence that 6-desaturation, the first step in the metabolism of both linoleic acid and ALA to the derived EFAs, is rate-limiting. Stimulated by this observation, the last ten years have seen increasing interest in the idea that there may be a place for the direct administration of derived EFAs of both the n-6 and n-3 series. Brenner and colleagues have shown that not only is 6-desaturation rate limiting in normal individuals, but also that it can be further inhibited by such factors as aging, diabetes, high intakes of alcohol and cholesterol, or by catecholamines, ACTH and adrenal steroids released during stress.

This laboratory and animal work has been followed by human studies which have demonstrated low levels of GLA, DGLA, arachidonic acid, EPA or DHA in certain diseases, even when dietary intake of linoleic and alpha-linolenic acids appears satisfactory. Such low levels of derived EFAs have been found in individuals with atopic disorders (6,7), diabetics (8,9), alcoholics (10), in bottle-fed as opposed to breast-fed infants (11), and in normal individuals who will later develop strokes (12) or heart disease (13,14). Some racial groups may also have abnormalities of desaturation, of 6-desaturation in the Vancouver Island Indians (15), and of 5-desaturation in Eskimos (16). Cats also appear to have little or no ability to 6-desaturate (17).

High dietary intakes of linoleic acid or alpha-linolenic acid have proved unable to change plasma concentrations of DGLA, arachidonic acid, or EPA (18). Because of this, and because of provocative epidemiological observations in Eskimos, there has been a surge of interest in the direct administration of EPA and DHA in the forms of marine oils. It is not always clear that those carrying out such studies are aware of the probability that Eskimos have low arachidonic acid levels and high DGLA levels in blood because of an inability to 5-desaturate, as well as because of their high intakes of marine oils. Many studies have now shown that administration of such marine oils can raise plasma levels of DHA and EPA in humans and there is currently great interest in the idea that administration of such oils may treat or prevent certain diseases such as coronary heart disease, psoriasis and rheumatoid arthritis. The roles of the derived EFAs from marine oils are discussed in detail in several other chapters and will not be addressed further in this paper.

To date there has been less interest in possible therapeutic uses of the derived n-6 EFAs, although the biochemical considerations which have led to the interest in n-3 EFAs apply equally to the n-6 series. All the derived n-6 EFAs are found in modest quantities in organ meats such as liver, kidney and adrenals. Arachidonic acid is found in substantial amounts in egg yolks and in fish oils, especially those from warmer waters in the Southern Hemisphere (19). Contrary to a commonly held view, arachidonic acid is not found in peanut oil: it is arachidic acid which occurs in peanuts.

DGLA has not yet been identified in substantial amounts in readily available natural oils, although it may possibly be produced by some fungi. GLA, in contrast, is found in the oils from Spirulina, from various fungi such as strains of Mortierella and Rhizopus, from oats, barley, hops, hemp and certain other plants (1,20). The most abundant sources are the seed oils from evening primrose (Oenothera spp), borage (Borago officinalis), and blackcurrant and other Ribes species (21,22). These oils are now beginning to be tested clinically although to date almost all published work has related to evening primrose oil. This has now been found effective in treating such conditions as atopic eczema (23), rheumatoid arthritis (24), diabetic neuropathy (25), alcoholic liver disease (26) and premenstrual syndrome and premenstrual breast pain (cyclical mastalgia (27).

The remainder of this paper will be devoted to exploring three main topics: the evidence in humans for restricted conversion of dietary linoleic acid to its metabolites: the evidence for the therapeutic effects of evening primrose oil in atopic eczema, diabetic neuropathy, rheumatoid

arthritis and premenstrual breast pain: and the biological activities of
derived EFAs in oils from different sources.

RESTRICTED METABOLISM OF LINOLEIC ACID IN HUMANS

There are at least five published studies concerning the effects of
dietary linoleic acid supplementation on the plasma levels of the linoleic
acid 6-desaturated metabolites, DGLA and arachidonic acid (28-32). These
studies are all consistent in failing to demonstrate any rise in plasma
levels of either DGLA or arachidonic acid following dietary supplementation
with additional amounts of linoleic acid ranging from 2.5g/day in children
(32) up to 36g/day in adults (29,30). These studies therefore show that
it is not possible to modulate plasma DGLA and arachidonic acid concentrations in humans by changing linoleic acid intake. Thus in those situations
mentioned earlier (atopic eczema, diabetes, alcholism, stroke, heart disease) in which plasma levels of DGLA and arachidonic acid have been found
to be below normal, the concentrations cannot be restored to normal by increasing linoleic acid intake.

The two main possible explanations for these observations are as follows.
1. Linoleic acid added to the diet in excess of the amounts normally present
is not effectively metabolized to DGLA and arachidonic acid, probably because the 6-desaturation step, the conversion of linoleic acid to GLA, is
severely rate-limiting. 2. Linoleic acid added to the diet is metabolized
to DGLA and arachidonic acid, but the concentrations of these derived EFAs
do not change because they are maintained by a homeostatic mechanism which
metabolizes the additional DGLA and arachidonic acid and so maintains their
concentrations constant.

It is possible to test these two hypotheses by supplying GLA directly
in order to by-pass the rate-limiting 6-desaturation. GLA will be rapidly
converted to DGLA, since elongation is not a rate-limiting step (3-5), and
then more slowly to arachidonic acid since 5-desaturation is again known to
be rate-limiting (3-5). If direct administration of GLA fails to raise
the plasma levels of DGLA and arachidonic acid, then the second explanation
is likely to be correct: increases in DGLA and arachidonic acid formation
must be balanced by increases in their rate of metabolism or disposal, so
maintaining blood levels constant. If administration of GLA does elevate
plasma concentrations of DGLA and arachidonic acid, then the first explanation is likely to be correct: DGLA and arachidonic acid do not rise after
linoleic acid administration because 6-desaturation is so rate-limiting
that significant quantities of the derived EFAs cannot be formed.

The question has now been conclusively answered (33,34). Administration of relatively small amounts of GLA is able to raise the concentrations of both DGLA and arachidonic acid in human plasma. GLA given in a
dose of around 360mg/day as Efamol evening primrose oil to 20 groups of
humans consistently produced significant elevations of both DGLA and
arachidonic acid in plasma phospholipids (34). The rise in DGLA was always greater than the rise in arachidonic acid, leading to an increase in
the DGLA/arachidonic acid ratio in all twenty studies. This last observation is consistent with a rate-limiting 5-desaturation step.

As reviewed in the Introduction, there are now several examples of disease states in which although dietary intake of linoleic acid is normal,
concentrations of DGLA and arachidonic acid are below normal. Except in
those few situations where low concentrations of DGLA and arachidonic acid
are due to subnormal intake of linoleic acid (e.g. in total parenteral
nutrition without lipids), administration of increased amounts of dietary

linoleic acid will not raise DGLA and arachidonic acid concentrations to or towards normal. In order to achieve this result it is necessary to by-pass the rate-limiting 6-desaturation step and to administer GLA directly.

THERAPEUTIC EFFECTS OF EFAMOL EVENING PRIMROSE OIL

The seeds of the evening primrose, Oenothera spp., are unusual in containing GLA. The natural varieties of Oenothera contain very variable amounts of GLA ranging from as low as 2% up to 12% or more (21). It is to be expected that oils of such different compositions will have different biochemical and therapeutic effects. It is therefore necessary to be specific about the precise source of the oil used. Efamol is the oil from certain selected and/or hybridised varieties of Oenothera, chosen because of their ability to produce a constant and stable level of GLA (8.5-9.0% of total fatty acids present) and of other constituents. The major part of the rest of the oil consists of linoleic acid (70-72%). Efamol has been subjected to a complete toxicological and carcinogenicity evaluation (35,36). No important adverse effects could be generated even at doses as high as 10ml/kg/day.

Efamol has now been tested in double-blind, placebo-controlled trials in several situations. Statistically significant and clinically important therapeutic effects have been observed, indicating the likely value of by-passing the 6-desaturation block. The most complete documentation to date as been in the areas of atopic eczema, diabetic neuropathy, rheumatoid arthritis and premenstrual breast pain.

Atopic eczema

In adults with atopic eczema, plasma phospholipid linoleic acid levels are normal or elevated, while concentrations of all the 6-desaturated metabolites, including DGLA and arachidonic acid, are significantly below normal (6). A similar but more marked deviation from the normal blood EFA picture is found in children with atopic eczema (7): in children the linoleic acid concentrations are strikingly elevated indicating a block in 6-desaturation. Linoleic acid levels are also elevated in the umbilical cord blood from babies with elevated IgE concentrations, a marker of atopic eczema risk (7). Eczema is not present in neonates, indicating that the EFA abnormality precedes the development of the abnormal skin condition and is not a secondary consequence of the skin inflammation. These observations are consistent with the idea that people with atopic eczema have a lifelong modest reduction in their ability to convert linoleic acid to GLA and its metabolites.

These biochemical observations provide a rationale for administering GLA directly to test the hypothesis that it may have beneficial effects. Nine double blind, randomised, placebo-controlled trials involving over 300 adults and children with atopic eczema have now been performed using Efamol (23,37,38). The results are unequivocal. Efamol has a highly significant effect in improving the eczema as compared to the baseline state (p<0.0001) and the effect of Efamol is highly significantly better than that of placebo (p<0.01). The effects of Efamol on itch (pruritus) are particularly striking. Itch is an important feature of atopic eczema which is not confined to the obviously inflamed areas of skin. Itch is central to the production of symptoms, since the scratching exacerbates the inflammation and may precipitate outbreaks of eczema in non-inflamed areas. There are no effective treatments for the itch of atopic eczema. The antihistamines, although widely used, have not been found effective in placebo-controlled trials and have central sedative effects. This sedation is a major problem for people who must drive, operate machinery at work or in the home or perform intellectually demanding work. Efamol was highly significantly better than placebo at relieving itch (p<0.0001) and was effective without causing any drowsiness (38). The use of Efamol is therefore a substantial development in the treatment of atopic eczema and is the first treatment available which is logically based on understanding of a biochemical abnormality.

Diabetic neuropathy

Diabetes impairs the conversion of linoleic acid to its 6-desaturated metabolites in both animals and humans (3,8,9,39,40). It has been suggested that this reduced ability to desaturate linoleic acid may account for many of the long term cardiovascular, renal, retinal and neurological complications of diabetes (40). Reduced levels of derived EFAs would be expected to lead to elevated cholesterol and triglyceride concentrations, increased platelet aggregation, and increased blood viscosity, as well as changed membrane properties leading to increased permeability of capillary and other membranes. All these factors have been thought to contribute to the long term complications of diabetes.

It is probable that the major part of conversion of linoleic acid to GLA within the body occurs within the liver. When insulin is normally secreted from the pancreas into the hepatic portal system, the concentration of insulin in the blood perfusing the liver is many times higher than its concentration in arterial blood perfusing the peripheral tissues. This differential concentration may be important in the maintenance of normal EFA metabolism. When insulin is injected in the treatment of diabetes, it usually enters the general circulation so that no differential concentration between the hepatic blood and the rest of the blood is maintained. This may help to explain why insulin treatment of diabetes may not correct abnormal EFA metabolism and may not prevent the long term development of complications.

Reduced levels of linoleic acid metabolites could account for some of the peripheral neuronal degeneration in diabetes. The metabolites are important in the structures of axonal membranes, and also in the maintenance of a normal neuronal microcirculation by reducing blood viscosity, inhibiting platelet aggregation, and maintaining normal permeability of the capillaries (40). If this is the case, then by-passing the impaired 6-desaturation by the direct administration of GLA might be expected to prevent or reverse diabetic neuropathy. In streptozotocin diabetes in rats, Julu demonstrated that the administration of Efamol could prevent the development of neuropathy (40). This occurred when treatment with Efamol was started at or before the time of induction of diabetes. Julu has now shown that administration of Efamol several weeks after the onset of diabetes is successful in reversing established neuropathy (P.O. Julu, University of Zimbabwe, Harare, personal communication).

To date only one double-blind, placebo-controlled randomised trial of Efamol has been carried out in humans with diabetic neuropathy although several more are in progress. This trial showed over a period of six months that in the Efamol group motor and sensory nerve conduction and thermal sensitivity to heat and cold in both the upper and lower limbs all improved. The same eight parameters all deteriorated in the placebo group. For six of the eight parameters, the difference between the results in the Efamol and placebo groups was statistically significant (41).

These investigations raise the exciting possibility that direct administration of GLA may be able to prevent and/or reverse other diabetic complications and studies are under way to test that possibility.

301

Premenstrual breast pain

Women with premenstrual syndrome have a pattern of plasma phospholipid n-6 fatty acids which is similar to that seen in atopic eczema (42). This is perhaps not surprising since women with severe premenstrual syndrome are known to demonstrate an increased incidence of atopic disorders. Linoleic acid levels are normal or slightly elevated, while concentrations of 6-desaturated metabolites are reduced. However, the plasma phospholipid EFA patterns in atopic eczema and premenstrual syndrome are not identical. In atopic eczema concentrations of 6-desaturated n-3 EFAs are also significantly below normal (6,7), whereas the n-3 EFAs tend to be elevated in premenstrual syndrome (42).

The most consistent symptom of premenstrual syndrome in many women is premenstrual breast pain and tenderness (cyclical mastalgia). The only current treatments for mastalgia are potent hormone-modifying drugs such as bromocriptine and danazol. These have numerous side effects and the hormonal modifications they produce make them unacceptable to many women. In a major comparative study, surgeons at the Breast Clinic at the University of Wales compared the clinical efficacy of Efamol, bromocriptine and danazol. They found that the three agents were approximately similar in efficacy but that side effects with Efamol were few and trivial (27). The risk/benefit ratios of the products therefore made Efamol the best first line treatment (27).

Four other double-blind, placebo-controlled studies of Efamol in premenstrual syndrome have now been performed, in addition to open studies (43-48). All have shown Efamol to be superior to placebo, not only in relation to breast pain but also with regard to other features of the syndrome such as bloating, depression and anxiety. Efamol did not produce any changes in hormone levels in women with premenstrual syndrome (43).

Rheumatoid arthritis

Many investigators have shown that prostaglandin E_1 (PGE_1) has a wide variety of anti-inflammatory actions. PGE_1 itself is too unstable and short-lived to be employed therapeutically and its stable analogues do not always have the same pharmacological profile. GLA, via DGLA, is a precursor of PGE_1 and can elevate the levels of PGE_1 in the body. Efamol was found to be effective in treating adjuvant arthritis, an animal model of arthritis in which PGE_1 had previously been shown to be effective (49). A double-blind, placebo-controlled, randomised trial of Efamol in human rheumatoid arthritis was therefore set up (24).

Either placebo, or Efamol or Efamol Marine (80% evening primrose oil + 20% fish oil) was administered to patients with stable rheumatoid arthritis for a period of 12 months on a double-blind basis. This was followed by a further 6 month single blind period during which, unknown to the patients, all were switched to placebo. All the patients to start with were taking non-steroidal anti-inflammatory drugs (NSAIDs) for relief of pain and inflammation. NSAIDs are the most widely used drugs in rheumatoid arthritis but cause many side effects, notably gastrointestinal ulceration and bleeding.

The results were clear cut. By the end of the first 12 months, 56% of the patients in the placebo group, but only 6% in the Efamol group and 13% in the Efamol Marine group had dropped out because of worsening rheumatoid arthritis. Of the remaining patients, 30% in the placebo group, 76% in the Efamol group and 82% in the Efamol Marine group were able to stop or substantially reduce their intake of NSAIDs. Over 90% of the patients in the Efamol and Efamol Marine groups, but only 31% of the patients in the placebo group rated themselves as substantially improved by the end of the 12 months. During the single blind phase of the study when all the patients were switched to placebo, almost all the patients on Efamol or Efamol Marine experienced a relapse to their original condition and had to increase their use of NSAIDs to its previous level.

EFAs do not have dramatic short term effects in rheumatoid arthritis. Their action is slow and prolonged. It is therefore important that studies are conducted over substantial periods of time. On the basis of the results in this study, at least six months active treatment should be involved in clinical trials. Otherwise the results are likely to be less than optimum and the therapeutic efficacy of the EFAs may be underestimated. Because of their freedom from important side effects, the ability to replace use of NSAIDs with substances such as Efamol or Efamol Marine represents an important advance in rheumatoid arthritis treatment.

In an earlier, open study of Efamol in rheumatoid arthritis, twenty patients on NSAIDs were tested. Their NSAIDs were abruptly stopped and replaced by Efamol for a period of three months (50). Two patients dropped out for reasons unrelated to arthritis or Efamol but of the remaining 18, only one had to use NSAIDs during the period of the study. An objective measure of inflammatory response was obtained by testing the effect of ultra-violet radiation on the skin. Efamol inhibited this measure of inflammation but did not produce other improvements in the clinical state. This study demonstrated that in the short term Efamol was as effective as NSAIDs in controlling rheumatoid arthritis in most patients. Longer periods of study are likely to be required to demonstrate actual improvements in clinical status.

BIOLOGICAL EFFECTS OF DERIVED EFAS IN DIFFERENT FORMS

A major emerging consideration relates to the precise chemical configuration of the derived EFAs in relation to their biological activity. It used to be somewhat naively thought that any configuration would be active, the only determining factor being the actual amount of the particular fatty acid administered. It is now clear that this is not true. For example, the ethyl ester forms of the EFAs may be less efficiently absorbed from the intestines than triglyceride or other forms (51). Even within a particular class of compound, such as triglycerides, the details of chemical structure may influence biological and therapeutic actions to a major degree. Perhaps the first demonstration of this phenomenon concerned peanut oil. The natural oil is a relatively atherogenic material when added to the diet, but the randomized oil has little atherogenic activity even though the fatty acid compositions of the two materials are absolutely identical (52). The precise positions of the individual fatty acids and the nature of their companions on the triglyceride molecule are clearly very important determinants of biological activity.

Some years ago when our research group was first investigating the clinical effects of GLA-containing oils, we also made the naive assumption that the clinical effect of the oil in humans would be directly proportional to the amount of GLA in the oil. We therefore thought that borage, blackcurrant and fungal oils would be much more clinically effective than Efamol evening primrose oil and we took steps to procure large commercial quantities of these other oils. Unfortunately we were sadly disappointed by our first pilot studies in which it became apparent that our assumption was untrue. In certain circumstances, not only were the other oils not any more effective than Efamol, they actually appeared to produce poorer clinical effects even though they contained more GLA. This may partly be because the other fatty acids present are different (table 1) and partly because the triglyceride configurations of the four oils are also different (53).

Table 1. Major fatty acids present in various GLA-containing oils. Results are expressed as a percentage of the total fatty acids present in the oil.

	PRIMROSE	FUNGAL	BORAGE	BLACKCURRANT
Saturates	7.3	27.1	15.5	7.4
Oleic	9.0	40.4	15.8	11.6
Linoleic	72.6	10.4	39.1	43.9
GLA	9.1	18.9	18.7	18.7
ALA	-	-	4.7	14.5
Stearidonic	-	-	-	2.9
Erucic	-	-	3.5	-

We have therefore conducted a series of experiments in which we have given identical amounts of GLA/day in the form of different oils to rats. After two weeks on the oil supplemented diets, the animals were killed and the outputs of n-6 EFAs and prostaglandins from perfused mesenteric vascular beds were measured (54). As can be seen from tables 2 and 3, the four oils had very different effects on the outputs of both 6-desaturated n-6 EFAs and of prostaglandins.

Table 2. Effects of different oils at doses providing the same daily intake of GLA on the outflow of EFA metabolites of GLA from the mesenteric vascular bed. Results are shown as the mean outflow per 30 minutes in the effluent from six animals. The amounts are in micrograms.

GROUP	GLA	DGLA	ARACHIDONIC	ADRENIC	22:5n-6
Borage	16.2	4.2	32.4	1.5	1.9
Blackcurrant	26.1	3.7	28.6	0.9	1.1
Fungal	12.2	3.6	15.6	0.6	1.2
Primrose	19.8	8.6	40.6	5.4	5.9

These results supported our clinical impression that the biological effects of these oils cannot simply be related to the crude GLA content. Primrose oil was significantly more effective at increasing the outflow of DGLA, arachidonic acid, adrenic acid, 22:5n-6, and PGE_1 than any of the other oils. These biochemical effects may possibly be related to its favourable clinical actions. The biochemical effects produced by the other oils may also be therapeutic in specific circumstances. However, only formal testing in randomised, double-blind, placebo-controlled trials will enable a decision to be made as to whether or not a particular oil is effective in a particular clinical situation. It is impossible to extrapolate from one to the other on the simple basis of the GLA content of the oil.

Table 3. Effects of different oils at doses providing the same daily intake of GLA on the outflow of prostaglandin metabolites of GLA from the mesenteric vascular bed. Results are shown as the mean concentration (pg/ml) seen in the results from six animals.

GROUP	PGE_1	PGE_2	PGI_2	TxA_2	PGI_2/TxA_2
Borage	2	315	4811	46	105
Blackcurrant	7	281	4885	129	38
Fungal	39	652	3043	28	109
Primrose	82	784	4812	26	185

Abbreviations: PGE_1 = prostaglandin E_1. PGE_2 = prostaglandin E_2. PGI_2 = prostacyclin, measured as 6-keto-PGF_1alpha, the stable degradation product. TxA_2 = thromboxane A_2, measured as TxB_2, the stable degradation product. PGE_1 and PGI_2 inhibit platelet aggregation and are vasodilators, whereas TxA_2 promotes aggregation and is a powerful vasoconstrictor.

Although our work has been on the oils containing derived n-6 EFAs, we suspect that exactly the same considerations will apply to n-3 EFAs. This may help to account for some of the confusion and the contradictory results obtained in the EPA/fish oil field. Many different EPA preparations from many different sources are being used by many different investigators. It is usually assumed that the total EPA content of the oil being used is the only determinant of biological activity. Our research group has serious doubts as to whether this is true. We suspect that it will prove to be just as invalid to extrapolate from one EPA preparation to another in which the precise configuration is different as is the case in the GLA field.

CONCLUSIONS

There are now consistent studies which show the presence of abnormal patterns of EFA concentrations of the blood in certain diseases. For the most part these abnormalities cannot be explained on the basis of differences in dietary EFA intake: they appear to be caused by differences in the rates of endogenous metabolism. As a general rule, except in the rare case of truly inadequate dietary intake of the parent EFAs, administration of linoleic acid and ALA cannot correct the observed abnormalities in concentrations of the 6-desaturated EFAs.

In contrast, double-blind, placebo-controlled trials have now shown that certain oils rich in derived EFAs can correct these abnormal EFA blood patterns and can produce therapeutic effects. Each oil seems to have its own specific spectrum of activity, based partly on its content of a derived EFA such as GLA or EPA, and partly on other fatty acids present and on the precise triglyceride configuration. Only double-blind, placebo-controlled trials, and not theoretical arguments, can demonstrate whether or not a particular oil is effective in a particular situation.

REFERENCES

1. Hudson, B.J.F. Human Nutrition: Food Sci. Nutr. 41F: 1 (1987).
2. Kummerow, F.A. J. Environ. Path. Toxicol. Chem. 6: 123 (1986).
3. Brenner, R.R. Progr. Lipid Res. 20: 41 (1982).
4. Garcia, P.T. and Holman, R.T. J. Am. Oil Chem. Soc. 42: 1137 (1965).

5. Sprecher, H. Progr. Lipid Res. 20: 13 (1982).
6. Manku, M.S., Horrobin, D.F., Morse, N.L., Wright, S. and Burton, J.L. Br. J. Dermatol. 110: 643 (1984).
7. Strannegard, I.L., Svennerholm, L. and Strannegard, O. Int. Arch. Allergy Appl. Immunol. 82: 423 (1987).
8. Jones, D.B., Carter, R.D., Haitas, B. and Mann, J.I. Br. Med. J. 286: 178 (1983).
9. Tilvis, R.S., Helve, E. and Miettinen, T.A. Diabetologia 29: 690 (1986).
10. Glen, I., Skinner, F., Glen, E. and MacDonell, L. Alcoholism: Clin. Exp. Res. 11: 37 (1987).
11. Putnam, J.C., Carlson, S.E., DeVoe, P.W. and Barness, L.A. Am. J. Clin. Nutr. 36: 106, 1982.
12. Miettinen, T.A., Huttunen, J.K., Naukkarinen, V. Monogr. Atherosclerosis 4: 19 (1986).
13. Salonen, J.T., Salonen, R. and Penttila, I. Am. J. Cardiol. 58: 226, (1985).
14. Wood, D.A., Butler, S. and Riemersma, R.A. Lancet 2: 117 (1984).
15. Bates, C. van Dam, C. and Horrobin, D.F. Prostaglandins Leukotr. Med. 17: 77 (1985).
16. Horrobin, D.F. Med. Hypotheses 22: 421 (1987).
17. Frankel, T.L. and Rivers, J.P.W. Br. J. Nutr. 39: 227 (1978).
18. Manku, M.S., Morse-Fisher, N. and Horrobin, D.F. Eur. J. Clin. Nutr. 42: 55 (1988).
19. Gibson, R.A., Sinclair, A.J. Comp. Biochem. Physiol. 78C: 325 (1984).
20. Qureshi, A.A., Schnoes, H.K., Din, Z.Z. and Peterson, D.M. Fed. Proc. 43: 2626 (1984).
21. Hudson, B.J.F. J. Am. Oil Chem. Soc. 61: 540 (1984).
22. Traitler, H. and Winter, H. Progr. Lipid Res. 25: 255 (1986).
23. Wright, S. and Burton, J.L. Lancet 2: 1120 (1982).
24. Belch, J.J.F., Ansell, D., Madhok, R., O'Dowd, A. and Sturrock, R.D. Ann. Rheum. Dis. 47: 96 (1988).
25. Jamal, G.A., Carmichael, H. and Weir, A.J. Lancet 1: 1098 (1986).
26. Glen, E., MacDonnell, L., Glen, I. and MacKenzie, J. Possible pharmacological approaches to the prevention and treatment of alcohol-related CNS impairment: results of a double blind trial of essential fatty acid, in Pharmacological Treatments for Alcoholism, G. Edwards and J. Littleton, eds., Croom Helm, London.
27. Pye, J.K., Mansel, R.E. and Hughes, L.E. Lancet 2: 373 (1985).
28. Dayton, S., Hashimoto, S. and Dixon, W. J. Lipid. Res. 7: 103 (1966).
29. Singer, P., Berger, I. & Wirth, M. Prostaglandins Leukotr. Med. 24: 173 (1986).
30. Singer, P., Jaeger, W. and Voigt, S. Prostaglandins Leukotr. Med. 15: 159 (1984).
31. Lasserre, M., Mendy, F. and Spielman, D. Lipids 20: 227-233 (1985).
32. Gibson, R.A. Proc. Nutr. Soc. Australia 10: 196 (1985).
33. Vericel, E., Lagarde, M., Mendy, F., Courpron, P. and Dechavanne, M. Nutr. Res. 7: 569 (1987).
34. Manku, M.S., Morse-Fisher, N. and Horrobin, D.F. Eur. J. Clin. Nutr. 42: 55 (1988).
35. Everett, D.J., Greenough, R.J., Perry, C.J., McDonald, P. and Bayliss, P. Med. Sci. Res. in press.
36. Perry, C.J., Everett, D.J., Macnaughten, F., Howroyd, P. and Bayliss, P. Med. Sci. Res. in press.
37. Schalin-Karrila, M., Mattila, L., Jansen, C.T. and Uotila, P. Br. J. Dermatol. 117: 11, 1987.
38. Morse, P.F., Horrobin, D.F., Manku, M.S. and Stewart, J.C.M. Br. J. Dermatol. in press.
39. Poisson, J.P. Enzyme 34: 1 (1985).
40. Horrobin, D.F. Prostaglandins Leukotr. EFAs 31: 181 (1988).
41. Jamal, G.A., Carmichael, H. and Weir, A.J. Lancet 1: 1098 (1986).

42. Brush, M.G., Watson, S.J., Horrobin, D.F. and Manku, M.S. Am. J. Obstet. Gynecol. 150:363 (1984).
43. Puolakka, J., Makarainen, L., Viinikka, L. and Ylikorkala, O. J. Reprod. Med. 30: 149 (1985).
44. Massil, H., O'Brien, P.M.S. and Brush, M.G. presented at 2nd International Symposium on Premenstrual Syndrome, Kiawah Island, Sept. 1987.
45. Casper, R. presented at 2nd International Symposium on Premenstrual Syndrome, Kiawah Island, Sept. 1987.
46. Oeckermann, P.A., Bachrack, I., Glans, S. and Rassner, S. Rec. Adv. Clin. Nutr. 2: 404 (1986).
47. Horrobin, D.F. J. Reprod. Med. 30: 149 (1985).
48. Horrobin, D.F. and Manku, M.S. presented at 2nd International Symposium on Premenstrual Syndrome, Kiawah Island, Sept. 1987.
49. Kunkel, S.L., Ogawa, H., Ward, P.A. and Zurier, R.B. Progr. Lipid Res. 20: 885, 1981.
50. Hansen T.M., Lerche, A., Kassis, V., Lorenzen, I. and Sondergaard, J. Scand. J. Rheumatol. 12: 85 (1983).
51. Lawson, L.D. and Hughes, B.G. Biochem. Biophys. Res. Comm. 152: 328 (1988).
52. Myher, J.J., Marai, A., Kuksis, A. and Kritchevsky, D. Lipids 12: 775 (1977).
53. Lawson, L.D. and Hughes, B.G. Lipids 23: 313 (1988).
54. Jenkins, D.K. Mitchell, J.C., Manku, M.S. and Horrobin, D.F. Med. Sci. Res. in press.

BIOCHEMICAL AND BIOCLINICAL ASPECTS OF BLACKCURRANT SEED OIL

ω3-ω6 BALANCED OIL

D. Spielmann*, H. Traitler*, G. Crozier*, M. Fleith*,
U. Bracco*, P.A. Finot*, M. Berger, and R.T. Holman **

* Nestlé Research Centre, Nestec Ltd, Vers-chez-les-Blanc,
Ch-1000 Lausanne 26 (Switzerland)
** Hormel Institute, University of Minnesota, Austin (USA)

INTRODUCTION

The essential nature of the parent fatty acids of the n-6 and n-3 families, namely linoleic acid and α-linolenic acid, is now well established.
However, since they must undergo successive desaturations and elongations (figure 1) to develop most of their various biological actions, the efficiency of their transformation in different dietary and pathological situations has been questioned.

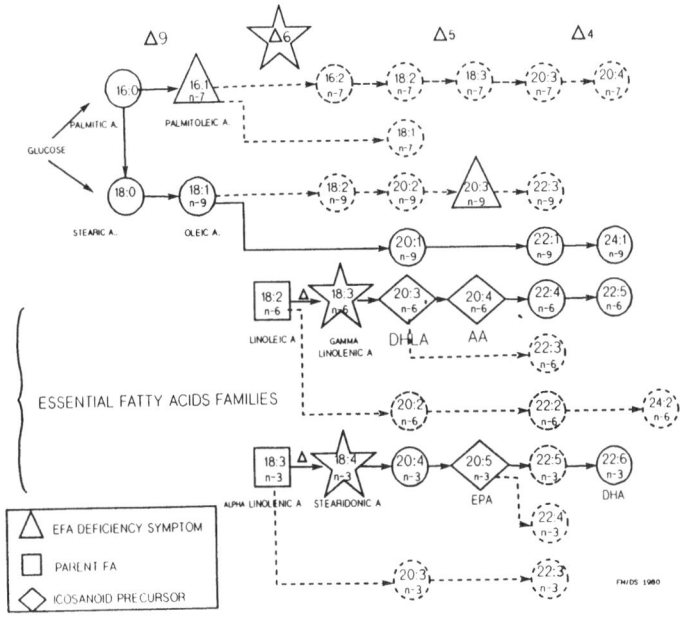

Fig. 1. Biosynthesis of polyunsaturated fatty acids in man.
From F. Mendy, D. Spielmann (18).

Already in 1958, Mead[1] hypothesized that Δ6 desaturation was the limiting step in this enzymatic pathway. This was confirmed by Brenner[2,3], by Marcel, Holman, Kristiansen[4] (in the rat) and by Sprecher[5].
Moreover Crawford and Rivers[6] showed that Δ6 and Δ5 desaturases do not exist in felines, which are thus obligatory carnivores. They also showed the relative low rates of transformation in both families (figure 2).

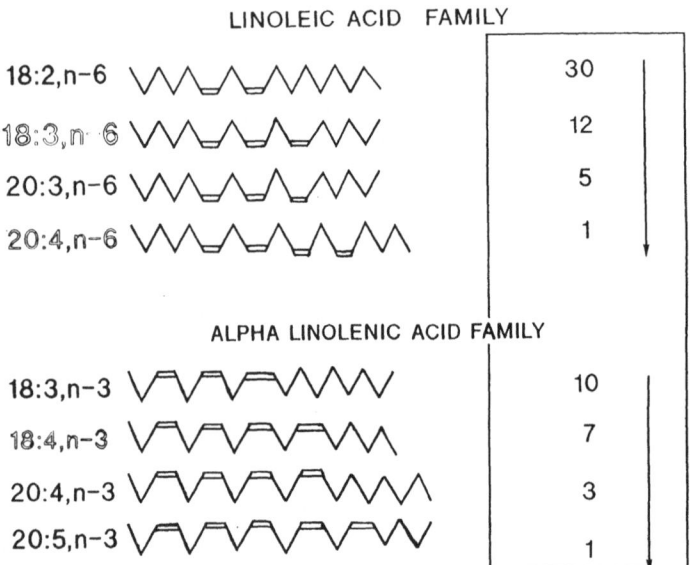

Fig. 2. Conversion coefficients in the two essential families. From M. Crawford.

Δ6 DESATURASE AND Δ5 DESATURASE REGULATION

The importance of substrates and of various hormones in Δ6 and Δ5 regulation was also well studied. There is now a growing body of evidence leading to the conclusion that high supplies of linoleic acid inhibit Δ6 desaturase (Iacono[7]; Mendy[8]; Hansen[9]; Nouvelot[10]; Agradi and Galli[11]; Friedman[12]).
Moreover, Brenner[13,14,15,16], Mendy et al.[17,18,19], Blond, Lemarchal, Spielmann[20] have shown that Δ6 desaturase is very sensitive to metabolic and endocrine regulation. The fatty acid, protein and carbohydrate content of the diet influence its activity. It is stimulated by insulin and inhibited by age, epinephrin, cortisol, thyroxin and glucagon (figure 3).

Presently more and more data are being gathered on Δ5 desaturase leading from 20:3,n-6 (dihomo-γ-linolenic acid to 20:4,n-6 (arachidonic acid). It appears that it is also regulated by cortisol, epinephrin, ACTH (Alaniz[21]; Mandon[22,23]) and insulin (El Boustani[24]).

The various regulations on both Δ6 and Δ5 desaturases lead to the possibility of manipulating the amounts of 20 carbons precursors of icosanoids in phospholipids (figure 4).

By-passing Δ6 desaturase with γ-linolenic acid (GLA) would lead to increased dihomo-γ-linolenic acid (DHLA), the precursor of series 1 eicosanoids, and to a lesser increase of arachidonic acid, the precursor of series 2 eicosanoids, if Δ5 desaturase is partially inhibited.

Thus increasing the 20:3,n-6 / 20:4,n-6 ratio could modify the synthesis of icosanoids playing a role in various cell-cell interactions,

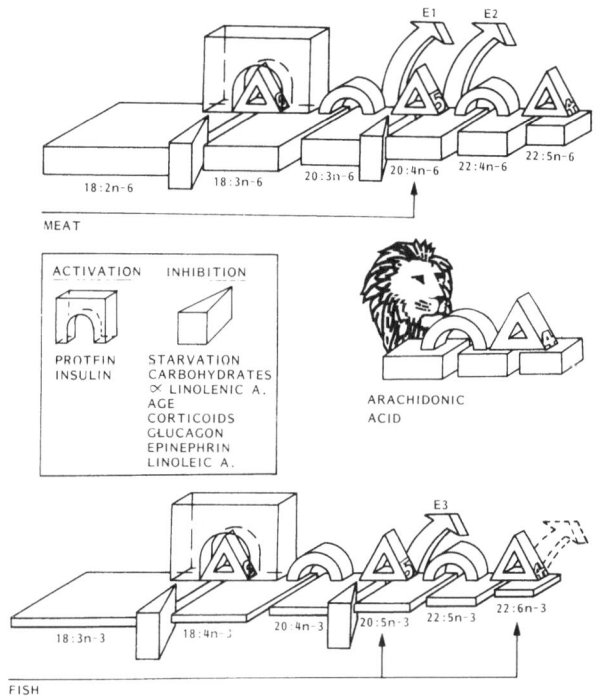

Fig. 3. The physiological situation in omnivores compared to carnivores. Regulation of Δ6 desaturase. (Δ5 desaturase seems to almost share the same factors of regulation).

like platelet-endothelium interaction or immune functions. DHLA, for instance, is a very bad substrate for 5 lipoxygenase and cannot be transformed into leukotrienes.

Stearidonic acid 18:4,n-3, (18:4,Δ6,9,12,15), also by-passing Δ6 desaturase, could compete for Δ5 desaturase through its elongation product 20:4,n-3, and lead to an increased 20:5,n-3 synthesis (figure 5).

THE "ω3 / ω6" RATIO

The importance of both the n-3 and n-6 families is being more and more recognized, as well as their mutual interaction and the importance of their respective metabolisms from the parent fatty acids to their various long chain derivatives.

The optimal intakes of n-3 and n-6 fatty acids have also been studied. According to Lasserre et al.[25] the optimal ω3 / ω6 ratio has been evaluated at 1 to 5-6. In general, linoleic acid should contribute to 5 to 6% of total calories, and α-linolenic acid: 0.5 to 1%. According to Bjerve[26] the minimal daily requirements of n-3 fatty acids has been estimated at 0.54% of energy, and the minimal absolute amount of long chain n-3 acids to 100-200 mg daily.

Which nutritional strategy should be followed in order to provide the

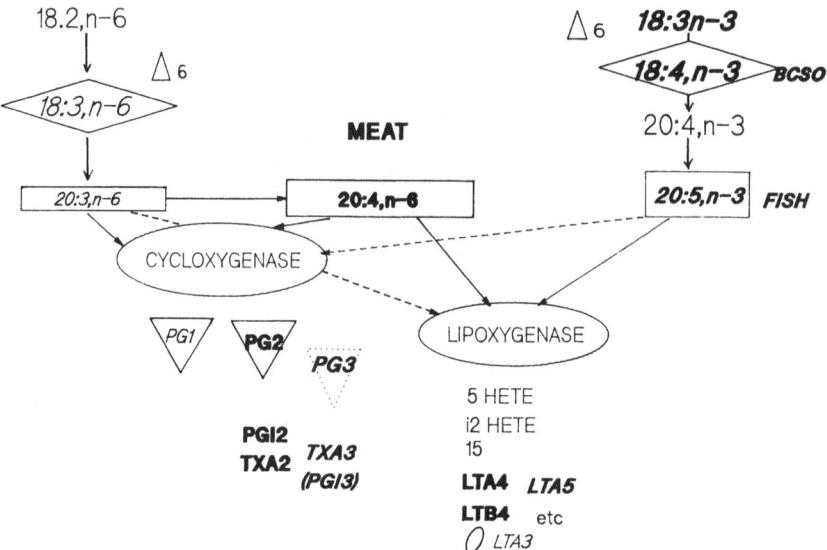

Fig. 4.   Eicosanoids precursors of the 1, 2 and 3 families.

adequate amounts and ratios of parent fatty acids and long chain derivatives?

The new low erucic acid rape seed oil answers one aspect of the ω3 / ω6 ratio. Evening primrose oil and borage oil provide the linoleic acid metabolite, γ-linolenic acid, in case of Δ6 desaturase deficiency. But with its particular composition black currant seed oil (BCSO) covers all the aspects of the search for optimal fatty acid intakes: it contains both n-3 and n-6 families with a ω3 / ω6 ratio of 1/5, but also γ-linolenic acid and stearidonic acid (18:4,n-3), the products of Δ6 desaturase in both families (table 1) (Traitler[27]).

Animal studies

This balanced supply of fatty acids has been studied in rats (Traitler et al.[28,29]) guinea pigs (Crozier et al.[30]) and monkeys (Ward[31]). A study in the rat is described in this book. It gives a first example of the possible manipulation of prostaglandin production in the heart by BCSO.

Guinea pigs have a more similar Δ6 desaturase activity to humans than rats, and so were chosen for a further study.
Three groups of guinea pigs were fed semi-synthetic diets containing

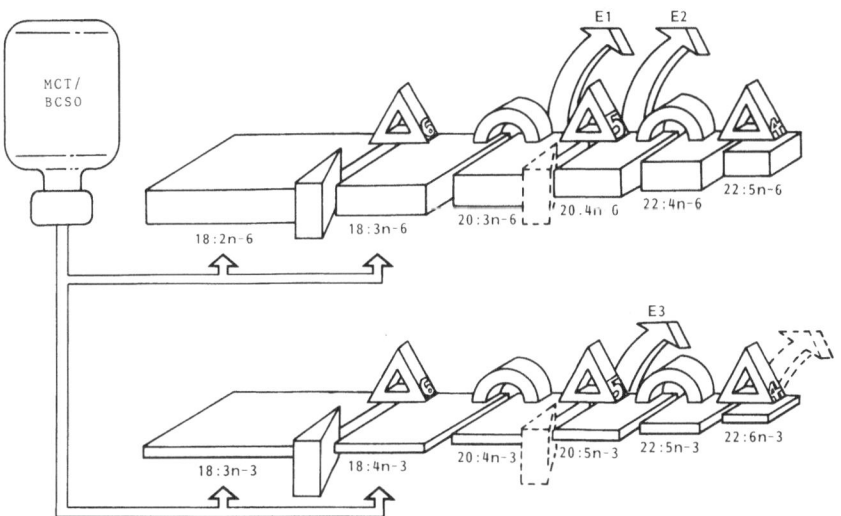

Fig. 5. How to reach an optimal eicosanoids precursors ratio (20:3,n-6 + 20:5,n-3 / 20:4,n-6): optimal amounts of linoleic and α-linolenic acids and their Δ6 desaturase products, γ-linolenic and stearidonic acids.

10% BCSO, walnut oil or lard for 40 days.
Livers were frozen clamped and the fatty acid compositions of triglycerides (TG), free fatty acids (FFA), cholesterol esters (CE), phosphatidylinositol (PI), phosphatidylcholine (PC), phosphatidylserine (PS), phosphatidylethanolamine (PE) and cardiolipin (CL) were determined by gas chromatography.
The 20:3,n-6 to 20:4,n-6 ratio was significantly increased in all classes and particularly in PI and CE in the BCSO groups (figure 6).
A very recent study by Ward et al. reported the antiaggregatory properties of a BCSO emulsion compared to a fish oil emulsion in 10 African green monkeys. After a 6 hour infusion of 125 mg/kg of either GLA or EPA, there was a significant decrease of thromboxane B_2 and of platelet aggregation to ADP (not to collagen) with both emulsions. These studies need further confirmation.

Table 1. BLACKCURRANT SEED OIL COMPOSITION

	% of total fatty acids	Total ω 3	Total ω 6
16:0	6-7		
18:0	1-2		
18:1,n-9 (cis)	9-10		
18:1,n-9 (trans)	0.5		
18:2,n-6 (cis)	47-49		
18:3,n-6 (cis) (γ-linolenic acid)	15-19	....	...⟩ x 2*
18:3,n-3 (cis)	12-14	15-18	77-87
18:4,n-3 (cis) (stearidonic acid)	3-4		

Ratio ω-3 / ω-6 : 1/5

* linoleic acid equivalent given the transformation rate.

Human studies

A human metabolic study was performed at the Hormel Institute (R. Holman).
A BCSO concentrate was administered to a young male volunteer (aged 25) over 6 weeks. The capsules contained 450 mg of a concentrate (74% GLA, 25% stearidonic acid, ie the same ratio of GLA to stearidonic acid as in the oil).
The following protocol was followed: 3 capsules/day during 20 days (1 g of γ-linolenic acid/day), 10 capsules/day during 6 days (approx. 3 g of γ-linolenic acid/day), washing out period of 12 days, 10 capsules/day during 4 days.
Plasma lipid classes were analyzed and compared to a control population. Table 2 shows results obtained in plasma phospholipids: a significant decrease of palmitoleic acid (16:1,n-7) and linoleic acid, a significant increase of 18:3,n-6 and 20:3,n-6 (100%) and a slight increase of 20:4,n-6 (10%) leading to a very interesting increase of the 20:3,n-6/20:4,n-6 ratio from 0.26 to 0.51. Eicosapentaenoïc acid (20:5,n-3) was slightly but significantly increased.
Figure 7 shows the dose responses of different individual fatty acids in cholesterol esters and phospholipids: 18:3,n-6 is quickly incorporated into all plasma lipid fractions in a dose related manner. Plateau levels appear to be attained in about one week. 18:3,n-6 is also quickly eliminated from all lipid fractions when capsule supplementation is discontinued. It appears to be metabolized, quickly accumulating especially as 20:3,n-6 and to a lesser degree as 20:4,n-6; 18:4, n-3 is incorporated and slightly transformed into long chain derivatives. Similar results are seen with cholesterol esters. Given the very high intakes of γ-linolenic acid, (much higher than the threshold proposed by Willis[32] to avoid an excessive increase of 20:4,n-6, the hypothesis of a relative blocking of Δ5 desaturase due to the composition of the oil should be put forward and further studied.

Given the influence of age on Δ6 desaturase and the possible advantage of supply by Δ6 desaturase products, a clinical trial has been undertaken in elderly patients (Dillon[33]).

Sixty elderly patients under institutional care, aged 61 to 83 years (mean 72 years) were studied for 3 months. Twenty were free of vascular disease on clinical grounds, 20 had vascular disorders such as coronary heart disease, stroke, or peripheral vascular disease, and 20 were senile patients. During the study they took no medication such as antiinflammatory drugs. All the subjects received the same diet, supervised by a dietician. The subjects were randomly assigned either to blackcurrant seed oil supplementation or to a control oil (a mixture of

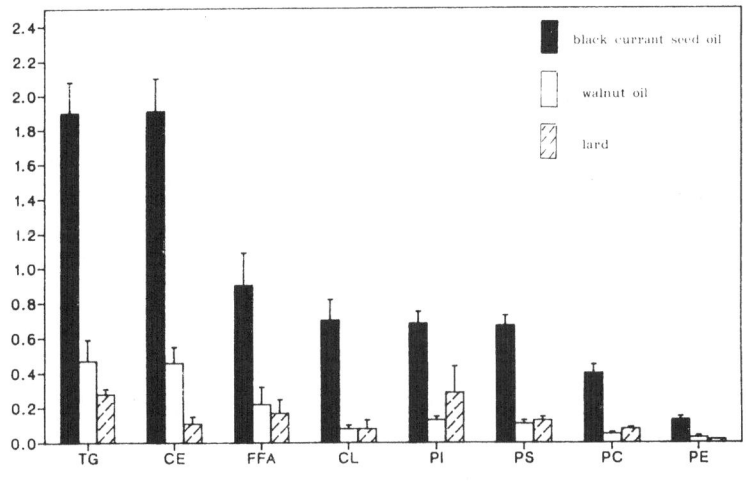

Fig. 6.   20:3,n-6 / 20:4,n-6 ratio in different lipidic classes in the liver after feeding guinea pigs with different oils.

linseed oil and corn oil). Each subject received 10 g of oil daily, thus 1.5 g of γ-linolenic acid in the BCSO group.

Both groups received the same amount of PUFA and the same ratio ω-3 / ω-6 thus the only difference was the γ-linolenic content.

At the end of the trial there was a slight but significant decrease in diastolic blood pressure (from 95 ± 11 mmHg to 83 mm ± 10 mmHg; $p < 0.01$ in the group of elderly subjects who were clinically free of cardiovascular symptoms. Biologically there was a significant decrease of α 2 and β globulins and a slight increase in serum albumin levels.

Several blood lipid fractions and related parameters were measured: total cholesterol; triglycerides; phospholipids; LDL-, HDL-, and very low density lipoprotein (VLDL)-cholesterol, LDL-, HDL-, and VLDL-triglycerides; α-lipoprotein, pre-β-lipoprotein, and β-lipoprotein concentrations. Stable plateau levels were obtained for all these lipid fractions, with the exception of significantly decreased values of VLDL-cholesterol ($p < 0.01$) in the γ-linolenate group, and increased values in the placebo group.

This led us to check the cholesterol lowering effect in well-defined hypercholesterolemic patients, in collaboration with M. Berger: twelve patients, 7 women and 5 men, with type II_A and II_B hyperlipidemia were randomized to receive either BCSO or grape seed oil capsules (GSO) in a double blind study. Five patients received 6 capsules of BCSO daily during 12 weeks, corresponding to 2,7 g of oil containing 450 mg of γ-linolenic acid, 200 ppm ascorbyl palmitate and 600 ppm α-tocopherol. Seven patients received the same amount of grape seed oil

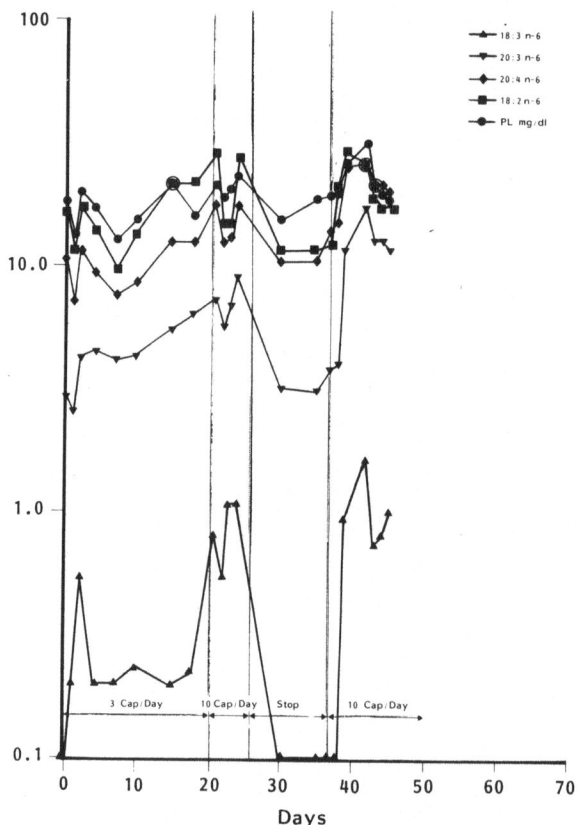

Fig. 7. Variation of FA composition in
PL (percentage of total fatty
acids) - 3 capsules/day
correspond to 1 g of γ-linolenic
per day - 10 capsules to approx.
3 g. From R. Holman.

(with the same antioxidants). The 2 groups were similar in terms of age (means 61 and 55 years respectively) and cholesterol values at the beginning of the study.
Both groups followed dietary advice. Blood samples were taken at days 0,28,42,84. Total cholesterol (mg/dl), HDL-cholesterol (mg/dl) LDL-cholesterol (mg/dl) and triglycerides (mg/dl) were measured.
Statistical analysis were performed using Student's t test.
Total cholesterol decreased significantly after 11 weeks, 8 weeks and 12 weeks in the BCSO groups (from 309.2 to 261.0) but not in the GSO group (from 304.4 to 293.7). The difference between the two groups was significant (p < 0.02).

HDL cholesterol increased significantly already 8 weeks in the BCSO group, the difference between the two groups being significant after 8 and 12 weeks (p < 0.001) (BCSO: from 42.8 to 61.4, GSO from 40,9 to 39.6). LDL cholesterol decreased significantly in both groups after 8 weeks but the variation was much more intense in the BCSO groups. At week 12 the difference was significant (p < 0.02) (figure 8).

Table 2.   METABOLIC STUDY

Variations of fatty acids composition of human plasma lipids
FA concentrations are expressed as percent of total fatty acids
p < * 0,05, ** 0.01, *** 0.001

	PL		FFA		TG		CE	
	CONTROL	TREATMENT	CONTROL	TREATMENT	CONTROL	TREATMENT	CONTROL	TREATMENT
20:3 n-6 DHLA	3.83	8.35***	0.48	0.93***	0.38	1.60***	0.88	1.97***
20:4 n-6 AA	14.92	16.30*	1.29	1.61*	1.49	3.12***	9.68	11.75*
20:3 n-6 / 20:4 n-6	0.26	0.51***	0.02	0.58***	0.25	0.51***	0.09	0.17***
20:5 n-3	0.68	0.94*	0.00	0.00	0.27	0.58**	0.65	0.91*
22:5 n-3	1.28	1.52*	0.19	0.87**	0.50	0.90*	0.00	0.18***
22:6 n-3	2.76	2.52	0.15	0.57**	0.19	0.36***	0.43	0.45

p < * 0.05,   ** 0.01,   *** 0.001

Triglycerides were higher in the GSO group at the beginning. There was no significant decrease at day 84 between the two groups.
These results lead to the following LDL/HDL ratios, which decreased from 6.34 to 3.48 in the BCSO group (p < 0,001) and was practically unchanged with GSO.
The usual effect of polyunsaturated oil is to lower both LDL and HDL cholesterol. Evening primrose oil apparently has no effect on HDL cholesterol (D. Horrobin[34,35]), and this is the first time that a vegetable oil is shown to decrease LDL but increase HDL cholesterol.
Further studies are needed to find out which HDL fraction is modified by blackcurrant seed oil but these results already seem very promising.

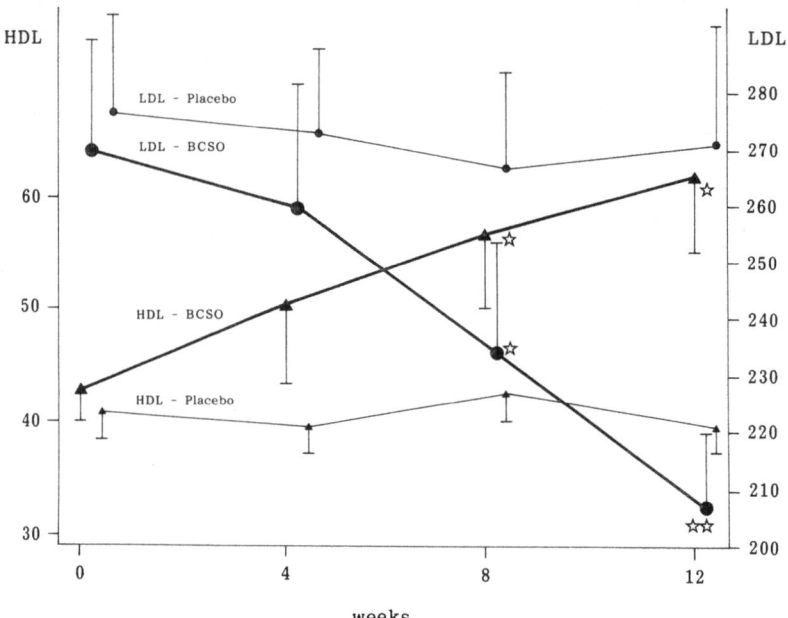

Fig. 8. Blackcurrant seed oil and plasma lipids.
(BCSO n = 5   Placebo n = 7)

In conclusion, the composition of blackcurrant seed oil alone can raise realistic hopes about its potential. The quite interesting n-3 / n-6 ratio, γ-linolenic acid and stearidonic acid, can lead to efficient manipulations of membrane biochemistry and cell-cell inter-actions, and thus pathological conditions like immune or cardiovascular diseases. These preliminary animal and clinical trials seem to vali-date the hypotheses we could suggest after originally analysing the composition of blackcurrant seed oil.

ACKNOWLEDGEMENT

The authors thank Dr. I. Horman for correcting the English text and Georgette Guex for assisting in the preparation of this manuscript.

REFERENCES

1. J. F. Mead, The metabolism of the essential fatty acids, Am. J. Clin. Nutr. 6:656 (1958).

2. R. R. Brenner, M. E. Detomas, and R. O. Peluffo, Effect of poly-unsaturated fatty acids on the desaturation in vitro of linoleic to γ-linolenic acid, Biochim. Biophys. Acta. 106:640 (1965).

3. R. R. Brenner, Regulatory function of Δ6 desaturase-key enzyme of polyunsaturated fatty acid synthesis, in: "Function and Biosyn-thesis of Lipids Advances in Experimental Medecine and Biology", Plenum Press ed., (1983).

4. Y. L. Marcel, K. Kristiansen, and R. T. Holman, The preferred metabolic pathway from linoleic acid to arachidonic acid in vitro, Biochim. Biophys. Acta. 164:25 (1968).

5. H. Sprecher, and L. Chui Long, A reevaluation of optional pathways for the metabolism of linoleic and linolenic acids, Biochim. Biophys. Acta. 398:113-125 (1975).

6. M. A. Crawford, The relationship of dietary fats to the chemistry and morphological development of muscle liver and brain, Riv. Ital. Sostanze Grasse. 6,9:302-309 (1975).

7. J. M. Iacono, J. F. Muller, and D. C. Zellner, Changes in plasma erythrocytes lipids during short term administration of intravenous fat emulsion and its subfraction, Am. J. Clin. Nutr. 16:165-171 (1965).

8. E. Lebreton, P. Lemarchal, and F. Mendy, Intérêt des acides gras polydésaturés essentiels dans la pratique quotidienne de la médecine d'enfants, Revue de Pédiatrie. 1:43 (1966).

9. A. E. Hansen, Influence of diet on blood serum lipids in pregnant women and newborn infants, Am. J. Clin. Nutr. 15:11-19 (1964).

10. A. Nouvelot, P. Dewailly, and J. C. Fruchart, Etude métabolique de 3 émulsions lipidiques utilisées dans l'alimentation parentérale, Nutr. Métab. 19:307-326 (1975).

11. E. Agradi, E. Tremoli, C. Colombo, and C. Galli, Influence of short term dietary supplementation of different lipids on aggregation and arachidonic acid metabolism in rabbit platelets, Prostaglandins. 16:973-984 (1978).

12. Z. Friedman, C. Fröhlich, Essential fatty acids and the major urinary metabolites of the E-prostaglandins in thriving neonates and infants receiving parenteral fat emulsions, Ped. Res. 13:932-936 (1979).

13. C. A. M. Marra, J. T. de Alaniz, and R. R. Brenner, Dexamethasone blocks arachidonate biosynthesis in isolated hepatocytes and cultured hepatoma cells, Lipids. 21:212-219 (1966).

14. I. N. T. Gomez Dumm de, M. J. T. de Alaniz, and R. R. Brenner, Effect of epinephrine on the oxidative desaturation of fatty acids in the rat, J. Lipid. Res. 17:616-621 (1976).

15. I. N. T. Gomez Dumm de, M. J. T. de Alaniz, and R. R. Brenner, Effect of catecholamines and β-blockers on linoleic acid desaturation activity, Lipids. 13:649-652 (1978).

16. I. N. T. Gomez Dumm de, M. J. T. de Alaniz, and R. R. Brenner, Effect of glucocorticoids on the oxidative desaturation of fatty acids by rat liver microsomes, J. Lipid. Res. 20:834-839 (1979).

17. F. Mendy, J. Hirtz, R. Berret, B. Rio, and A. Rossier, Etudes de la composition en acides gras polydésaturés des lipides sériques de nourrissons soumis à des régimes différents, Ann. Nutr. Alim. 22:264-284 (1968).

18. F. Mendy and D. Spielmann, "De l'acide linoléique à ses dérivés supérieurs: linéaires et cycliques, un chemin difficile, une étape clef pour un équilibre, L'acide γ-linolénique", Sopharga, ed., Paris (1980).

19. F. Mendy, M. Lasserre, P. Piganeau, D. Spielmann and B. Jacotot, Facts about linoleic acid metabolic fate during PUFA rich diets in human, in: "Lipid Metabolism and its Pathology", Halpern MJ, 213-223 (1986).

20. J. P. Blond, P. Lemarchal, and D. Spielmann, Comparative desaturation of linoleic and dihomo-γ-linolenic acids by homogenates of human liver in vitro, CR Acad. Sci (D) Paris. 292 (16):911-914 (1981).

21. M. J. T. Alaniz de, I. N. T. de Gomez Dumm, and R. R. Brenner, Effect of epinephrine, dibutyryl cyclic AMP and dexamethasone on fatty acid Δ5 desaturation activity of isolated cells, An. Assoc. Quim. Argent. 70:815-822 (1982).

22. E. C. Mandon, I. N. T. de Gomez Dumm, and R. R. Brenner, Effect of epinephrine on the oxidative desaturation of fatty acids in the rat adrenal gland, Lipids. 21:401-404 (1986).

23. E. E. Mandon, I. N. T. de Gomez Dumm, H. de Alaniz, A. Marra, and R. R. Brenner, ACTH depresses Δ6 and Δ5 desaturation activity in rat adrenal gland and liver, J. Lipid. Res. 28:1377-1383 (1987).

24. S. El Boustani, B. Descomps, L. Monnier, J. Warnant, F. Mendy, and A. Crastes de Paulet, In vivo conversion of dihomo-γ-linolenic acid into arachidonic acid in man, Prog. Lipid. Res. 25:67-72 (1986).

25. M. Lasserre, F. Mendy, D. Spielmann, and B. Jacotot, Effects of different dietary intake of essential fatty acids on 20:3 ω6 and 20:4 ω6 serum levels in human adults, Lipids. 20,4:277-233 (1985).

26. K. S. Bjerve, I. Lovoldmostad, and L. Thoresen, α-linolenic acid deficiency in patients on long term gastric tube feeding, Estimation of linolenic acid and long chain unsaturated ω3 fatty acid requirement in man, Am. J. Clin. Nutr. 45:66-77 (1987).

27. H. Traitler, H. Winter, U. Richli, and Y. Ingenbleek, Characterization of γ-linolenic acid in ribes seed, Lipids. 19,12:923-928 (1984).

28. H. Traitler, and H. Winter, Fatty acid patterns in organ lipids in response to dietary blackcurrant seed oil rich in γ-linolenic acid, Prog. Lipid. Res. 25:255-261 (1986).

29. H. Traitler, U. Richli, A. M. Kappeler, H. Winter, R. Munoz-Box, N. C. Fournier, Blackcurrant seed oil and prostaglandin development in rat heart tissues, in: "Dietary ω3 and ω6 Fatty Acids: Biological EFfects and Nutritional Essentiality", C. Galli, Plenum Press, New York. (in press).

30. G. L. Crozier, M. Fleith, P. A. Finot, Effect of feeding blackcurrant seed oil on fatty acid composition of lipid classes in the guinea pig liver, Internat. J. Vit. Nutr. Res. 57;343 (1987).

31. M. Ward, T. Pavlina, M. Truckenbrod, W. B. Rowe, and R. Cotter, Comparison of the metabolism and platelet aggregation effects of IV-lipid emulsions rich in n-3 and n-6 lipids in African green monkeys, Am. J. Clin. Nutr. 47,4:770 (1988).

32. A. L. Willis, K. Comai, D. C. Kuhn, and J. Paulsrud, Dihomo-γ-linolenate suppresses platelet aggregation when administered in vitro or in vivo, Prostaglandins. 8:509-519 (1974).

33. J. C. Dillon, Essential fatty acid metabolism in the elderly: effects of dietary manipulation, in: "Lipids in Modern Nutrition", Raven Press. New York.

34. D. F. Horrobin, and M. F. Manku, How do polyunsaturated fatty acids lower plasma cholesterol ?, Lipids. 18,8:558-562 (1983).

35. D. F. Horrobin, and Y. S. Huang, The role of linoleic acid and its metabolites in the lowering of plasma cholesterol and the prevention of cardiovascular disease, Int. J. Cardiology. 17,(3):241-255 (1987).

QUANTITATIVE DETERMINATION OF PROSTANOIDS BY STABLE ISOTOPE DILUTION GAS CHROMATOGRAPHY / MASS SPECTROMETRY

H. Traitler, U. Richli, A.M. Kappeler, H. Winter
R. Munoz-Box and N. Fournier

Nestlé Research Centre, Nestec Ltd.
Vers-chez-les-Blanc,
CH-1000 Lausanne 26 (Switzerland)

1.   INTRODUCTION

Since a couple of years there is increasing interest in particular fatty acids first of the n-6 series and later on also of the n-3 series. It was generally accepted that fatty acids such as γ - linolenic acid, stearidonic acid, eicosapentaenoic acid and docosahexaenoic acid have important implications in the general context of health, either prevention or intervention. Lipids, from a quantitative point of view but more and more from a qualitative point of view have become a centre of focus and play an important role in the strategy "health by nutrition".

With our findings on black currant seed oil (BCO)[1] and with all present knowledge or pretended knowledge on fish oils it was clear that we also became interested in the biochemical and nutritional implications of dietary fatty acids. It was also clear that biochemical parameters which were of interest were not only to know about the fate of the fatty acids as such in various tissues[2] but also to have as close as possible insight into the metabolic scenario and in particular into the products which were generated by the cyclooxygenase pathway. An animal feeding trial (rats) was set up by one of us which was initially aimed towards other objectives in which heart homogenates were analysed for some prostaglandins. The exact experimental parameters of this trial will be discussed later on. In the context of prostanoïds this feeding trial which compared dietary oils in the diet (sunflower (SFO) and black currant seed) at two different levels (5 % and 20 %, respectively) aimed at the hypothesis that oils containing γ-LA such as BCO would possibly modulate the PGE_1/PGE_2 ratio with all its biological consequences by increasing PGE_1 in the concerned tissues.

2.   EXPERIMENTAL PART

2.1 All experiments were carried out with male Sprague-Dawley rats which we received at the age of 21 days. They had a mean body weight of 50g and they were divided into 4 groups receiving the following diet:

Diet composition (WT %)

Diet	Oil	Starch	Casein	Cellulose	Vitamins	Minerals
SFO-R5 BCO-R5	5	60	25	3.75	1.25	5
SFO-R20 BCO-R20	20	38.79	29.4	4.38	1.44	5.88

SFO: Sunflower oil; BCO: black currant seed oil, from Nestlé, Switzerland. Casein (according to Hammarsten) was obtained from Merck, Germany. Cellulose, sold as Solka Floc BW-100, was from Christ, Aesch Switzerland. Vitamin diet fortification mixture to which 4.1 % choline bitartrate was added was from Eurobio, Paris. Minerals were Salt Mix XVII from ICN, Nutritional Biochemical, Cleveland OH. Starch was from corn.

After 30, 122, 255, 427 and 560 days of feeding with the four different diets as described above, 3 rats of each group were sacrificed. After rapid excisison on ice, the hearts were weighed, cut into small pieces and subsequently homogenized (Polytron) with 10 volumes of a 0.025 M barbital buffer pH 8.4.

All homogenates were then stored at - 22°C.

2.2 EXTRACTION OF HOMOGENATES[3]

2.2.1 The homogenate is acidified with citrate/HCl-buffer to around pH 3.5 and 20 μ l of an internal standard is added. This standard contains 20 ng of each: 6-keto-PGF$_1$ α -d$_4$, PGE$_2$-d$_4$ and PGF$_2$ α -d$_4$. The homogenate is thouroughly mixed and then centrifuged for 5 min. at 2000 rpm.

2.2.2 PREPARATION OF RP-18 CARTRIDGES

All manipulations can be done under suction on a manifold. Pass 6 ml of methanol, immediately followed by 6 ml of water; before the water meniscus reaches the bottom add the above (2.2.1) prepared sample, wash with 6 ml of water and elute without letting the cartridge run dry with 4 ml ethylacetate.

Collect the ethylacetate phase in a conical vial (polypropylene), let the possibly present water separate and suck it out with a syringe; centrifuge again at 2000 rpm for 5 min and separate residual water as above. Blow off ethylacetate with nitrogen or by vacuum centrifugation and take up the sample in 150 μ l of solvent mixture II (for details see paragraph 2.5).

2.5.3 CLEAN-UP WITH CC4-CARTRIDGES

In order to further clean up dirty samples but also for a possible preseparation of fatty acids, hydroxy fatty acids and prostanoids the above prepared sample (2.2.2.) may be eluted through acidic silica gel cartridges (CC4-material). For this purpose prepare a pasteur pipette in such a way that a small amount of prewashed cotton (as described under 2.5) or silanized glass wool - the latter has turned out to give more reliable results - is put into the pipette followed by 0.3 g acidic silica gel (CC4) material. The column is prewashed with 12 ml of solvent II (described under 2.5). Sample (2.2.2) is passed through this column which is subsequently washed twice with 150 μ l solvent II which has been used to rince the sample vial[4,5]. The prostanoid fraction is eluted with 8 ml of solvent mixture III. The solvents are evaporated (see 2.2.2) and the sample is now ready for derivatization.

2.3   DERIVATIZATION

2.3.1 Esterification of the carboxyl group

To   the dry sample 10 μ l of diisopropyl ethylamine and 40 μ l of a 7 % pentafluorobenzylbromide solution in acetonitrile are added. After shaking well the mixture is left at 40°C during 10 min. All solvents are evaporated[6].

2.3.2 METHOXIMATION OF THE CARBONYL GROUP

70 μ l of a 2 % solution of 0-methylhydroxylamine hydrochloride in pyridine (MOX) is added, shaken well, and left at ambient temperature overnight. The mixture is then evaporated to dryness[7].

2.3.3 INTERMEDIARY CLEAN-UP

The sample is taken up in 1 ml of water and extracted twice with 1 ml of hexane. The hexane is evaporated and the sample is thoroughly dried[8].

2.3.4 SILYLATION OF HYDROXY GROUPS

50 μ l of a 2:1 solution of NO-bis(trimethylsilyl)trifluoroacet-amide (BSTFA) in pyridine are added to the sample. After shaking well the mixture is left at room temperature for 1 hour. All solvents are evaporated and the sample is taken up in 25 μ l n-decane. The sample is ready for injection into the GC either in on-column or splitless mode.

2.5   MATERIALS AND SOLVENTS

2.5.1   RP-18 cartridges
        Baker - 10 SPE No. 7020-6

2.5.2. Baker 10 extraction system

2.5.3   75 ml reservoirs Baker No. 7120-3 all from J.T. Baker Chemical Co., g, NJ 08865

2.5.4   CC4 acidic silica gel
        Mallinckrodt No. 7086

2.5.5   Titrisol buffer pH 3.0 citrate/HCl, ready for use, Merck No. 9883

2.5.6   Phosphate/citrate buffer pH 3.8 for the deactivation of the cotton-wool; 13.68g citric acid monohydrate and 12.42g $Na_2HPO_4.2 H_2O$ are diluted to 1000 ml. The cotton-wool is put into the buffer for ca. 10 min and then dried at 120°C for 30 min.

2.5.7   Pasteur capillary pipettes short size 150 mm

2.5.8   Solvent I: diethyl ether: petrol ether 25:75 (v/v)

2.5.9   Solvent II: diethyl ether: petrol ether 75:25 (v/v)

2.5.10 Solvent III: ethyl acetate: methanol 90:10 (v/v)

2.5.11 Pentafluorobenzyl bromide (PFBB) 7 % in acetonitrile

2.5.12 N,N-Diisopropylethylamine Fluka No. 03440

2.5.13 0-Methyl hydroxylamine hydrochloride 2 % in pyridine, MOX, Pierce Nr. 45950

2.5.14 N,0-bis(trimethylsilyl)trifluoroacetamide   2:1   in   pyridine   (v/v) BSTFA; Pierce No. 38827.

2.6.   GC/MS

2.6.1   CAPILLARY GAS CHROMATOGRAPHY

        Gas chromatography was carried out on a Hewlett Packard (HP) 5880A. Injections were made in the splitless or on-column mode. Helium was used as carrier gas. The used columns were directly coupled to a deactivated fused silica interface capillary with an internal diameter of 0.16mm (SGE Infochroma AG, Zug) and a length of about 70cm. This interface capillary was introduced directly into the mass spectrometer so that the end of the column was within 5mm of the ion source. This interface acts as a restriction and a relatively high pressure of 30 psi was applied in the injector in order to have acceptable chromatographic conditions with a gas flow of about 1.5ml/min into the mass spectrometer. With these conditions the pressure at the end of the chromatographic column is slightly higher than atmospheric pressure. The interface temperature was kept between 250 to 280°C.

        Conditions for splitless injection:
        Just before injecting 1 to 4 μ l of the sample the split valve was closed. 1 minute after the injection the valve was opened again.
        Injector temperature : 280°C

        Capillary column: DB-5 (J&W) 15m, 0.32mm I.D., 0.25 micron film thickness

        Oven temperature program :
        Oven temperature initial value : 160°C
        Oven temperature initial time   : 1 '
        Oven temperature program rate   : 30°/'
        Oven temperature final value   : 250°C
        Oven temperature final time   : 0 '
        Oven temperature 2 program rate: 5°/'
        Oven temperature 2 final value : 300°
        Oven temperature 2 final time : 0 '

2.6.3 MASS SPECTROMETRY

An AEI (KRATOS) MS-30 with a DS-55 data system (Data
General NOVA 3, preprocessor interface) was used.
This double focusing double beam instrument was equipped
with an EI/CI source and negative ion facilities.
Ion source temperature : 200°C

MIS mode (NICI):
mass range   : 512-630 AMU
masses   : 594.9663 (Lock mass)
524.3228
526.3384
528.3478
612.3572
614.3729
618.3979
electron energy   : 100 eV
emission current   : 2.5 mA
reagent gas   : CH_4 OR NH_3

3.   DISCUSSION AND RESULTS

3.1   RESULTS OF MEASUREMENTS IN THE HOMOGENATE SAMPLES

In the four different groups one can try to interpret the results
basically from two aspects:

a)   influence of the diet (sunflower oil or black currant seed oil diet)
and

b)   influence of the dose of each diet (5% or 20%)

These two aspects then can again be subdivided into the various time
intervals at which the samples were taken. This latter aspect can
obviously also be related to the capacity of the organ to produce prostaglandins either under normal physiological conditions or upon stimulation
by stress as in our case (sacrifice of the animal). Again it must be
emphasized that during sampling no further precautions like addition of
cyclooxygenase inhibitors were undertaken. Yet, due to the fact that all
manipulations have been carried out under strictly identical conditions we
can assume that comparative interpretation of the results is legitimate.
At least until the contrary is proved in a further study.

3.1.1 INFLUENCE OF THE DIET

If we regard at the 5% fat level, the two different diets BCO and
SFO, respectively show differences which are more or less significant.
For PGE_1 the differences are statistically significant at 255 days (at the
5% level) and at 427 days (at the 1% level) but neither at 122 nor at 560
days, respectively.

PGE_2 at no time shows any diet dependent significant difference
whereas 6-keto-PGF_1 α again shows significant differences (at the 1%
level) at 255 and 427 days.

It is interesting to observe that there are significant differences due to the diet as far as PGE_1 is concerned but there are no significant differences on the PGE_2 level.

This suggests that the dietary administration of γ-linolenic acid gives rise to an increase in dihomo-γ-linolenic acid and to PGE_1 synthesis whereas arachidonic acid levels are hardly influenced thus reflecting no enhancement of PGE_2 production.

3.1.2 TIME EVOLUTION

The time evolution over the whole assay shows a clear general decreasing tendency with increasing time of diet. This effect could be due to increasing age of the animals and further experiences are necessary. The Prostacyclin metabolite shows again a different behaviour.

The statistical treatment shows that the errors produced by the GC/MS method measurments are in the order of .1 to 17% for quantities bigger than 1ppb. For smaller quantities the error rapidly increase.

3.1.3 INFLUENCE OF DOSE

Comparing the two doses 5% and 20%, respectively shows significant differences for PGE_1, PGE_2 and 6-keto-PGF_1 $_\alpha$ at 427 days (5% level) and for 6-keto-PGF_1 $_\alpha$ at 255 days (at the 1% level).

In the case of PGE_1 this dose dependence in very pronounced for BCO at 255 days and 427 days, respectively, whereas at 560 days there seems to be no dose response at all. This dose response cannot be seen with sunflower oil; irrespective of the dose there are always low levels of PGE_1 in this diet group.

For PGE_2 all groups in both diets and at all time intervals respond positively to an increased dose with the exception of the BCO diet group at 560 days which even decreases (significant at the 5 % confidence level) slightly at the higher dose.

As far as 6-keto-PGF_1 $_\alpha$ (PGI_2) is concerned there is very little dose dependence for the black currant seed oil. Pronounced influence can be observed for the sunflower oil where at 255 days and at 427 days there is a marked decrease for the higher dose whereas at 560 days there is an increase with the higher dose.

The following table (Tab. 1) and figures (Fig. 1-6) give an overview on the development of the three prostanoids during the study.

Note: In Table 1 and Figs. 1-6:
S ... sunflower seed oil
C ... black currant seed oil

Table 1

		122 days		255 days		427 days		560 days	
		5%	20%	5%	20%	5%	20%	5%	20%
PGE_1	S	0.0		3.57	10.30	1.04	4.53	.03	.48
		0.0		1.20	1.55	.88	1.23	.01	.68
	C	32.64	73.89	24.68	60.41	12.21	43.67	5.07	2.75
		10.24	0.0	7.91	41.38	2.28	16.69	4.71	.64
PGE_2	S	378.26		81.56	92.86	17.98	74.56	6.34	37.43
		76.47		8.21	9.56	.68	22.73	.57	13.22
	C	170.24	154.04	63.93	85.51	43.50	86.15	16.54	11.80
		75.60	0.0	7.24	14.50	18.35	23.30	10.52	5.69
PGI_2	S	48.27		383.31	111.32	256.43	149.88	128.99	187.85
		5.46		90.50	33.37	39.42	33.77	8.36	64.39
	C	41.00	44.21	123.88	77.99	134.36	140.62	127.92	105.16
		7.44	0.0	16.22	15.04	10.40	14.43	52.92	7.29

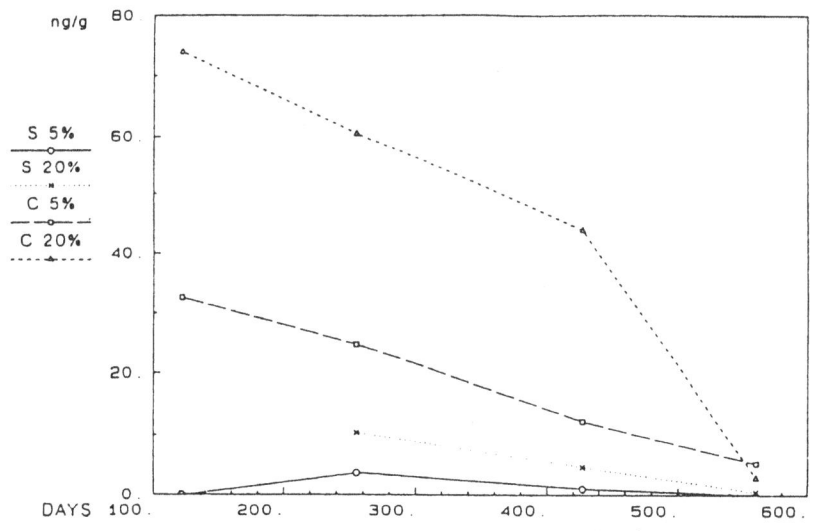

PGE 1 (MASS 526)

Figure 1. Evolution of PGE_1 with time.

329

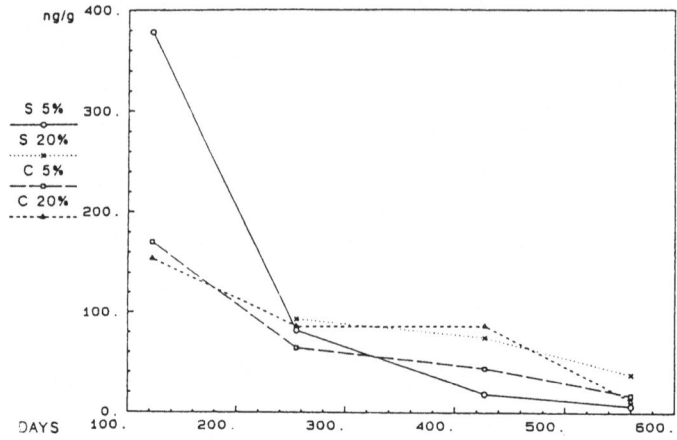

Figure 2. Evolution of 6 keto-PGF$_{1\alpha}$ with time.

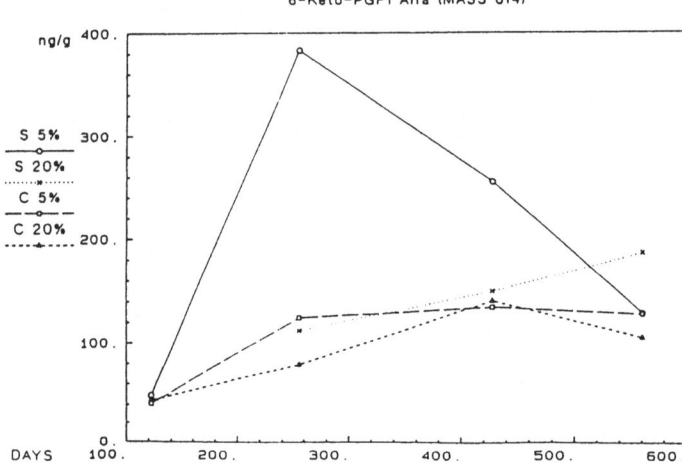

Figure 3. Evolution of PGE$_2$ with time.

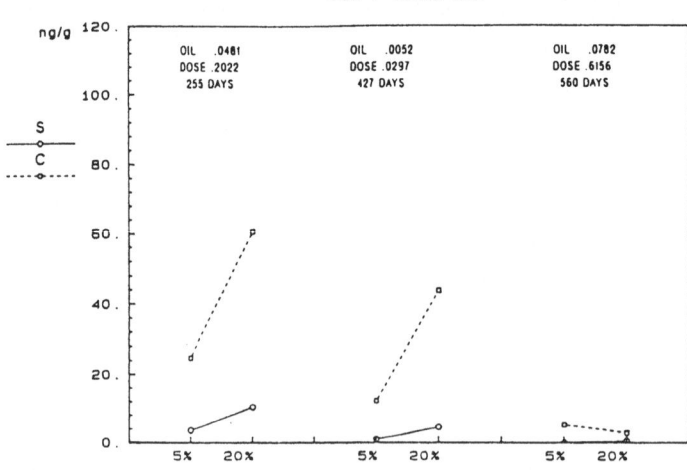

Figure 4. Dependence of oil and dose in the diet on the evolution of PGE$_1$

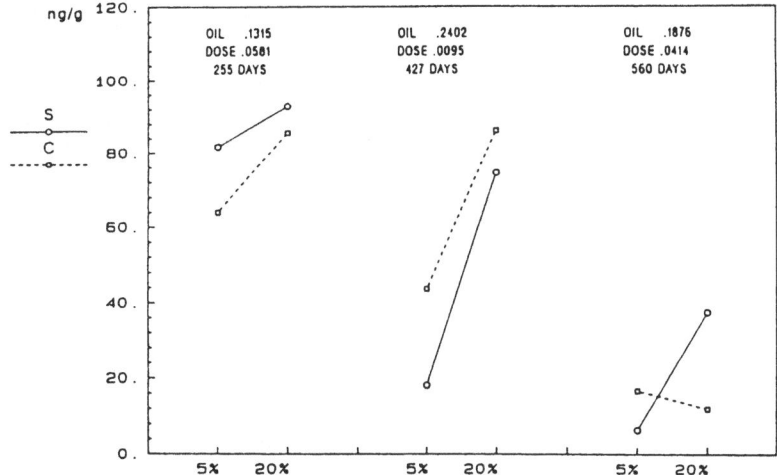

Figure 5. Dependence of oil and dose in the diet on the evolution of PGE_2

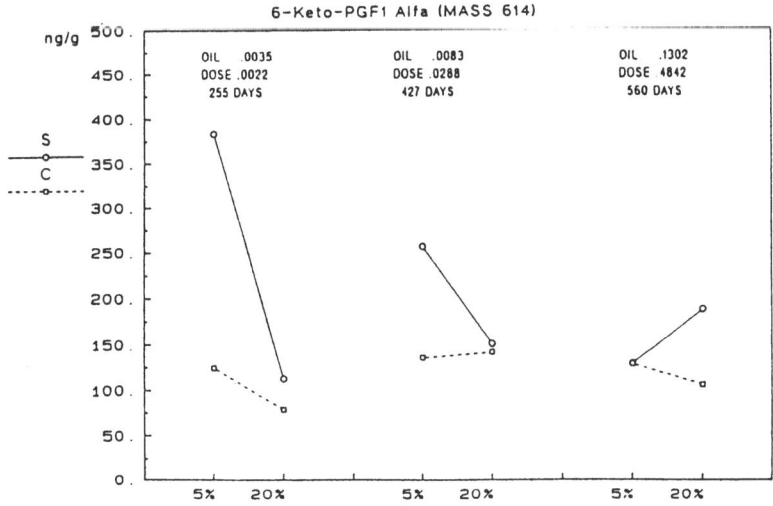

Figure 6. Dependence of oil and dose in the diet on the evolution of 6 Keto- $PGF_{1\alpha}$

4. CONCLUSION

1) Dietary administration of black currant seed oil containing γ - linolenic acid in rats compared to another polyunsaturated fatty acid oil such as sunflower seed oil leeds to an increased ratio PGE_1/PGE_2.

2) Overall abundances of PGE_1 in the BCO group and PGE_2 in both groups are significantly reduced with increasing age of the animals.

3) GC/MS with isotope dilution is an appropriate method for the quantitative determination of prostanoids on a physiological level in biological specimen. Much care has to be taken during sampling and work up of the samples in order to have control over the labile prostaglandins in the biological environment and to keep recovery high during the various work-up and derivatization procedures.

5.   BIBLIOGRAPHY

1.   Traitler, H., Winter, H., Richli, U. and Ingenbleek, Y., Lipids 19
      (1984) 223-228.

2.   Traitler, H., Richli, U., Kappeler, A.M. and Winter, H., in Biology
      of Icosanoids and Related Substances in Blood and Vascular Cells, M.
      Lagarde Ed., Editions Inserm Vol. 152, Paris 1987, p. 129-142.

3.   Traitler, H. and Winter, H., Progr. Lipid. Res. 25 (1986) 255-261.

4.   Mayer, B. (1985) private communication.

5.   Vesterquist, O. and Gréen, K., Prostaglandins (1984) 139-154.

6.   Waddell, K.A., Wellby, J. and Blair, I.A., Biomed. Mass Spectrom. 10
      (1983), 83-88.

7.   Gréen, K., Chem Phys. Lipids 3 (1969), 254-272.

8.   Gleispach, H. (1985) private communication.

LINOLEIC ACID AND EPIDERMAL WATER BARRIER

Harald S. Hansen

Biochemical Laboratory, Royal Danish School of Pharmacy
Universitetsparken 2, DK-2100 Copenhagen Ø
Denmark

INTRODUCTION

The (n-6)-fatty acids appear to be essential dietary components for mammals and birds (Holman, 1971; Mead and Fulco, 1976), whereas their nutritive importance in fish (Greene and Selvivonchick, 1987) and insect nutrition (de Renobales et al., 1987) are not yet clear. Some insects seem to be able to grow and develop without any polyunsaturated fatty acids'(Sohal et al., 1984; Rapport et al., 1984).

When Burr and Burr (1929; 1930) discovered a fat-deficiency disease in laboratory rats which could be cured by dietary linoleic acid, some of the disease symptoms were associated with the skin, i.e. scaliness of the skin, dandruff, and loss of yellow pigmentation. Another deficiency symptom, increased water consumption, was later found to be caused by an excessive transepidermal water loss (Sinclair, 1952). The excessive evaporation of water over the skin may also account - at least in part - for the decreased growth seen in essential fatty acid-deficient rats (Houtsmuller, 1975; Hansen, 1986).

EPIDERMAL STRUCTURE AND LIPIDS OF TERRESTRIAL MAMMALS

An important function of the epidermis is to generate a tough sheet that protects the organism from desiccation and external assault. Histological studies of mammalian epidermis indicate a structure in which cells in the basal layer divide and provide several layers of living cells, which migrate towards the skin surface, where they become keratinized and finally form the dead, flattened cells of the horny layer. Named from the lowest part of the epidermis the cells are histologically divided into strata: stratum basale, stratum spinosum, stratum granulosum, stratum lucidum, and stratum corneum. It is now generally accepted that the stratum corneum contains the most efficient barrier property of the different epidermal strata (Scheuplein, 1976; Elias, 1981; Wertz and Downing, 1982; Bowser and White, 1985) and that the permeability barrier is primarily confined to neutral lipids in the intercellular space between the cornocytes in the stratum corneum (Elias, 1981; Wertz and Downing, 1982; Landmann, 1984; Golden et al., 1987). Elias (1983) has suggested that the permeability barrier may be compared to a brick wall, where the horny cells represent the bricks and the extracellular lipids represent the motar. The lipid composition of the epidermis changes from the stratum basale to the stratum corneum in association with migration and differentiation of the epidermal cells, i.e. in the stratum corneum phospholipids are virtually absent, and ceramides, cholesterol and free fatty acids are the major lipids (Yardley and Summerly, 1981; Wertz et al., 1987). The ceramides are important constituents in the extracellular membrane sheets which seem to constitute a major part of the barrier function (Elias, 1981; Wertz and Downing, 1982). Gray et al. (1978) isolated an acylglucosylceramide from pig and human epidermis in which linoleate amounted to 77% and 56% of the esterlinked fatty acids, respectively. Gray et al. (1978) proposed that linoleate was esterified to the glucosyl-group, but more recent studies have shown that it is esterified to the long chain hydroxy fatty acid in amide linkage with sphingosine (Bowser et al., 1985; Abraham et al., 1985).

Glucosylceramides including acylglucosylceramide are found especially in the stratum granulosum, whereas they are less abundant in the stratum corneum (Yardley and Goldstein, 1976; Cox and Squier, 1986). Two other linoleate-containing lipids, which may be metabolites og acylglucosylceramide, have been found in epidermis, i.e. acylceramide (Wertz and Downing, 1983) and acyl-acid (Bowser et al., 1985) (Fig. 1). In stratum spinosum and stratum granulosum the cells contain many lamellar bodies (also called Odland bodies, keratinosomes or membrane coating granules) which contain polysaccharides, hydrolytic enzymes, and neutral lipids (Elias, 1981). The lipids within the lamellar bodies appear to be arranged in stacks of membraneous disks, which are assumed to be arranged into broad extracellular bilayers after secretion of the lamellar bodies (Elias, 1981; Wertz and Downing, 1982; Landmann et al., 1984). Available data indicate that acylglucosylceramide is formed in the lamellar bodies, and that during or after secretion acylglucosylceramide is hydrolyzed to acylceramide and

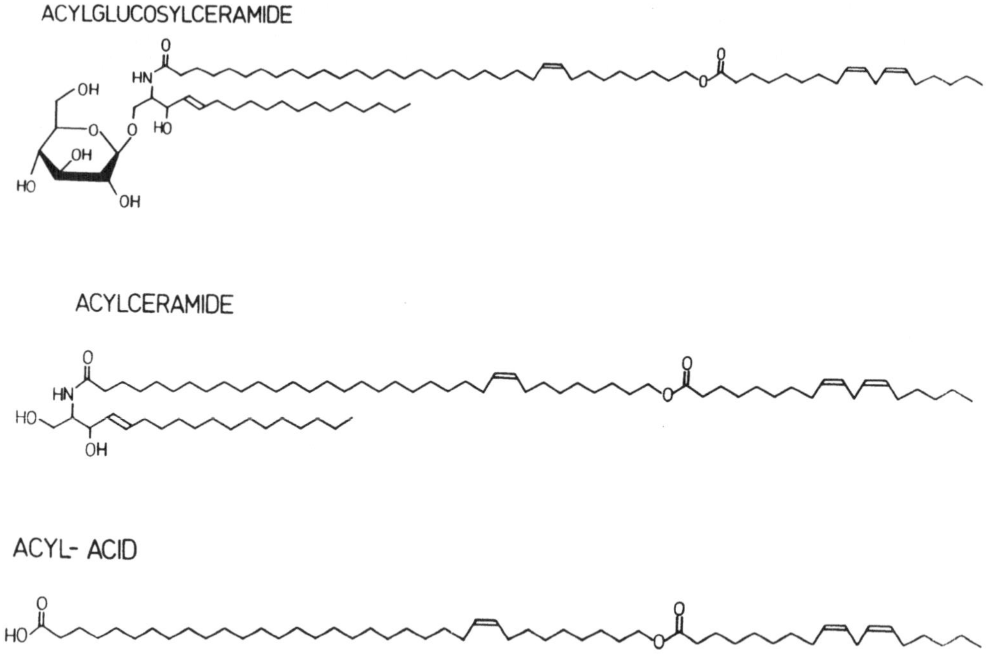

ACYLGLUCOSYLCERAMIDE

ACYLCERAMIDE

ACYL- ACID

Fig. 1   Structure og linoleic acid-rich epidermal lipids

and acyl-acid (Nugteren et al., 1985). Nugteren has also suggested that linoleate must be oxygenated by epidermal lipoxygenases in order to fulfil its role in the barrier (Nugteren et al., 1985; Nugteren and Kivits, 1987). However, for such a role of lipoxygenases evidence is not obvious from his data nor from the litterature references presented (Nugteren et al., 1985). On the contrary Elias et al. (1980) have reported that intraperitoneal injections of a well-known lipoxygenase inhibitor, eicosatetraynoic acid, did not inhibit the formation of the epidermal water permeability barrier. Nugteren has provided this author (October 1987) with unpublished data produced by U. M. T. Houtsmuller and A. van der Beek, demonstrating that eicosatetraynoic acid applied to the skin of essential fatty acid-deficient rats delayed or prevented the decrease in transepidermal water loss induced by either topical linoleate or dietary sunflower seed oil. However, nor-dihydroguaiaretic acid, which is also a wellknown inhibitor of lipoxygenases, did not prevent the effect of topical linoleate (U. M. T. Houtsmuller and A. van der Beek, unpublished). It is unknown whether nordihydroguaiaretic acid can inhibit epidermal lipoxygenases to the same extent as eicosatetraynoic acid. Thus, there is some evidence in favour of involvment of lipoxygenases as being important in formation of the barrier, but this has to be more substantiated.

In humans the epidermal water permeability barrier is formed around term, and preterm babies have a very high transepidermal water loss associated with a considerable heat loss (Hammarlund et al., 1977; Hammarlund et al., 1986). The newborn mouse has a low linoleate percentage (10%) in acylglucosylceramide. The percentage increases to 45% within 50 days (Wertz and Downing, 1986). Although the data above originate from two different species, respectively, it may be hypothesized that the high transepidermal water loss in preterm infants is caused by a low amount of linoleate-containg ceramides in the epidermis.

The skin surface consists of keratinized epithelium. In the mouth there is keratinized and non-keratinized eithelia. The latter is the most permeable (Squier and Hall, 1985). Lipid analysis of oral epithelia revealed that the keratinized epithelia, gingiva and palate epithelium, were more similar to the epidermis, i.e. containing acylglucosylceramide and acyl-ceramide, whereas the non-keratinized epithelia, floor of mouth and buccal epithelium, contained no acylated ceramides, but a high content of glucosylceramides (Wertz et al., 1986). These results support the view, that acylglucosylceramide and acylceramide are important for water barrier function. Free fatty acids are a major lipid class in stratum corneum, but analyses of fatty acid composition of these fatty acids have shown that linoleic acid amounts to a maximum of 12.5%, and that the remaining acids seem to be saturated and monoene fatty acids (Yardley and Summerly, 1976; Lampe et al., 1983; Bowser et al., 1985). In its free form linoleic acid does not appear to be important to the barrier function, since topical application of linoleic acid to the skin of essential fatty acid-deficient rats only cured the transepidermal water loss after a lack-period of several days (Houtsmuller and van der Beek, 1981). However, fatty acids may be important for the formation of lamellar structures in epidermis (Friberg and Osborne, 1985). Preliminary studies of epidermal lipids of marine mammals, a harbor porpoise and a West Indian manatee, have indicated similarities as well as dissimilarities with lipids of terrestrial mammals, i.e. acylglucosylceramides were found in the manatee but not in the porpoise (Elias et al., 1987).

Two of the symptoms of essential fatty acid-deficiency are increased transepidermal water loss and scaliness of the skin. Reversal of these two symptoms by feeding essential fatty acid-deficient rats a linoleic acid-rich diet revealed a difference in time scale, i.e. transepidermal water loss was corrected much faster than scaliness of the skin (Basnayake and Sinclair, 1956). Eliott et al. (1985a; 1985b) found that topical application of columbinic acid (5t,9c,12c-18:3) to the skin of essential fatty acid-deficient rats decreased transepidermal water loss as well as the degree of scaliness. The 13-hydroxy-derivative of columbinic acid only decreased scaliness, whereas transepidermal water loss was unchanged. This indicate that these two deficiency symptoms may occur as two independent phenomena.

LINOLEIC ACID ESSENTIALITY

A linoleate-rich ceramide identified in pig and human epidermis (Gray et al., 1978) has been suggested to be associated with the defect of the epidermal water permeability barrier seen in essential fatty acid-deficient animals (Elias et al., 1980; Yardley and Summerly, 1981). Previously, Hartop and Prottey (1976) and Prottey (1977) had reported that topical application of arachidonate as triglyceride as well as intraperitoneal injection of arachidonic acid did not decrease transepidermal water loss in essential fatty acid-deficient rats, only linoleate could do this. This suggested specificity of linoleate has later proved to be wrong (Houtsmuller and van der Beek, 1981; Hansen and Jensen, 1985). Further, it was not in accordance with the fundamental results of Thomasson (1953, 1962) that all the (n-6)-fatty acids could increase growth in water-restricted essential fatty acid-deficient rats, nor with the general view among nutritionists that arachidonic acid is the essential metabolite, and linoleic acid, the essential nutrient, serves as precursor for arachidonic acid (Gurr, 1984). Ziboh and Hsia (1972) reported that topical application of prostaglandins to the skin of essential fatty acid-deficient rats could alleviate the scaliness. However, the application of prostaglandins could not improve the defective water barrier (Prottey, 1977). Administration of cyclooxygenase inhibitors (Elias et al., 1980) to essential fatty acid-deficient rodents had no effect on the recovery of the barrier function following supplementation with a linoleic acid-rich diet. All these results, i.e. 1) the erroneous claim of specificity of linoleate versus arachidonate for correction of the epidermal water barrier; 2) the lack of effect of cyclooxygenase and lipoxygenase inhibitors on recovery of barrier function; and 3) the discovery of linoleate-rich acylglucosylceramides in epidermis, were seen as evidence for a specific role of linoleate in epidermal barrier function (Elias et al., 1980; Wertz and Downing, 1982; Elias, 1985). This view was supported by earlier reports by Jelenko et al. (1972a; 1972b), i.e. a lipid extracted from the skin (identified as ethyl linoleate) could retard water loss across experimental wounds. However, data were lacking on direct comparison of fatty acids in epidermal lipids with function of the epidermal water barrier. Wertz et al. (1983) analyzed the epidermal ceramides of essential fatty acid-deficient rats and found that the amounts of acylglucosylceramide and acylceramide increased relative to other ceramides, but linoleic acid was replaced by oleic acid. No arachidonic acid was found in these O-acylated epidermal ceramides (Gray et al., 1978; Wertz et al., 1983), but reports of Houtsmuller and van der Beek (1981) and of Thomasson (1953, 1962) indicate that arachidonic acid could correct the transepidermal water loss in essential fatty acid-deficient rats.

DIETARY FATTY ACIDS AND EPIDERMAL BARRIER FUNCTION

We measured fatty acid composition of epidermal acylglucosylceramide and acylceramide as well as transepidermal water loss in essential fatty acid-deficient rats fed different pure fatty acid esters, i.e. ethyl arachidonate, ethyl linoleate, ethyl oleate, and methyl columbinate (Hansen and Jensen, 1985). The three (n-6)-fatty acids - including columbinic

acid - all decreased evaporative water loss (Fig. 2), and we found high
levels of linoleate in these epidermal ceramides (Hansen and Jensen, 1985).
The rats fed methyl columbinate had columbinic and linoleic acid both in
the O-acylated ceramides (Hansen and Jensen, 1985). This linoleate may
originate either from the dietary columbinate preparation, which contained
3 wt% linoleate, or it could be mobilized from other tissues to the skin
because of the columbinate feeding. O-acylated ceramides from our control
rats contained small amounts of arachidonic acid only (Hansen and Jensen,
1985), as has also been reported in recent papers (Bowser et al., 1985;
Wertz et al., 1986). Feeding ethyl arachidonate to essential fatty acid-
deficient rats resulted in formation of O-acylated ceramides with high
amounts of linoleate and very little arachidonate only (Hansen and Jensen,
1985). A significant part of the linoleate seems to be the result of
retroconversion of some of the dietary arachidonate (Hansen et al., 1986).
Feeding ethyl α-linolenate did not result in incorporation of this acid
into epidermal O-acylated ceramides (Hansen and Jensen, 1985), and the
rats had elevated evaporative water loss (Hansen and Jensen, 1983; Hansen
and Jensen, 1987). This very weak effect of dietary α-linolenate on evaporative water loss is in accordance with results of Thomasson (1953; 1962),
who found α-linolenate to have little growth-promoting activity in water-
restricted essential fatty acid-deficient rats*. However, this weak activity of dietary α-linolenate seems to be contrary to results of Houtsmuller
and van der Beek (1981), who found considerably higher potency of α-linolenate regarding correction of transepidermal water loss after topical

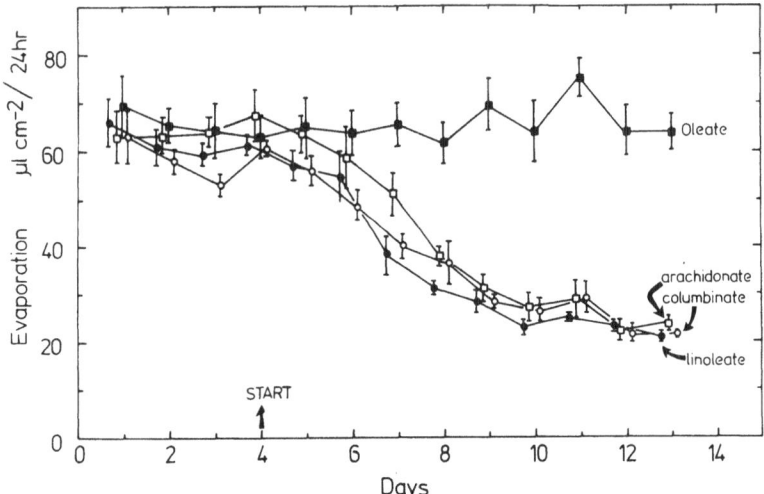

Fig. 2 Evaporative water loss of four groups of essential fatty acid-
        deficient rats, before and during dietary supplementation for ten
        days with fatty acid esters, i.e. ethyl oleate, ethyl arachido-
        nate, ethyl linoleate, and methyl columbinate, respectively.
        (from: Hansen and Jensen, 1985).

---

* α-Linolenate may have a growth-promoting effect which is independent of
  changes in the transepidermal water loss (Hansen and Jensen, 1983).

application. These data may indicate the existence of a selective mechanism for providing dietary fatty acids (linoleate) to the epidermis for synthesis of acylglucosylceramide, a mechanism which may be bypassed by topical applications. A similar observation was made with arachidonic acid. Topical application of arachidonate decreased the elevated transepidermal water loss (Houtsmuller and van der Beek, 1981). Application of $1-^{14}C$-arachidonic acid to the skin of essential fatty acid-deficient rats resulted in incorporation of radioactivity into epidermal O-acylated ceramides, to the same extent as that found with $1-^{14}C$-linoleic acid (Nugteren et al., 1985). Thus, the selectivity of fatty acids for formation of the acylglucosylceramide does not seem to be due to specificity of the acyl-glucosylceramide-forming enzymes, it must in some way be related to a selective availability of fatty acids, i.e. linoleate and oleate, for the synthesis of acylglucosylceramide. A recent report of Melton et al. (1987) confirms that also in young pigs, increased transepidermal water loss is associated with decreased linoleate content of epidermal O-acylated ceramides, apparently with a linear correlation.

CONCLUSION

The available data indicate that the water permeability barrier of mammalian skin may be related to intercellular membraneous lipid layers in stratum corneum of the epidermis. These lipid layers comprise primarily cholesterol, fatty acids, and ceramides, including some unusual linoleate-rich O-acylated ceramides. Linoleic acid is the only polyunsaturated fatty acid found in significant amounts in the O-acylated ceramides. Essential fatty acid-deficient mammals show high transepidermal water loss and the linoleate is replaced by oleate in these unusual ceramides. At present it is unclear whether the function of linoleate, as esterified to the ceramides, is to stabilize the intercellular membraneous structure, or whether linoleate serves as a substrate for epidermal lipoxygenases, resulting in polyoxygenated linoleate-derivatives which are important for the integrity of the water barrier in some unknown way.

There is strong evidence for a direct role of linoleate, i.e. whitout being converted to arachidonate, in formation of the epidermal water permeability barrier. By topical application of arachidonate to rat skin, it appears that arachidonate can substitute for linoleate. The same may be true for α-linolenate when applied to the skin but not when given in the diet. Results from feeding radioactive labelled arachidonate to essential fatty acid-deficient rats indicate, that it can be retroconverted to linoleate. This retroconversion may be the reason for the curative effect of dietary arachidonate on transepidermal water loss of essential fatty acid-deficient rats. It appears resonable to assume that transepidermal water loss in essential fatty acid-deficient rats may in part account for the growth retardation seen in the essential fatty acid-deficient rats, due to excessive loss of energy caused by evaporation. Future studies of lipid metabolism in keratinocyte-cultures (Madison et al., 1988) may help elucidating the role of linoleic acid in formation of the epidermal extracellular lamellar structures.

REFERENCES

Abraham, W., Wertz, P.W., and Downing, D.T., 1985, Linoleate-rich acylglucosylceramides of pig epidermis. Structure determination by proton magnetic resonance. J. Lipid Res., 26:761.
Basnayake, V., and Sinclair, H.M., 1956, The effect of deficiency of essential fatty acids upon the skin, in: "Biochemical Problems of Lipids", G. Popjak and E. Le Bretonm eds.,

Butterworth Scientific Publisher, London.

Bowser, P.A., Nugteren, D.H., White, R.J., Houtsmuller, U.M.T., and Prottey, C., 1985, Identification, isolation and characterization of epidermal lipids containing linoleic acid, Biochim. Biophys. Acta 834:419.

Bowser, P.A., and White, R.J., 1985, Isolation, barrier properties and lipid analysis of stratum compactum, a discrete region of stratum corneum, Brit. J. Dermatol. 112:1.

Burr, G.O., and Burr, M.M., 1929, A new deficiency disease produced by rigid exclusion of fat from the diet, J. Biol. Chem. 82:345.

Burr, G.O., and Burr, M.M., 1930, On the nature and role of the fatty acids essential in nutrition, J. Biol. Chem. 86:587.

Cox, P., and Squier, C.A., 1986, Variations in lipids in different layers of porcine epidermis, J. Invest. Dermatol. 87:741.

de Renobales, M., Cripps, C., Stanley-Samuelson, D.W., Jurenka, R.A., and Blomquist, G.J., 1987, Biosynthesis of linoleic acid in insects, Trends Biochem. Sci. 12:364.

Elias, P.M., 1981, Lipids and the epidermal permeability barrier, Arch. Dermatol. Res. 270:95.

Elias, P.M., 1983, Epidermal lipides, barrier function, and desquamation, J. Invest. Dermatol. 80:44s.

Elias, P.M., 1985, The essential fatty acid deficient rodent - evidence for a direct role of intercellular lipid in barrier function, in: "Models in Dermatology" vol. 1, H.I.Maibach and N.J.Lowe, eds., Karger, Basel.

Elias, P.M., Brown, B.E., and Ziboh, V.A., 1980, The permeability barrier in essential fatty acid deficiency. Evidence for a direct role for linoleic acid in barrier function. J. Invest. Dermatol. 74:230.

Elias, P.M., Menon, G.K., Grayson, S., Brown, B.E., and Rehfeld, S.J., 1987, Avian sebokeratinocytes and marine mammal lipokeratinocytes: structural, lipid biochemical, and functional considerations, Am. J. Anatomy 180:161.

Elliott, W.J., Morrison, A.R., Sprecher, H.W., and Needleman, P., 1985a, The metabolic transformations of columbinic acid and the effect of topical application of the major metabolites on rat skin, J. Biol. Chem. 260:987.

Elliott, W.J., Sprecher, H., and Needleman, P., 1985b, Physiologic effects of columbinic acid and its metabolites on rat skin. Biochim. Biophys. Acta 835:158.

Friberg, S.E., and Osborne, D.W., 1985, Small angle X-ray diffraction patterns of stratum corneum and a model structure for its lipids, J. Dispersion Sci. Technol. 6:485.

Golden, G.M., Guzek, D.B., Kennedy, A.H., McKnie, J.E., and Potts, R.O., 1987, Stratum corneum lipid phase transitions and water barrier properties, Biochemistry 26:2382.

Gray, G.M., White, R.J., and Majer, J.R., 1978, 1-(3´-O-acyl)-β-glucosyl-N-dihydroxypentatriacontadienoylsphingosine, a major component of the glucosylceramides of pig and human epidermis, Biochim. Biophys. Acta 528:127.

Greene, D.H.S., and Selivonchick, D.P., 1987, Lipid metabolism in fish, Prog. Lipid Res. 26:53.

Gurr, M.I., 1984, "Role of Fat in Food and Nutrition", Elsevier Appl. Sci. Publ., London.

Hammarlund, K., Nilsson, G.E., Öberg, P.Å., and Sedin, G., 1977, Transepidermal water loss in newborn infants. Relation to ambient humidity and site of measurement and estimation of total transepidermal water loss. Acta Pædiatr. Scand. 66: 553.

Hammarlund, K., Strömberg, B., and Sedin, G., 1986, Heat loss from the skin of preterm and fullterm newborn infants during the first weeks after birth, Biol. Neonate 50:1.

Hansen, H.S., 1986, The essential nature of linoleic acid in mammals, Trends Biochem. Sci. 11:263.

Hansen, H.S., and Jensen, B., 1983, Urinary prostaglandin E$_2$ and vasopressin excretion in essential fatty acid-deficient rats: effect of linolenic acid supplementation, Lipids 18: 682.

Hansen, H.S., and Jensen, B., 1985, Essential function of linoleic acid esterified in acylglucosylceramide and acylceramide in maintaining the epidermal water permeability barrier. Evidence from feeding studies with oleate, linoleate, arachidonate, columbinate and α-linolenate, Biochim. Biophys. Acta 834:357.

Hansen, H.S., and Jensen, B., 1987, Why is linoleic acid essential ? in: "Proc. AOCS Short Course on Polyunsaturated Fatty Acids and Eicosanoids", W.E.M.Lands, ed., American Oil Chemists' Society, Champaign, Illinois.

Hansen, H.S., Jensen, B., and von Wettstein-Knowles, P., 1986, Apparent in vivo retroconversion of dietary arachidonic to linoleic acid in essential fatty acid-deficient rats, Biochim. Biophys. Acta 878:284.

Hartop, P.J., and Prottey, C., 1976, Changes in transepidermal water loss and the composition of epidermal lecithin after applications of pure fatty acid triglycerides to the skin of essential fatty acid-deficient rats, Br. J. Dermatol. 95:255.

Holman, R.T., 1971, Essential fatty acid deficiency, Prog. Chem. Fats Lipids 9:275.

Houtsmuller, U.M.T., 1975, Specific biological effects of polyunsaturated fatty acids, in: "The Role of Fat in Human Nutrition", A.J.Vergroesen, ed., Academic Press, London.

Houtsmuller, U.M.T., and van der Beek, A., 1981, Effects of topical application of fatty acids, Prog. Lipid Res. 20:219.

Jelenko, C., Wheeler, M.L., and Scott, T.H., 1972a, Studies in burns, X. Ethyl linoleate: the water holding lipid of skin, J. Trauma. 12:968.

Jelenko, C., Wheeler, M.C., and Scott, T.H., 1972b, Studies in burns, XI. Ethyl linoleate: effects on in vivo burn eschar, J. Trauma. 12:974

Lampe, M.A., Burlingame, A.L., Whitney, J., Williams, M.L., Brown, B. E., Roitman, E., and Elias, P.M., 1983, Human stratum corneum lipids: characterization and regional variations, J. Lipid Res. 24:120.

Landmann, L., 1984, The epidermal permeability barrier. Comparison between in vivo and in vitro lipid structures, Eur. J. Cell. Biol. 33:258.

Landmann, L., Wertz, P.W., and Downing, D.T., 1984, Acylglucosylceramide causes flattening and stacking of liposomes. An analogy for assembly of the epidermal permeability barrier. Biochim. Biophys. Acta 778:412.

Madison, K.C., Swartzendruber, D.C., Wertz, P.W., and Downing, D.T., 1988, Lamellar granule extrusion and stratum corneum intercellular lamellae in murine keratinocyte cultures. J. Invest. Dermatol. 90:110.

Mead, J.F., and Fulco, A.J., 1976, "The Unsaturated and Polyunsaturated Fatty Acids in Health and Disease", Charles C. Thomas Publisher, Springfiled, U.S.A.

Melton, J.L., Wertz, P.W., Swartzendruber, D.C., and Downing, D.T., 1987, Effects of essential fatty acid deficiency on epidermal O-acylsphingolipids and transepidermal water loss in young pigs, Biochim. Biophys. Acta 921:191.

Nugteren, D.H., Christ-Hazelhof, E., van der Beek, A., and Houtsmuller, U.M.T., 1985, Metabolism of linoleic acid and other essential fatty acids in the epidermis of the rat, Biochim. Biophys. Acta 834:429.

Nugteren, D.H., and Kivits, G.A.A., 1987, Conversion of linoleic acid and arachidonic acid by skin epidermal lipoxygenases, Biochim. Biophys. Acta 921:135.

Prottey, C., 1977, Investigation of functions of essential fatty acids in the skin, Br. J. Dermatol. 97:29.

Rapport, E.W., Stanley-Samuelson, E., and Dadd, R.H., 1984, Ten generations of drosophila melanogaster reared axenically on a fatty acid-free holidic diet, Arch. Insect Biochem. Physiol. 1:243.

Scheuplein, R.J., 1976, Permeability of the skin: a review of major concepts and some new developments, J. Invest. Dermatol. 67:672.

Sinclair, H.M., 1952, Essential fatty acids and their relation to pyridoxine, in: "Lipid Metabolism", R.T.Williams, ed., The University Press, Cambridge, UK.

Sohal, R.S., Bridges, R.G., and Howes, E.A., 1984, Relationship between lipofuscin granules and polyunsaturated fatty acids in the house fly, Musca Domestica, Mechanism Age. Develop. 25:355.

Squier, C.A., and Hall, B.K., 1985, The permeability of skin and oral mucosa to water and horseradish peroxidase as related to the tickness of the permeability barrier, J. Invest. Dermatol. 84:176.

Thomasson, H.J., 1953, Biological standardization of essential fatty acids (a new method), Int. Rev. Vitamin Res. 25:62

Thomasson, H.J., 1962, Essential fatty acids, Nature 194:973.

Wertz, P.W., Cho, E.S., and Downing, D.T., 1983, Effect of essential fatty acid deficiency on the epidermal sphingolipids of the rat, Biochim. Biophys. Acta 753:350.

Wertz, P.W., Cox, P.S., Squier, C.A., and Downing, D.T., 1986, Lipids of epidermis and keratinized and non-keratinized oral epithelia, Comp. Biochem. Physiol. 83B:529.

Wertz, P.W., and Downing, D.T., 1982, Glycolipids in mammalian epidermis: structure and function in the water barrier, Science 217:1261.

Wertz, P.W., and Downing, D.T., 1983, Ceramides of pig epidermis. Structure determination, J. Lipid Res. 24:759.

Wertz, P.W., and Downing, D.T., 1986, Linoleate content of epidermal acylglucosylceramide in newborn, growing and mature mice, Biochim. Biophys. Acta 876:469.

Wertz, P.W., Donald, C., Swartzendruber, C., Abraham, W., Madison, K.C., and Downing, D.T., 1987, Essential fatty acids and epidermal integrity, Arch. Dermatol. 123:1381.

Yardley, H.J., and Goldstein, D.J., 1976, Changes in dry weight and projected area of human epidermal cells undergoing keratinization as determined by scanning interference microscopy, Br. J. Dermatol. 95:621.

Yardley, H.J., and Summerly, R., 1981, Lipid composition and metabolism in normal and diseased epidermis, Pharmac. Ther. 13:357.

Ziboh, V.A., and Hsia, S.L., 1972, Effects of prostaglandin E_2 on rat skin: Inhibition of sterol ester biosynthesis and clearing of scaly lesions in essential fatty acid deficiency, J. Lipid Res. 13:458.

DIFFERENT DOSES OF FISH - OIL FATTY ACID INGESTION IN ACTIVE RHEUMATOID

ARTHRITIS: A PROSPECTIVE STUDY OF CLINICAL AND IMMUNOLOGICAL PARAMETERS

Joel M. Kremer, David A. Lawrence, and William Jubiz

Division of Rheumatology, Department of
Microbiology/Immunology
Albany Medical College and Albany Veteran's Administration
Hospital, Albany, New York

INTRODUCTION

Lee, et al, have demonstrated that fish - oil ingestion leads to decreased production of leukotriene B_4 (LTB_4) derived from arachidonate through the 5-lipoxygenase pathway with the new production of leukotriene B_5 (LTB_5) from EPA(1). Since LTB_4 is a potent inflammatory and chemotactic compound, a decrease in its production could favorably affect the clinical manifestations of an inflammatory disease like rheumatoid arthritis. It was not surprising, then, when we observed improvement in certain clinical manifestations of rheumatoid arthritis which were significantly correlated with decreased production of neutrophil LTB_4 in patients receiving fish - oil(2).

Leukotrienes are also potent modulators of immune reactivity(3-6) which can significantly affect T and B cell activity through modulation of the production of certain cytokines including interleukin-1 (IL-1)(7). IL-1 could potentially influence the course of an autoimmune inflammatory disease like rheumatoid arthritis through several different mechanisms including effects on synovial tissue and cartilage metabolism(8) as well as immunomodulation(9).

Previous studies of fish - oil fatty acid ingestion in patients with rheumatoid arthritis have employed fixed daily dietary supplements regardless of body weight(2,10,11). Additionally, it is not certain whether the observed benefits would be sustained as omega-3 fatty acid dietary supplements have not been administered for periods longer than 14 weeks. We thus measured clinical and immunological parameters in patients with rheumatoid arthritis consuming different doses of fish - oil dietary supplements over a period of 24 weeks.

METHODS

Study Design - This was a prospective, randomized, double - blinded, placebo - controlled parallel study. Three groups were studied including 2 different doses of fish - oil dietary supplementation and a third group which received olive - oil. Patients were randomized for age, sex, treatment with slow - acting anti - rheumatic drug and disease severity. For the purposes of randomization the following system of rating disease activity was employed: total joint count 6 - 10 = Category 1; total joint count 11 - 20 = Category 2; total joint count $\geq$ 21 = Category 3. Patients were placed into 3 study groups: 1) A control group which ingested 9 olive - oil capsules/day containing a total of 6.84 grams of oleic acid, .93 grams of palmitic acid, .53 grams of linoleic acid and .13 grams of stearic acid; 2) A "low - dose" fish - oil group which ingested 27 mg/kg/day EPA and 18 mg/kg/day DHA and 3) a "high dose" fish - oil group which ingested 54 mg/kg/day EPA and 36 mg/kg/day DHA. All fish - oil patients ingested capsules containing 330 mg EPA and 240 mg DHA per capsule in an ethyl ester form provided by Pharmacaps , Elizabeth, NJ. The total daily caloric intake for the placebo group was 54 KCal (9 capsules) in all patients. The mean daily caloric supplement was 52 ± 9.7 KCal in the low dose group (range 36 to 72) and 103 ± 20.5 KCal in the high dose group (range 81 to 153).

Patients - Forty - nine patients with definite or classical rheumatoid arthritis from the outpatient clinic population of the Division of Rheumatology at Albany Medical College successfully completed the study. All patients had active disease as previously defined.

Any change in medications for rheumatoid arthritis including non - steroidal anti - inflammatory drugs, slow - acting anti - rheumatic drugs, oral corticosteroids or intraarticular steroid injections during the entire duration of the study was considered a reason to withdraw a patient from the data analysis.

Clinical Assessment - Clinical evaluations were performed by the same rheumatologist at baseline and every 6 weeks through 30 weeks (6 visits). Patients on fish - oil ingested these supplements through 24 weeks. All 3 groups ingested olive - oil during weeks 24 - 30. Patients were not aware of the washout period. Physicians were aware of the washout, but remained blinded through the 30 week visit. Patients received no dietary instructions. A 3 day detailed dietary history was performed on all patients at baseline and at the 30 week visit and the information obtained was analyzed for consistency of nutrient intake by an IBM AT computer using a Short Report software package.

Clinical evaluations were performed at the time of each visit as previously described.

Compliance - Compliance was monitored by pill counts and gas chromatographic analysis of plasma fatty acids at the time of each visit. Any patient in either fish - oil group who failed to demonstrate a rise in plasma EPA was eliminated from the data analysis.

Laboratory and Immunological Evaluations - Laboratory determinations included a complete blood count with platelets and Westergren erythrocyte sedimentation rate at the time of each visit. In addition, the following laboratory tests were performed at baseline and after 24 weeks (maximum duration) of fish - oil ingestion: ionophore - stimulated neutrophil LTB_4 and LTB_5; interleukin-1 and interleukin-2 production; T - cell proliferation after stimulation with Concanavalin-A (ConA) and Phytohemagluttinin (PHA); T - cell dependent B - cell proliferation after stimulation with Pokeweed mitogen (PWM) and B - cell proliferation after stimulation with Staph A; immunoglobulin production (IgG, IgM and IgA) after PWM stimulation; and rheumatoid factor (latex test).

Laboratory Methodology -

Measurement of Leukotriene - were performed as previously described(2,18).

<u>Lymphoproliferation Assay</u> - was performed as previously described(13,14).

<u>IL-1 Production</u> - PBMC (1 X 10^6 / ml) were cultured for 24 hours in the presence or absence of 1 mg / ml LPS. Cell - free supernatants were stored at -70°C prior to quantitation of IL-1 in the C3H / HeJ thymocyte co - stimulation assay(15).

<u>IL-2 Production</u> - PBMC (1 X 10^6 / ml) were cultured for 24 hr. in the presence or absence of 5 mg / ml Con A. Cell - free supernatants were stored at -70°C prior to quantitation of IL-2 in a HT-2 bioassay(16).

<u>Thiol quantitation</u> - Plasma and cellular thiols were quantitated in a previously described fluorometric assay(17).

<u>Iq Production</u> - PBMC were cultured for 7 days in the presence or absence of PWM (1:200). Cell - free supernatants were stored at -70°C prior to quantitation of IgG, IgA, or IgM by an ELISA system(2).

RESULTS

<u>Within - Group Changes in Clinical Parameters From Baseline</u> - Changes in clinical parameters with time can be seen in Table 1. Statistically significant decreases from baseline measurements in the number of tender joints in the group ingesting low - dose fish - oil were observed at the time of the 24 week visit (-1.85 ± 4.0, p = .05) and in the high - dose fish - oil group at the time of the 18 week (-2.6 ± 4.9, p = .04) and 24 week visit (-1.7 ± 2.9, p = .02). Decreases in tender joints after 12 weeks reached borderline significance (-2.4 ± 4.8, p = .06) in the high - dose fish - oil group.

Swollen joints decreased from baseline in a statistically significant way in the group ingesting low - dose fish - oil after 12 weeks (-2.7 ± 3.6, p = .003), 18 weeks (-3.6 ± 4.4, p = .002) and 24 weeks (-4.1 ± 4.9, p = .001). The significant decrease in swollen joints was also noted at the time of the 30 weeks washout visit (-3.6 ± 5.4, p = .007).

Significant decreases in swollen joints in the high - dose group were noted after 12 weeks (-2.9 ± 2.2, p = .0001), 18 weeks, (-2.3 ± 3.1, p = .008) and 24 weeks (-2.8 ± 4.2, p = .02). No significant decreases in tender or swollen joints were noted at any time in the patients ingesting the olive - oil supplement.

Significant decreases in morning stiffness from baseline were noted only in the high dose group (Table 1). They occurred after 18 weeks (-41.8 ± 62.5 minutes, p = .01) and 24 weeks (-46.5 ± 58.1 minutes, p = .004) with a tendency to a carry-over benefit noted at the time of the 30 week visit (-30.0 ± 63.3 minutes, p = .07). Significant improvements in morning stiffness did not occur in the low - dose fish - oil or olive - oil patients.

Grip strength increased in the low dose group at 18 weeks (+12.3 ± 2.0 mm Hg, p = .01) and 24 weeks (+13.0 ± 21.9 mm Hg, p = .02). In the high dose group and 24 weeks (+21.1 ± 28.9, p = 008) and a carry-over benefit was seen after 30 weeks (+19.6 ± 32.1, p = .02). Significant improvements in grip strength were not observed in the olive - oil group.

Physician evaluations of global arthritis activity decreased at 24 weeks in the group ingesting olive – oil (-.41 + .5, p = .02) and the low – dose fish – oil group (-.42 + .7, p= .03). Significant improvements in physician evaluation of global arthritis activity were seen in the high – dose fish – oil group after week 6 (-.35 + .61, p = .03), week 12 (-.35 + .70, p = .05), week 18 (-.35 + .70, p = .05) and week 24 (-.52 + .71, p = .008) with a carry-over improvement noted at 30 weeks (-.31 + .47, p = .02)(Figure 2C).

A statistically significant improvement in patient evaluation of global arthritis activity was noted only in the olive – oil group after 6 weeks (-.41 + .51, p = .02), 18 weeks (-.41 + .66, p = .05), 24 weeks (k- .41 + .51, p = .02) and 30 weeks (-.41 + .51, p = .02).

Physician evaluation of pain improved in the high – dose group after 12 weeks (-.41 + .61, p = .01), 18 weeks (-.41 + .51, p = .004) and 24 weeks (-.41 + .51, p = .004) (Figure 2D).

Improvement in patient evaluation of pain was noted only once at 24 weeks in the high – dose fish – oil group (-.29 + .46, p = .02). No significant improvements were noted at any time in any group in the interval to the onset of fatigue.

A reduction in systolic blood pressure was noted in the low – dose group after 12 weeks (-8.1 mm Hg + 17.4, p = .05), 24 weeks (-10.5 mm Hg + 21.1, p = .03) and 30 weeks (-9.9 mm Hg + 18.9, p = .03). No other significant changes in systolic blood pressure or any changes in diastolic blood pressure were observed.

Between Group Comparisons – Only grip strength improved significantly in all patients ingesting fish – oil when compared with the olive – oil control patients (p = .04).

1, grip strength improved after 12 weeks (+12.2 + 23.1, p = .04), 18 weeks (+17.8 + 30.7, p = .02)

Laboratory – No significant changes were observed in hemoglobin, Westergren erythrocyte sedimentation rate or rheumatoid factor in any group. Ionophore stimulated neutrophil LTB$_4$ decreased by 19% in the low – dose fish – oil patients (p = .0003) and 20% in the high – dose fish – oil patients (p = .02) (Table 2 and Figure 1).

Immunological Parameters

Interleukin-1 and 2 Changes from Baseline Measurements

Interleukin-1 production and release decreased by 239.9 + 503 units / ml (40.6%, p = .05) in the low dose fish – oil group after 24 weeks and decreased by 416.2 + 402.1 units / ml (54.7%, p = .0005) in the high – dose fish – oil group (Table 2 and Figure 2). Although IL-1 decreased by 243.1 + 133.4 in the olive – oil group (38%) the change was not statistically significant. Interleukin-2 increased by 33.8% from baseline after 24 weeks of fish – oil consumption in the low – dose group, and increased by 2% from baseline in the high – dose group. Neither of these changes from baseline measurements were significant. There was, however, a significant correlation between changes in interleukin-1 and interleukin-2 within individual subjects which was observed in all patients ingesting fish – oil (R = .40, p = .05).

Results of mitogen stimulation, immunoglobulin production and thiol determinations will not be presented due to space considerations, but may be seen in Table 2.

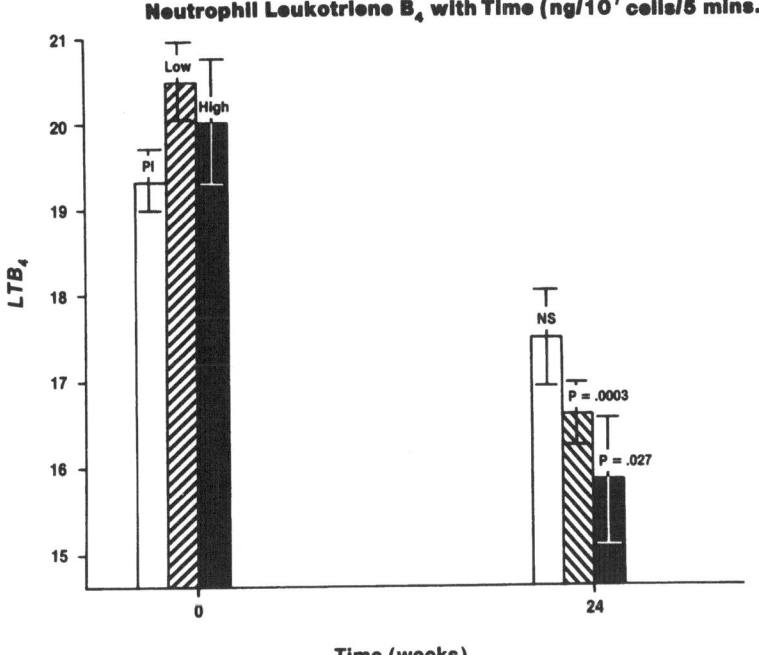

Figure 1. Ionophore stimulated neutrophil Leukotrine B₄ with time

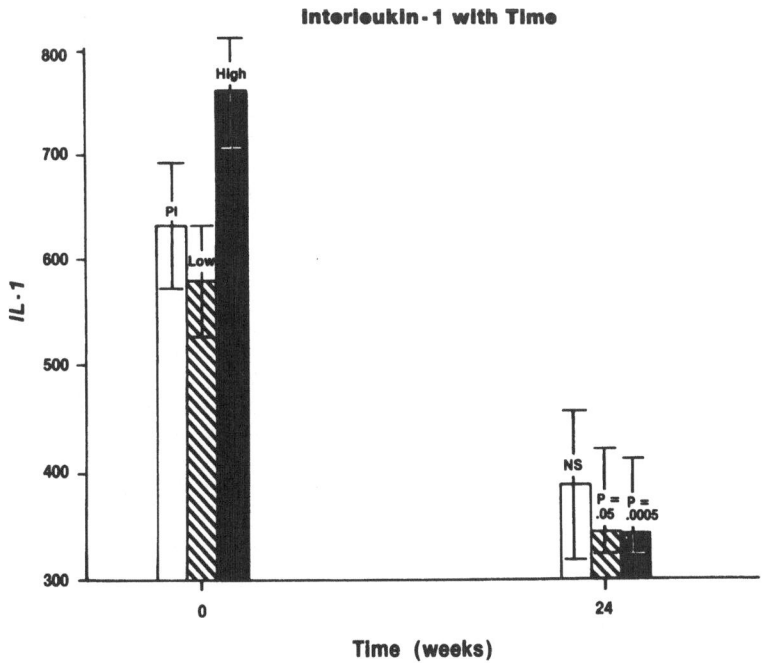

Figure 2. Macrophage interleukin-1 with time (u/ml)

DISCUSSION

        We have previously demonstrated significant improvements in many
clinical variables in patients with rheumatoid arthritis ingesting omega-3
fatty acids for 12 - 14 weeks(10,2). We report here that the beneficial
effects of omega-3 fatty acids on the clinical manifestations of
rheumatoid arthritis including fewer tender and swollen joints, decreased
duration of morning stiffness and improvements in grip strength and
assessments of pain and disease activity are sustained and more common
after 18 - 24 weeks of treatment.

        The improvements from baseline in the patients ingesting fish - oil
were usually not statistically significant when compared with the olive -
oil control subjects. This may be because clinical parameters in patients
ingesting olive - oil often improved from baseline, although not in a
statistically significant way. It should be noted, however, that the
clinical outcomes in the olive - oil patients were biased towards more
favorable results because of the withdrawal of 10 of the original 22
patients in this arm of the study. Six of these 10 were withdrawn for
reasons relating to increased disease activity and the concomitant need
for increased oral steroid, change in disease modifying anti - rheumatic
drug or intraarticular steroid injection. The distribution of dropouts
for increased disease activity in the olive - oil group vs. both fish -
oil groups was highly significant (p = .008).

        Olive - oil itself could have potential beneficial effects in
patients with RA. These changes are due to cell membrane changes in
lymphocytes resulting in altered immune function through a variety of
mechanisms(19-22). Altered immunological reactivity has been observed in
mice fed a diet high in polyunsaturated fatty acid(22). We did observe
significant increased reactivity in lymphocytes from patients ingesting
olive - oil after both ConA and PWM stimulation. Whether or not these
effects would have clinical significance is unclear.

        We had previously noted a correlation between the magnitude of the
decrease in LTB_4 and the reduction in the number of tender joints on
physical examination(2) and it would be tempting to link the decrease in
this potent inflammatory substance to clinical improvements. In the
current investigation there was a highly significant decrease in the
production of interleukin-1 after fish - oil ingestion. Leukotriene B_4
exerts a positive modulating effect on the genetic control of IL-1,
probably at the translational level within the cytoplasm(7). Decreased
IL-1 production has recently been observed in normal subjects ingesting
dietary supplements of omega-3 fatty acids(23).

        IL-1 has many varied biological actions in promoting rheumatoid
arthritis disease activity. It induces the production of acute phase
reactants and promotes a negative nitrogen balance and anorexia(24,25).
IL-1 has potent stimulatory effects on polymorphonuclear leukocytes(26)
and fibroblasts(27) and can stimulate synovial cells to release
prostaglandins and collagenase(28). Rheumatoid synovial tissue is thought
to produce enhanced amounts of IL-1(29) and recent observations correlate
synovial IL-1 production to inflammatory changes in the synovium observed
by arthroscopy(30). The inflammatory response in gouty arthritis is
completely abrogated by IL-1 inhibition(31).

        IL-1 also has potent biological effects on connective tissue
including the promotion of bone resorption via osteoclast stimulation(29)
and the induction of increased secretion of collagenase and neutral
proteases(33). In addition, IL-1 enhances the release of
glucosaminoglycans from cultured cartilage and is a potent mediator of

cartilage destruction(33). Suppression of these biological effects would be expected to result in amelioration of the clinical manifestations of rheumatoid arthritis. In summary, patients with active rheumatoid arthritis ingesting fish - oil for a period of 24 weeks had significant improvements in multiple clinical parameters which was accompanied by significant decreases in the production of neutrophil LTB_4 and macrophage IL-1 production.

We did not observe clear dose dependent clinical effects between fish oil groups when analyzed by regression analysis. Forty - five measures of clinical efficacy were performed during the course of this investigation (9 measures at each of 5 follow - up visits). Twenty - one of 45 showed significant improvement ($p \leq .05$) from baseline in the high dose group, 8 / 45 in the low - dose group and 5 / 45 in the olive - oil group. The likelihood that this distribution of significant clinical results occur by change is 2 in 10,000 ($p = .0002$). It should also be noted that IL-1 decreased by 54.7% from baseline in the patients on high - dose fish - oil vs. a 40.6% decrease in the low - dose group.

In summary, patients with active rheumatoid arthritis ingesting fish - oil for a period of 24 weeks had significant improvement in multiple clinical parameters which was accompanied by significant decreases in the production of neutrophil LTB_4 and macrophage IL-1 production.

REFERENCES

1. Lee TH, Hoover RL, Williams JD, et al. Effect of dietary enrichment with eicosapentaenoic and docosahexaenoic acids on in vitro neutrophil and monocyte leukotriene generation and neutrophil function. N Engl J Med 1985; 321:1217-1224
2. Kremer JM, Jubiz W, Michalek A, Rynes RI, Bartholomew LE, Bigaouette J, Timchalk M, Beeler D, Lininger L. Fish - oil fatty acid supplementation in active rheumatoid arthritis. Ann Int Med 1987; 106:497-503
3. Payan DG, Goetzl EJ. Specific suppression of human T lymphocyte function by leukotriene B_4. J Immunol 1983; 131:551-553
4. Gualde N, Durgaprasadarao A, Goodwin JS. Effect of lipoxygenase metabolites of arachadonic acid on proliferation of human T cells and T cell subsets. J Immunol 1985; 134:1125-1129
5. Rola-Pleszczynski M, Bouvrette L, Gingras D, Girard M. Identification of interferon- as the lymphokine that mediates leukotriene B_4 induced immunoregulation. J Immunol 1987: 139:513-517
6. Blomgren H, Hammarstrom S, Wasserman J. Synergistic enhancement of mitogen responses of human lymphocytes by inhibitors of cyclo-oxygenase and 5,8,11-eicosatriynoic acid, an inhibitor of 12-lipoxygenase and leukotriene biosynthesis. Int Archs Allergy Apply Immun 1987; 83:247-255
7. Dinarello CA. Biology of interleukin-1. FASEB J 1988; 2:108-115
8. Dayer JM, de Rochemonteix B, Burrus B, Demczuk S, Dinarello CA. Human recombinant interleukin 1 stimulates collagenase and prostaglandin E_2 production by human synovial cells. J Clin Invest 1986; 77:645-648
9. Dinarello, CA. An update on human interleukin-1: from molecular biology to clinical relevance. J Clin Immunol 1985; 5:287-297
10. Kremer JM, Bigauoette J, Michalek AV, Timchalk MA, Lininger L, Rynes RI, Huyck C, Zieminski J, Bartholomew LE. Effects of manipulation of dietary fatty acids on clinical manifestations of rheumatoid arthritis. Lancet 1985; Jan. 26:184-187
11. Sperling RI, Weinblatt M, Robin JL, Ravalese J III, Hoover RL, Hoase

F, Coblyn JS, Fraser PA, Spur BW, Robinson DR, Lewis RA, Austen KF. Effects of dietary supplementation with marine fish oil on leukocyte lipid mediator generation and function in rheumatoid arthritis. Arthritis Rheum 1987; 30:988-997

12. Jubiz W, Draper RE, Gale J, Nolan G. Decreased leukotriene B$_4$ synthesis by polymorphonuclear leukocytes from male patients with diabetes mellitus. Postaglandins Leukotrienes Med 1984; 14:305-311

13. Mendelsohn J, Skinner A, Kornfeld S. The rapid induction by PHA of increased r-aminoisobutyric acid uptake by lymphocytes. J Clin Invest 1971; 50:818-826

14. Noelle RJ, Lawrence DA. Determination of glutathione in lymphocytes and possible association of redox state and proliferative capacity of lymphocytes. Biochem J 1981; 571-579

15. Rosenwasser LJ, Dinarello CA. Antibody of leukocyte pyrogen to enhance phytohemagglutinin induced murine thymocyte proliferation. Cell Immunol 1981; 63:134-142

16. Gillis S, Germ MM, Ou W, Smith KA. T-cell growth factor, parameters of production and a quantitative microassay for activity. J Immunology 1978; 120:2027-2032

17. Ayers FC, Warner GL, Smith KL, Lawrence DA. Fluorometric quantitation of cellular and nonprotein thiols. Anal Biochem 1986; 154:186-193

18. Check IJ, Piper M. Quantitation of immunoglobulin. In: Rose NR, Friedman H, Fahey JL, eds. Manual of clinical laboratory immunology; Washington, DC: ASM Press, 1986:138-151

19. Traill KN, Wick G. Lipids and lymphocyte function. Immunol Today 1984; 5:70-75

20. Johnston PV. Dietary fat, eicosanoids, and immunity. Adv Lipid Research 1985; 21:103-141

21. Del Buono BJ, Williamson PL, Schlegel RA. Alterations in plasma membrane lipid organization during lymphocyte differentiation. J of Cell Physiology 1986; 126:379-388

22. Erickson KL. Dietary fat modulation of immune response. Int J Immunopharmac 1986; 8:529-543

23. Endres, S. Personal communication.

24. Dinarello CA. Interleukin-1. Rev Infect Dis 1984; 6:51-90

25. Dinarello CA. Interleukin-1 and the pathogenesis of the acute-phase response. N Engl J Med 1984; 311:1413-1418

26. Luger TA, Charon JA, Colot M, Micksche M, Oppenheim JJ. Chemotactic properties of partially purified human epidermal cell-derived thymocyte-activating factor for polymorphonuclear and mononuclear cells. J Immunol 1983; 131:816-820

27. Schmidt JA, Mizel SB, Cohen D, Green I. Interleukin-1: a potential regulator of fibroblast proliferation. J Immunol 1982; 128:2177-2182

28. Mizel SB, Dayer JM, Krane SM, Mergenhagen SE. Stimulation of rheumatoid synovial cell collagenase and prostaglandin production bypartially purified lymphocyte-activating factor (interleukin 1). Proc Natl Acad Sci USA 1981; 78:2474-2477

29. Goto M, Sasano M, Yamanaka H, Miyasaka N, Kamatani N, Inoue K, Nishioka K, Miyamoto T. Spontaneous production of an interleukin-1-like factor by cloned rheumatoid synovial cells in long-term culture. J Clin Invest 1987; 80:786-796

30. Miyasaka N, Sato K, Goto M, Sasano M, Natsuyama M, Inoue K, Nishioka K. Augmented interleukin-1 production and HLA-DR expression in the synovium of rheumatoid arthritis patients. Arthritis Rheum 1988; 4:480-486

31. Di Giovine FS, Malawista SE, Nuki G, Duff GW. Interleukin-1 (IL-1) as a mediator of crystal arthritis. J Immunol 1987; 138:3213-3218

32. Gowen M, Wood DD, Ihrie EJ, et al. An IL-1 like factor stimulates bone resorption in vitro. Nature 1983; 306:378-380

DIETARY w-3 AND w-6 FATTY ACIDS IN CANCER

Rashida A. Karmali

Rutgers University, New Brunswick, NJ, and
Memorial Sloan-Kettering Cancer Center, New York, NY

INTRODUCTION

The evidence that dietary fat relates to cancers of the breast and colon, and probably other types of cancer, has been built on descriptive and metabolic epidemiology, correlation studies, migrant studies, case-control studies, and experimental animal studies.[1] This overall evidence has led to recommendations that dietary fat intake be reduced to decrease the risk of developing certain types of cancer.[2,3]

Data correlating age-adjusted mortality rates from breast and colon cancer show a high correlation with per capita supply of animal fat and total fat intake, but none with vegetable fat.[4] In evaluating the vegetable fat data, it is now recognized that the composition of different vegetable oils varies considerably in fatty acid composition. Some outliers in correlation studies of breast cancer need explanation; for example, Greece and Spain have relatively high fat intakes for their rates of breast cancer. Could this be related to olive oil, rich in oleic acid, predominantly used in these countries?[5]

Studies of Japanese in Japan and migrants to Hawaii provide important clues. In Japan, a gradual increase in the incidence of breast, colon, and prostate cancers, particularly in more affluent, urban segments of the population, has been recorded during a period when consumption of fat in the Japanese diet has increased from 10-15% to 22-23% of total calories.[6] Studies of Japanese migrants to Hawaii show a marked increase in the incidence of breast, colon, and prostate cancers compared with that of Japanese in their native country.[7] These data are explained in part by the westernization of the diet of native Japanese and emigrants to the United States. Not only has this change resulted in increased intake of total fat calories, but the type of fat increased in the diet is animal fat. The traditional Japanese diet includes large quantities of fish and some seaweed so that total fat is comprised of an appreciable portion of long-chain fatty acids of the w-3 family.[8]

The Greenland Eskimos are also known for eating large quantities of fish in the diet.[9] In the past, the Eskimo women were notable for their very low incidence of breast cancer. However, epidemiological evidence suggests that incidence of breast cancer among Eskimo women increased during a period of modernization that resulted in greater access to imported foods and an increased consumption of saturated fat and vegetable oil rich in w-6 polyunsaturated fatty acids (PUFA).[10]

Many western countries show a trend toward an increase in breast cancer, which may be correlated with increased use of vegetable oils that are good sources of linoleic acid (LA).[11]

EXPERIMENTAL CARCINOGENESIS

In experimental animals, both the amount and type of fat in the diet have clearly been shown to influence development of carcinogen-induced and transplanted tumors. These tumors include carcinomas of breast, colon, pancreas, and prostate.[12]

While there is some evidence to suggest that saturated fat may influence the initiation stage of mammary cancinogenesis,[13] the preponderance of evidence points to an effect of w-6 PUFA (specifically, linoleic acid) on the promotional stage.[4] The promotional phase in humans is of extended duration, and in animal studies it was found to be reversible in nature.[15] What is becoming apparent is the requirement for LA to facilitate tumor development by acting as promoter of mammary and colon carcinogenesis. Mammary tumor incidence and yield increased with increasing amounts of dietary LA, up to a level of 4% by weight of the diet, in rats challenged with dimethylbenz(a)anthracene (DMBA).[16] Additional increases up to 12% had no further effect. These observations in the DMBA model suggest that inhibition of tumor development by dietary modification of LA could only be achieved by reducing the level below 4% of total calories.

MECHANISMS

A number of mechanisms of dietary LA have been proposed to explain the enhancement of mammary tumorigenesis, e.g., suppression of cell-mediated immune response, alteration of membrane structure and properties, inhibition of intercellular communication, increase in secretion of mamotrophic hormones, and increase in production of dienoic prostaglandins. The tumor-promoting effects of LA on mammary cancer in experimental animals can be counteracted by cyclooxygenase inhibitors.[17]

It was therefore critical to determine whether dietary w-3 PUFA could inhibit arachidonic metabolism by cyclooxygenase and measure the effect on tumor development. This communication reviews reports on the effect of marine w-3 PUFA (eicosapentaenoic acid, EPA, + docosahexaenoic acid, DHA) on animal tumor systems. Preliminary studies of w-3 PUFA intervention in women at high-risk for breast cancer are discussed. In addition, recent studies on the role of gamma-linolenic acid (GLA) in mammary tumor development are presented.

EFFECTS OF w-3 PUFA IN ANIMAL TUMOR SYSTEMS

The effects of marine w-3 PUFA have been tested in a number of tumor models (Table 1). In studies of two transplantable mammary adenocarcinomas--the R3230AC growing in female Fischer 344 rats,[18] and the IX growing in mice,[19] feeding fish oil resulted in reduction of tumor size. Production of dienoic prostanoids and thromboxane was reduced in tumors taken from fish oil-fed rats. EPA + DHA were incorporated into tumor phosphoglycerides at the expense of LA acid and

Table 1. Studies of Marine w-3 PUFA in
Experimental Tumor Models

Mammary Cancer	Reference
R3230AC adenocarcinoma	18
IX adenocarcinoma	19
NMU-induced	20
DMBA-induced	21-23
Pancreatic Cancer	
L-azaserine-induced	24,26
Colon Cancer	
azoxymethane-induced	25
Prostatic Cancer	
DU-145	27
Metastatic Tumors	
CT-26 colon tumor	28
13762 MAT:B mammary tumor	29
BN472 mammary tumor	30

arachidonic acid.[18] Gabor et al.[19] suggested that the reduction of tumor size could be explained by an increased loss of tumor cells from the tumor site in the fish oil-fed mice.

A number of studies have been carried out to compare the effect of marine w-3 and LA in two carcinogen-induced mammary tumor models.[20-23] Compared to corn oil, inhibitory effects of menhaden oil were observed only at the high level of 20% by weight, both in the nitrosomethylurea and DMBA models.[20,21] In these long-term studies, it was not clear whether inhibition of mammary tumor development was a result of feeding w-3 PUFA or due to very low levels of LA in the diet. However, in two subsequent studies with the DMBA model, mixtures of corn and fish oil were tested. Inhibition of tumor incidence was reported at w-3/w-6 ratio equal to 0.7[22] and 1.[23]

Tumor models of other organs--colon, pancreas, and prostate--that respond to promoting effects of LA have been studied. Feeding menhaden oil resulted in significant inhibition of L-azaserine-induced pancreatic preneoplastic lesions[24] and azoxymethane-induced colon tumors[25] compared to feeding corn oil. When diets containing mixtures of corn and menhaden oil were used in subsequent studies with the L-azaserine model, inhibition was observed only at the high amounts of w-3 PUFA in the diet.[26]

We have tested the effect of fish oil (20.5%) + corn oil (3%) and corn oil alone (23.5%) on growth of the DU-145 human prostatic carcinoma in nude mice.[27] Tumors growing in the fish oil-fed mice were smaller, more differentiated, and produced significantly lower amounts of PGE_2 compared to those growing in corn oil-fed mice (Table 2).

A number of investigators have tested the effect of w-3 and w-6 PUFA on metastasis of tumor cells. Broitman et al.[28] compared the

Table 2. DU-145 Human Prostatic Tumors Grown in Nude Mice: Tumor Weight, Volume, and PGE_2

	Tumor Wt (g)	Tumor Vol (cm^3)	Tumor PGE_2 (ng/g tissue)
23.52% corn oil	1.8±0.7	1.4±0.5	104.1±20.0
20.52% fish oil + 3% corn oil	0.4±0.1 (p=0.045)	0.3±0.1 (p=0.039)	45.4± 2.2 (p=0.027)

effects of safflower oil with menhaden oil on growth and pulmonary colonization of CT-26 implanted into bowel of mice. Menhaden oil retarded colon tumor growth and lung colonization at the 20% fat level.

We are currently studying the effect of w-3 PUFA, in the presence of an adequate supply of LA, on metastasis in the artificial system using 13762 MAT:B mammary tumor cells and in the spontaneous system using 13762NF mammary tumor cells. Data obtained to date suggest that, in the artificial system, both dietary w-3 PUFA and a low-fat diet reduce the frequency and inhibit the growth of metastatic mammary tumor foci in the lung.[29] However, there was no significant effect on metastasis formation in the spontaneous model of 13762NF mammary tumor in 3-month-old Fischer rats (unpublished results).

Kort et al.[30] reported studies of w-3 PUFA in the spontaneous model of BN472 metastatic mammary adenocarcinoma in rats. Although growth of the primary tumor in rats receiving dietary w-3 PUFA was significantly inhibited, there was no effect on metastasis compared to rats receiving w-6 PUFA.

The problem with studying dietary effects of fat on metastatic processes in experimental mammary tumorigenesis is that most metastatic primary tumors are fast growing and quickly induce cachexia in young rats. In the 13762NF model, the growth and metastatic behavior of this tumor in young Fischer 344 rats does not appear to be influenced by the level of fat in the diet. It is only in the retired breeders that the level of fat appears to affect metastasis.[31]

A series of reports suggest that the intravascular balance between prostaglandin (an anti-aggregatory factor for platelets) and thromboxane A_2 (a pro-aggregatory factor for platelets) is disrupted in favor of platelet aggregation during tumor cell metastasis.[32] Since EPA and DHA inhibit TXA_2 synthesis and PGI_3 is produced from EPA, it is reasonable to test the hypothesis that selective inhibition of TXA_2 synthesis could be used in the control of hematogenous tumor metastasis under optimal experimental conditions.

Finally, modulatory influence of fatty acids on molecular events of initiation and cellular events of promotion associated with mammary tumor virus (MTV) have been demonstrated in preneoplastic mammary alveolar lesions from the RIII mouse. Linoleic acid and arachidonic acid, but not stearic acid, increased MTV expression (measured as reverse transcription activity), cell proliferation (^3H-thymidine uptake), and cytodifferentiation (secretory alveoli).[33] When these organ cultures were treated with equivalent amounts of LA and EPA, ras expression was inhibited in EPA-treated cultures (personal communication).

The carcinogenic activity of DMBA has been attributed to its ability to bind to DNA and thus to cause somatic mutations. The mutation in the DMBA-induced tumors is localized to codon 61 of H-ras-1 oncogenes.[34] Recently, Barbacid and coworkers[35] have shown that ras oncogenes may be involved in the initiation of carcinogenesis. Indirect evidence, such as presence of activated ras in benign tissues[36] and correlations with progression and stage of mammary tumor development,[37] suggest that elevated or mutated ras p21 may play an important role in mammary tumorigenesis. In light of these observations, we measured ras p21 in DMBA-induced carcinomas taken from rats that were fed 23.5% fat by weight provided by either corn or blackcurrant oil or 20.5% fish oil + 3% corn oil. In a small number of carcinomas, expression of H-ras was found to be in the order: corn > blackcurrant = fish + corn oil (relative area of ras p21:40 > 26 = 26, respectively).

HUMAN STUDIES

A pilot study exploring the potential chemopreventive effects of marine w-3 PUFA against breast cancer development has been carried out in women at enhanced risk. Changes in 16-α-hydroxyestrone, which is enhanced in breast cancer patients and in susceptible individuals,[38,39] were monitored to measure efficacy of w-3 PUFA intervention. Women in the w-3 PUFA group received nine capsules of fish oil per day (1.53 g EPA + 1.44 g DHA) and those in the placebo group received nine capsules of vegetable oil (corn/olive mixture = 2.48 g LA + 3.6 g oleic acid). Preliminary results of the trial demonstrated no perturbation in the

Table 3. Fatty acid Composition of Five Dietary Oil Mixtures Tested (%)[a]

	I BCO 23.5%	II CO 23.5%	III BCO 15.5% + FO 8%	IV FO 20.5% + CO 3%	V BCO 20.5% + FO 3%
14:0	---	---	2.361	5.241	0.957
14:1	---	---	---	---	---
16:0	6.875	11.840	9.966	13.170	8.029
16:1	---	---	2.885	6.690	1.119
18:0	1.295	1.353	1.825	3.140	1.607
18:1	9.199	25.140	9.636	13.609	9.138
18:2 w-6	43.785	61.043	31.659	11.684	38.010
18:3 w-6	17.538	1.281	12.482	---	14.856
18:3 w-3	15.547	---	11.426	3.561	13.175
18:4 w-3/20:1	5.285	---	5.211	6.056	4.655
20:3 w-6	---	---	---	---	---
20:4 w-6	---	---	---	1.844	0.371
20:5 w-3	---	---	6.122	14.101	2.682
22:5 w-3	---	---	0.722	1.628	0.311
22:6 w-3	---	---	3.386	8.310	1.244

[a]Percentage of total fatty acids.

extent of 16-α-hydroxylation of estradiol in women in the placebo
control group; there was, however, a reduction in 16-α-hydroxylation of
estradiol in women taking w-3 PUFA. The extent of this reduction was
consistent with the law of initial values in that those with the highest
baseline pre-intervention values had the most profound fall in the
biomarker.[40] In the long term, the use of dietary w-3 PUFA as an
intervention in enhanced-risk women is attractive, as compliance to
these PUFA can be measured in cell membrane lipids. However, the major
problem to be encountered is the large number of patients required in
the trial if changes in cancer incidence are to be evaluated by proper
statistical procedures.

GAMMA-LINOLENIC ACID

When studying the role of PUFA in cancer, the working hypothesis
widely tested is that w-6 PUFA act as promoters whereas w-3 PUFA may
have preventive or inhibitory effects in a number of tumor models. w-3
PUFA seem to prevent the exaggerated metabolism of arachidonic acid (AA)
in neoplastic tissues and to displace LA and AA from membrane lipids.
Until recently, there have been few studies of intermediate fatty acids
produced during the conversion of LA to AA. Some reports suggest that
feeding even small amounts of GLA has resulted in inhibition of tumor
development[41] and growth[42]. Some natural sources of GLA include:
evening primrose oil (9%), blackcurrant oil (18%), and borage oil (26%).

We have tested the effect of blackcurrant oil on development of
DMBA-induced mammary tumor development. The composition of the diet by
weight was: casein 23.5%, corn starch 32.9%, dextrose 8.3%,
DL-methionine 0.35%, choline bitartrate 0.24%, alphacel 5.9%, AIN-76
mineral mix 4.11% and vitamin mix 1.18%, and fat 23.52%. Three oils
were used: corn (CO), blackcurrant oil (BCO), and fish oil (FO). Table
3 describes fatty acid composition of five oil mixtures tested.
Analysis of tumor incidence data indicated that rats fed 23.52% CO
exhibited enhanced mammary tumor yields compared to rats on BCO and FO

Table 4. Tumor Incidence and Prostaglandin Production

Diets*	Total Tumors/Group	% Tumor-Bearing Rats/Group	PGE_1 ng/g Tissue	PGE_2 ng/g Tissue
I	84	74.3	40.9± 9.5	44.2±75.3
II	106[a]	80.6[a]	13.7± 1.7[c]	531.9±85.5[e]
III	66	76.5	65.3±17.6	268.4±38.6
IV	63[b]	67.9[b]	33.6± 4.3	87.0± 7.5[f]
V	76	74.2	86.5±19.7[d]	193.4±26.1[g]

[a,b] Significantly different from remaining groups (general linear
models procedure and Duncan's rest, p = 0.0107).
[c-g] Significantly different from Group I, student's t-test.
*Fatty acid composition described in Table 3.

diets: CO > BCO > FO + CO (Table 4). Fatty acid profiles of red blood cells and tumor phosphoglycerides reflected dietary fatty acid composition (unpublished observations). Levels of dihomo-gamma-linolenic acid were increased in BCO-fed groups. Tumor PGE_2 levels were lowest in FO + CO groups: FO + CO < BCO < CO; PGE_1 levels were higher in BCO-fed groups compared to CO- and FO + CO-fed groups (Table 4). These results support earlier findings that feeding marine w-3 PUFA with adequate LA_1 results in lower tumor yields and inhibition of tumor PGE_2 production.[18] In addition, data obtained from BCO-fed groups suggest that PGE_2 production was inhibited and PGE_1 production was increased in the tumors. Since levels of dihomo-gamma-linolenic acid (DHLA) were increased in tumor phospholipids, these data suggest that DHLA inhibited AA metabolism. However, the role of increased PGE_1 production in tumor development is still unclear.

In conclusion, knowledge linking carcinogenesis with fatty acid metabolism is strongly supported, and there is a potential for interaction between different types of w-3 and w-6 PUFA at the biochemical and metabolic level.

ACKNOWLEDGEMENTS

New Jersey Agricultural Experiment Station publication No. D-14501-3-88. Supported by State funds and the New Jersey Commission on Cancer Research (86-483-CCR).

We thank Barbara Hannon for typing the manuscript.

REFERENCES

1. E. L. Wynder, Amount and type of fat/fiber in nutritional carcinogenesis, Prev. Med. 16:451 (1987).
2. B. Armstrong and R. Doll, Environmental factors and cancer incidence and mortality in different countries, with special reference to dietary practices, Int. J. Cancer 15:617 (1975).
3. National Academy of Sciences, Committee on Diet, Nutrition and Cancer, Diet, Nutrition and Cancer, Natl. Acad. Press, Washington, D.C. (1982).
4. M. Kurihara, K. Aoki, S. Tominaga, eds., Cancer Mortality Statistics in the World, University of Nogoya Press, Nagoya, Japan (1984).
5. L. A. Cohen, D. O. Thompson, K. Choi, R. A. Karmali, and D. P. Rose, Dietary fat and mammary cancer. II. Modulation of serum and tumor lipid composition and tumor prostaglandins by different dietary fats: association with tumor incidence patterns, J. Natl. Cancer Inst. 77:43 (1986).
6. T. Hirayama, Epidemiology of prostatic cancer with special reference to the role of diet, Natl. Cancer Inst. Monogr. 53:149 (1979).
7. L. N. Kolonel, J. H. Hankin, J. Lee, S. Y. Chu, A. M. Y. Nomura, M. W. Hinds, Nutrition intakes in relation to cancer incidence in Hawaii, Br. J. Cancer 44:332 (1981).
8. A. Hirai, T. Hamazaki, and T. Terano, Eicosapentaenoic acid and platelet function in Japanese, Lancet 2:1132 (1980).
9. H. O. Bang, J. Dyerberg, and N. Hjorne, The composition of food consumed by Greenland Eskimos, Acta Med. Scand. 200:69 (1976).
10. N. H. Nielsen and J. P. H. Hansen, Breast cancer in Greenland-- selected epidemiological, clinical and histological features, J. Cancer Res. Clin. Oncol. 98:287 (1980).

11. E. L. Wynder, D. P. Rose, and L. A. Cohen, Breast cancer in causation and therapy, Cancer 58(Suppl.):1804 (1986).

12. K. K. Carroll and H. T. Khor, Dietary fat in relation to tumorigenesis, Prog. Biochem. Pharmacol. 10:308 (1975).

13. A. E. Rogers, B. H. Conner, C. L. Bonlanger, S. Y. Lee, F. A. Carr, and W. H. DuMouchil, Enhancement of mammary carcinogenesis in rats fed a high lard diet only before or before and after DMBA, in: Basic and Clinical Aspects of Dietary Fiber, G. Vahouny and D. Kritchevski, eds., Plenum Press (1985).

14. G. J. Hopkins, T. G. Kennedy, and K. K. Carroll, Polyunsaturated fatty acids as promoters of mammary carcinogenesis induced in Sprague-Dawley rats by 7,12,dimethylbenz(a)anthracene, J. Natl. Cancer Inst. 66:517 (1981).

15. R. Kalaneghan and K. K. Carroll, Reversal of the promotional effect of high-fat diet on mammary tumorigenesis by subsequent lowering of dietary fat, Nutr. Cancer 6:22 (1984).

16. C. Ip, C. A. Carter, and M. M. Ip, Requirement of essential fatty acid for mammary tumorigenesis in the rat, Cancer Res. 45:1997 (1985).

17. R. A. Karmali, Do tissue culture and animal model studies relate to human diet and cancer?, Prog. Lipid Res. 25:533 (1986).

18. R. A. Karmali, J. Marsh, and C. Fuchs, Effect of omega-3 fatty acids on growth of a rat mammary tumor, J. Natl. Cancer Inst. 73:457 (1984).

19. H. Gabor and S. Abraham, Effect of dietary menhaden oil on growth and cell loss of transplantable mammary adenocarcinoma in BALB/c mice, J. Natl. Cancer Inst. 76:1223 (1986).

20. J. J. Jurkowski and W. T. Cave, Jr., Dietary effects of menhaden oil on the growth and membrane lipid composition of rat mammary tumors. J. Natl. Cancer Inst. 74:1145 (1985).

21. K. K. Carroll and L. M. Braden, Dietary fat and mammary carcinogenesis, Nutr. Cancer 6:254 (1985).

22. C. Ip, M. M. Ip, and P. Sylvester, Relevance of trans fatty acids and fish oil in animal tumorigenesis studies, in: Dietary Fat and Cancer, E. Ip, D. F. Birt, A.E. Rogers, and C. Mettlin, eds., Alan R. Liss, Inc. (1986).

23. R. A. Karmali, R. U. Doshi, L. Adams, and K. Choi, Effect of n-3 fatty acids on mammary tumorigenesis, Adv. Prost. Thromb. Leukotr. Res. 17:886 (1987).

24. T. P. O'Connor, B. D. Roebuck, F. Peterson, and T. C. Campbell, Effect of dietary intake of fish oil and fish protein on the development of L-azaserine-induced preneoplastic lesions in the rat, J. Natl. Cancer Inst. 75:959 (1985).

25. B. S. Reddy and H. Maruyama, Effect of fish oil on azoxymethane-induced colon carcinogenesis in male F344 rats, Cancer Res. 46:3367 (1986).

26. T. P. O'Connor, F. Peterson, B. Lokesh, J. Kinsella, B. D. Roebuck, and T. C. Campbell, Effect of dietary omega-3:omega-6 ratio on prostaglandin metabolism, membrane fatty acid composition and preneoplastic pancreatic lesion development in rats, Fed. Proc. 46:582, Abs. 1548 (1986).

27. R. A. Karmali, P. Reichel, L. A. Cohen, T. Terano, A. Hirai, Y. Tamura, and S. Yoshida, The effects of dietary w-3 fatty acids on the DU-145 transplantable human prostatic tumor, Anticancer Res. 7:1173 (1987).

28. S. A. Broitman, F. Cannizzo, A. Rogers, and L. S. Gottlieb, Comparison of dietary marine oil and safflower oil on growth and pulmonary colonization of CT-26 implanted into bowel of mice, Fed. Proc. 46:437, Abs. 704 (1987).

29. L. M. Adams and R. A. Karmali, Inhibition of artificial metastasis in rat mammary tumor 13762 by dietary fish oil, Fed. Proc. 46:437, Abs. 703 (1987).

30. W. J. Kort, I. M. Weijma, A. M. Bijma, W. P. van Schalkwijk, A. J. Vegroesen, and D. L. Westbroek, Omega-3 fatty acids inhibiting the growth of a transplantable rat mammary adenocarcinoma, J. Natl. Cancer Inst. 79:593 (1987).

31. E. B. Katz and E. S. Boylan, Stimulatory effect of high polyunsaturated fat diet on lung metastasis from 13762 mammary adenocarcinoma in female retired breeder rats. J. Natl. Cancer Inst. 79:351 (1987).

32. K. V. Honn, R. S. Bockman, and L. J. Marnett, Prostaglandins and cancer: a review of tumor initiation through tumor metastasis, Prost. 21:833 (1981).

33. N. T. Telang, A. Basu, M. J. Modak, and M. P. Osborne, Modulatory influence of fatty acids on induction and promotion of preneoplasia in adult mouse mammary gland organ cultures, Fed. Proc 46:582, Abs. 1546 (1987).

34. H. Zarbl, S. Sukunar, A. V. Arthur, D. Martin-Zanca, and M. Barbacid, Direct mutagenesis of Ha-ras-1 oncogenes by N-nitro-N-methylurea during initiation of mammary carcinogenesis in rats, Nature (London) 315:382 (1985).

35. M. Barbacid, Mutagens, oncogenes and cancer, in: Oncogenes and Growth Factors, R. A. Bradshaw and S. Prentis eds., pp. 90-99, Elsevier Science Publishers (1987).

36. S. H. Reynolds, S. J. Stowers, R. R. Maronpot, M. W. Anderson, and S. A. Aaronson, Detection and identification of activated oncogenes in spontaneously occurring benign and malignant hepatocellular tumors of the B6C3F1 mouse, Proc. Natl. Acad. Sci. 83:33 (1986).

37. N. J. Agnantis, C. Petraki, P. Markoulatos, and D. A. Spandidas, Immunohistochemical study of the ras oncogene expression in human breast lesions, Anticancer Res. 6:1157 (1986).

38. J. Schneider, D. Kinne, A. Fracchia, V. Pierce, K. Anderson, H. L. Bradlow, and J. Fishman, Abnormal metabolism of estradiol in women with breast cancer. Proc. Natl. Acad. Sci. USA 79:3047 (1982).

39. M. P. Osborne, H. L. Bradlow, R. Hershkopf, and J. Fishman, A potential marker for the risk of breast cancer: abnormal estradiol 16-alpha-hydroxylation, (Abs.) 37th Annual Meeting of Soc. of Surgical Oncol., New York (1986).

40. M. P. Osborne, R. A. Karmali, R. J. Hershcopf, H. L. Bradlow, I. Kourides, W. Williams, P. P. Rosen, and J. Fishman, Omega-3 fatty acids, modulation of estrogen metabolism and potential for breast cancer prevention, Cancer Invest. (in press).

41. S. H. Abou-El-Ela, K. W. Prasse, R. Carroll, and O. R. Bunce, Effects of dietary primrose oil on mammary tumorigenesis by 7,12-dimethylbenz(a)anthracene, Lipids 22:1041 (1987).

42. R. A. Karmali, J. Marsh, C. Fuchs, W. Hare, and M. Crawford, Effects of dietary enrichment with gamma-linolenic acid upon growth of the R3230AC mammary adenocarcinoma, Nutr. Growth Cancer 2:41 (1985).

MARINE MAMMALS: ANIMAL MODELS FOR STUDYING THE DIGESTION AND TRANSPORT OF DIETARY FATS ENRICHED IN w-3 FATTY ACIDS. POSITIONAL ANALYSES OF MILK FAT TRIACYLGLYCEROL MOLECULES

Donald L. Puppione, R.J. Jandacek, S.T. Kunitake, and D.P. Costa

University of California at Santa Cruz and San Francisco;
The Procter and Gamble Co., Miami Valley Labs.,
Cincinnati , U.S.A.

Abstract

In the depot fat of both marine and terrestrial animals, the relative percent distribution of the major fatty acids at the three positions of the glycerol backbone are not the same. Interestingly, studies have revealed that the polyenoic omega-3 fatty acids in marine mammalian blubber and fish oils are distributed differently. In marine mammalian blubber, the polyenoic omega-3 fatty acids are located on the outer positions of the molecule, i.e. the 1- and 3-positions. In fish oils, the polyenoic omega-3 fatty acids are located almost exclusively in the 2-position.

Because most marine mammals eat either fish or plankton (killer whales, polar bears and leopard seals being exceptions), we were interested in determining if the fat of marine mammalian milk contained triacylglycerol molecules (TG) with the polyenoic fatty acids distributed as in fish oil or as in the blubber of the lactating female. In our studies, we have obtained data on the milk fat of two species of pinnipeds, the northern elephant seal (Mirounga angustirostris) and the antarctic fur seal (Artocephalus gazella). These data demonstrate that the polyenoic omega-3 fatty acids together with the long chain monoenoic fatty acids are located on the 1- and 3-positions.

Introduction

In studying the effects of omega-3 fatty acids on the dynamics of fat transport, we have been using marine mammals as our animal model (1-6). In contrast to other experimental animals, studies of marine mammals have several advantages.

361

1) **They are not coprophagous and they naturally have a diet enriched in omega-3 fatty acids**
2) **They are able to ingest a large bolus of fat in a very short time. In 10 minutes a 60 kg seal by eating 6 to 9 kg of fish ingests between 600 to 900 g of fat along with 3 to 9 g of cholesterol**
3) **In spite of their diet, reports of atherosclerosis in these animals are rare. Levels of atherogenic lipoproteins also are low**
4) **They readily develop chylomicronemia.**

We currently are studying developmental changes in the plasma lipoproteins of neonatal elephant seals. The plasmas of the suckling pups are milky. Although as lactescent as human plasmas with TG concentrations between 1000 and 5000 mg/dl, the plasma TG concentration of nursing pups rarely exceed values between 250 and 300 mg/dl. We have reported on the presence of large structures (100 to 1600 nm) in ultracentrifugal fractions isolated from the plasmas of nursing pups. These structures, initially interpreted by us to be chylomicra, (6), may also be chylomicron remnants.

To learn more about these particles and dietary fat, we have characterized the fatty acid distribution in elephant seal milk TG. In particular, we wanted to determine whether or not the positional distribution of omega-3 fatty acids on TG of marine mammalian milk was similar to what has been reported for marine mammalian blubber or for fish oil (7,8). Based on our studies of two different species of pinnipeds, we report here that the lactating female does not synthesize milk TG to mimic the distribution of fish oils with the omega-3 fatty acids in the 2-position. Instead, these acids are distributed primarily on the 1- and 3-positions.

Methods

Milk samples (30 to 120 ml) were collected from females immobilized with ketamine hydrochloride as described by Costa et al.(9) and then injected with 60 units of oxytocin. Samples were kept frozen at -20°C until lipid analyses were conducted.

The milk lipids were extracted with chloroform:methanol (2:1). Extracts were weighed to obtain the percent lipid in the milk. Analyses of lipid extracts by thin layer chromatography (TLC) confirmed that the milk fat consisted almost exclusively of TG. The total fatty acids in the lipid extracts were methylated with boron trifluoride after undergoing saponification (10). The distribution of FA methyl esters was determined by gas liquid chromatography. Methyl esters were separated on a durabond 225 column (.00025 x 30 m) in a Hewlett-Packard model 5890a equipped with a hydrogen flame detector column at a temperature of 220°C. The carrier gas was helium and the flow rate was 1.42 ml/min. The detector and injection port temperatures were 350°C. The resulting peak areas were measured electronically with a Hewlett-Packard 3392a and normalized to give values as weight percent of total FA.

To determine the positional FA distribution, the lipid extracts were first hydrolyzed with porcine pancreatic lipase (steapsin) in a 1.0 M Tris buffer (pH 8.0) as described by Mattson and Volpenhein (11). Following extraction, the enzymatic digestion products, 2-mono= glycerides and free fatty acid released from the 1- and 3-positions, were separated by thin layer chromatography (TLC). Using fatty acid and monoglyceride standards, the positions of the digestion products on the plate and the completion of the reaction could be determined visually in an iodine chamber. Following a subsequent TLC separation with only the standards being visualized, the monoglycerides and free fatty acid were scrapped from the plate and recovered. Methylation of fatty acid and analyses of the resulting methyl esters for each of the recovered regions were done as described above.

Results

Table 1 shows the distributions of major FA for the total extracted TG of pinniped milk and for the FA distribution in the 2-position and in the 1- and 3-positions of these TG. As can be seen from these data, the content of saturated FA in the 2-position, 68.2% in the fur seal and 43.8% in the elephant seal, is more than two fold higher than in the total milk extract of both species. The content of both long chain monoenoic and omega-3 FA (20:5, 22:5 and 22:6), on the other hand, is higher in the 1- and 3-positions than in the total extract. Interestingly, when the contents of the long chain monoenoic and omega-3 FA in the total extracts of the two species are comapred, fur seals are found to have much higher levels of omega-3 FA, particularly 20:5, whereas the elephant seals have much higher levels of 20:1. These differences may be due to a combination of several factors. In contrast to elephant seals which fast for four weeks during lactation lactating fur seals actively forage for food. The mammary gland of the fur seal is able to utilize circulating chylomicra as a source of fat whereas in the elephant seal circulating free fatty acids mobilized from depot fat would be the major source. Other major differences between these two species are the depths at which they obtain their food as well as the type of food which they eat. The principal food of fur seals consist of krill which they obtain near the surface. The elephant seals primarily eat fish at depths between 300 and 600 meters during dives which are 20 to 25 minutes in duration (12).

Discussion

As demonstrated by Mattson and Volpenhein (13), the selective hydrolysis of the primary ester bonds of ingested fat TG by pancreatic lipase results in 75% of enterically synthesized TG having the same fatty acids in the 2-position as were located originally in the ingested fat. These workers (13) also showed this to be true for the 2-position of chylomicron TG isolated from mesenteric lymph.

In the circulation, the core TG of chylomicra undergo hydrolysis once they come in contact with the enzyme lipoprotein (LPL) located on the surface of endothelial cells. Like pancreatic lipase, LPL has positional specificity for the primary ester bonds and has no

TABLE 1

Triacylglycerol fatty acid distribution in marine mammalian milk
Weight percent of major fatty acids

	14:0	16:0	18:0	16:1	18:1	20:1	22:1	20:5	22:5	22:6
Antartic fur seals										
Total	10.1	20.7	1.4	12.5	23.2	1.9	0.3	12.6	2.1	6.3
2-position	19.2	49.0		14.0	8.7			2.6		
1- @ 3- positions	7.1	10.3	2.2	13.7	32.0	2.7		14.6	2.9	8.4
Northern elephant seal										
Total	2.3	13.2	3.2	3.9	39.5	15.0	4.6	2.6	1.2	4.4
2-position	5.4	36.5	1.9	6.7	30.8	2.6	1.0	1.0	0.1	2.1
1- @ 3- positions	1.2	5.5	4.1	2.8	45.5	20.1	5.6	2.5	1.2	4.4

effect on 2- monoglycerides (14,15). Redgrave,Kodali and Small (16) have recently reported that the degree of saturation of fatty acids in the 2-position of dietary fats affects the metabolism of both chylomicra and the chylomicron remnants generated through the action of LPL. These workers (16) have observed that when saturated fatty acids are in the 2-position of dietary TG both the hydrolysis of core TG by LPL and the clearance of chylomicron remnants by the liver are retarded.

The observations of Redgrave et al.(16) together with the positional data presented here in Table 1 have provided us with an explanation to an unexpected observation, viz. that in contrast to lactating females and weaned pups (females fast for four weeks and pups fast for 2.5 months), nursing elephant seals had demonstrable levels of beta-LDL (unpublished results). In studies of the lipoprotein distribution of captive seals, which had been either fed or fasted overnight (4), no demonstrable levels of beta-lipoproteins were observed. Because hepatic TG-rich lipoproteins (also called VLDL or pre-beta lipoproteins) have been demonstrated to be the metabolic precursors of beta-LDL, we naturally assumed that during their periods of prolonged fasting beta-lipoproteins would be detected in their plasmas. Instead, beta-lipoproteins were detected in the chylomicronemic plasmas of nursing pups. We now propose that the presence of beta-lipoproteins in the plasmas of nursing elephant seal pups is related to the relatively high content of saturated fatty acids in the 2-position of milk TG. This would be consistent with the observations of Redgrave et al.(16) because incomplete hydrolysis of chylomicron core TG would result in relatively lipid enriched core remnants being taken up by the liver. Overloaded by this high intake of calories, the hepatocytes would be stimulated to secrete excessive amounts of VLDL.

Finally, the accumulation of saturated 2-monoglycerides on the surface of chylomicron core remnants may be giving rise to the conditions enabling us to detect the large light scattering structures which we have described (6). This being the case, these structures should not be observed once these animals are fed fish which contain fat with highly unsaturated fatty acids in the 2-position.

Future studies

Having an animal model which naturally ingests fats differing in terms of the location of omega-3 fatty acids on the glycerol backbone, our future studies will attempt to determine if the large light scattering structures isolated from the plasmas of nursing pups can also be isolated from the plasmas of these animals when they are fed fish.

In clinical trials involving the administration of ethylated omega-3 fatty acids to human subjects, enteric synthesis of TG will take place independent of the 2-monoglyceride pathway. As a result, physiological processes involved in transport of TG following the digestion and absorption of ethylated fatty acids will not mimic those taking place after the ingestion of fish or fish oil. Moreover, TG synthesized after ingestion of these ethylated fatty acids would contain the omega-3 fatty acids predominantly in the 1- and 3-positions. If the 2-position of these TG had a high content of saturated fatty acids, we would predict that the large light scattering structures, which we have isolated from the plasmas of nursing elephant seal pups, could then be isolated from the plasmas of human subjects on the test supplement as well. We intend to use our physicochemical techniques developed in studies of pinniped lipemic plasmas to analyze plasmas being obtained in these clinical trials.

References

1. Puppione D.L., A.V.Nichols (1970) Characterization of the chemical and physical properties of the serum lipoproteins of certain marine mammals. Physiol.Chem.Physics 2, 49

2. Puppione D.L., T.Forte, A.V.Nichols (1970) Partial characterization of serum lipoproteins in the density interval 1.04-1.06 gm/ml. Biochim.Biophys.Acta 202, 392

3. Puppione D.L., T.Forte, A.V.Nichols (1971) Serum lipoproteins of killer whales. Comp.Biochem.Physiol. 39, 673

4. Puppione D.L. (1978) Serum lipoproteins in two species of phocids (Phoca vitulina and Mirounga angustirostris) during alimentary lipemia. Comp.Biochem.Physiol. 59B, 239

5. Puppione D.L. (1982) Marine mammalian lipoproteins, in "Handbook of electrophoresis : Lipoprotein studies of nonhuman species", vol. IV, L.Lewis and H.Naito eds., Chemical Rubber Co., Boca Raton, FL, pp. 79-100

6. Puppione D.L., L.Corash, S.T.Kunitake, D.L.Smith, D.P.Costa (1987) Pinnipeds: animal models for studying the effects of dietary fats on lipoproteins and platelets, in "Polyunsaturated fatty acids and eicosanoids", W.E.M. Lands ed., American Oil Chemist Publication, Champaign, IL, pp. 352-357

7. Brocherhoff H., R.G.Ackman, R.J.Hoyle (1963) Specific distribution of fatty acids in marine lipids. Archives Biochem.Biophys. 100,9

8. Ackman R.G. (1988) Some possible effects on lipid biochemistry of differences in the distribution on glycerol of long chain n-3 fatty acids in the fats of marine fish and marine mammals. Atherosclerosis 70, 171

9. Costa D.P., B.J.LeBoeuf, A.C.Huntley, C.L.Ortiz (1986) The energetics of lactation in the Northern elephant seal, Mirounga angustirostris. J.Zool. London (A) 209, 21

10. Metcalfe L.D., A.A.Schmitz, J.R.Pelka (1966) Rapid preparation of fatty acid esters from lipids fro gas chromatographic analysis. Anal.Chem. 38, 514

11. Mattson F.H., R.A.Volpenhein (1961) The use of pancreatic lipase for determining the distribution of fatty acids in partial and complete glycerides. J.Lipid Res. 2, 58

12. LeBoeuf B.J., D.P.Costa, A.C.Huntley, S.D.Feldkamp (1988) Continuous, deep diving in female northen elephant seals, Mirounga angustirostris. Can.J.Zool. 66, 446

13. Mattson F.H., R.A.Volpenhein (1964) The digestion and absorption of triglycerides. J.Biol.Chem. 239, 2772

14. Nilsson-Ehle P., T.Egelrud, P.Belfrage, T.Olivecrona, B.Borgstrom (1973) Positional specificity of purified milk lipase. J.Biol. Chem. 248, 6734

15. Morley N.H., A.Kuksis, D.Buchnea, J.J.Myher (1975) Hydrolysis of diacylglycerols by lipoprotein lipase. J.Biol.Chem. 250, 3414

16. Redgrave T.G., D.R.Kodali, D.M.Small (1988) The effect of triacyl-sn-glycerol structure on the metabolism of chylomicrons and triacyl= glycerol-rich emulsions in the rat. J.Biol.Chem. 263, 5118 .

CHANGES IN THE POLYUNSATURATED FATTY ACID PROFILES IN ZELLWEGER SYNDROME

SUGGESTING A NEW ENZYMATIC DEFECT: DELTA-4 DESATURASE DEFICIENCY

Manuela Martinez

Laboratorio de Cromatografia
Hospital Infantil Vall d'Hebron
Barcelona, Spain

Zellweger (cerebro-hepato-renal) syndrome is a very severe peroxisomal disorder, leading to death within the first 6 months of life. A lack of peroxisomes in hepatocytes and renal proximal tubules has been detected in Zellweger patients (1). Deficiency of peroxisomal enzymes results in several biochemical alterations, such as a reduction in plasmalogen synthesis (2) and a decrease in the β-oxidation of saturated and monoinsaturated very long chain fatty acids (mainly 26:0 and 26:1), with a significant accumulation of these compounds in the affected tissues (3). To date, no changes in the polyunsaturated fatty acid (PUFA) patterns of Zellweger patients have been described. This report presents evidence of very drastic anomalies in the PUFA profiles of a case of Zellweger syndrome, strongly suggesting the existence of a new enzymatic defect in peroxisomal disorders involving the desaturase system of long polyunsaturated fatty acids.

Total fatty acids and plasmalogens were studied quantitatively in red blood cells, fibroblasts, forebrain, liver and kidney of a 3-month-old child who died from Zellweger syndrome with neurological and X-ray pictures very characteristic of the disease. Five neurologically normal (or 6, in the case of cells) infants, with ages ranging from 0 to 5 months, were also studied as controls. Given the enormous alterations found in the PUFA composition of the patient's tissues, the fatty acid patterns of ethanolamine and choline phosphoglycerides (EPG and CPG) were determined. Total fatty acid methyl esters (FAME) and aldehyde dimethyl acetals (DMA) were obtained by direct transesterification of the tissues and cells by the method of Lepage and Roy (4). After lipid extraction with chloroform-methanol-water 50:50:15 and separation of neutral from acidic lipids by DEAE Sephadex (5), EPG and CPG were separated by TLC and methanolyzed by the same procedure (4). Methyl 13:0 and 19:0 were used as internal standards to quantify all fractions. The FAME and DMA were separated by GLC on a 30 m long, SP-2330 capillary column (0.25 mm ID), working on a two-step temperature programme (140-180°C, at 4°C/min; 180-210°C, at 2°C/min). This programme allowed a very clean separation of 26:0 and 26:1 from the two artefactual peaks coming from cholesterol esters in total fatty acid analysis (see Fig. 1), without the need for using TLC for FAME purification, a step that always leads to losses of PUFA.

Table 1 presents the most significant data in absolute, quantitative terms. Besides the typical increase in the ratios 26:0/22:0 and 26:1/22:0 and the great reduction in the plasmalogen levels, confirming the diagnosis of Zellweger syndrome, some hitherto undescribed, striking changes in the

Table 1. Main total fatty acid and plasmalogen changes in Zellweger syndrome compared to controls.

	Cerebrum		Liver		Kidney		Erythrocytes		Fibroblasts	
	ZS	Controls (n = 5)	ZS	Controls (n = 5)	ZS	Controls (n = 5)	ZS	Controls (n = 6)	ZS	Controls (n = 6)
26:0/22:6ω3	0.008	0.02 ± 0.01	0.393	0.01 ± 0.002	5.28	0.03 ± 0.004	0.84	0.07 ± 0.02	0.25	0.02 ± 0.01
26:1/22:6ω3	0.035	0.011 ± 0.007	0.586	0.009 ± 0.002	2.13	0.041 ± 0.008	0.93	0.06 ± 0.01	0.196	0.021 ± 0.003
22:6ω3/22:5ω3	14.4	42.3 ± 6.7	0.73	21.6 ± 7.8	0.44	7.3 ± 2.1	0.73	5.9 ± 1.2	0.89	1.39 ± 0.13
22:5ω6/22:4ω6	0.09	0.49 ± 0.06	0.08	1.40 ± 0.19	0.05	0.38 ± 0.04	0.18	0.32 ± 0.02	0.06	0.14 ± 0.02
20:4ω6/20:3ω6	3.07	10.49 ± 1.59	3.89	5.15 ± 1.14	5.17	10.01 ± 0.56	1.86	9.14 ± 0.94	11.21	12.66 ± 1.16
26:0/22:0	0.09	0.19 ± 0.08	0.47	0.06 ± 0.02	0.66	0.05 ± 0.01	0.39	0.20 ± 0.03	1.04	0.05 ± 0.01
26:1/22:0	0.41	0.32 ± 0.21	0.70	0.05 ± 0.009	0.27	0.06 ± 0.01	0.43	0.17 ± 0.02	0.83	0.07 ± 0.01
22:6ω3	1478	4943 ± 633	183	3188 ± 281	58	933 ± 174	32	283 ± 31	16.8	22.2 ± 2.1
22:5ω6	181	1823 ± 170	32	619 ± 106	26	225 ± 50	28	58 ± 6	2.4	2.0 ± 0.13
20:4ω6	7314	6802 ± 736	7174	9880 ± 954	5768	8060 ± 816	619	853 ± 49	85.4	91.9 ± 5.6
20:4ω3	26.0	6.0 ± 3.8	23.6	16.1 ± 5.9	37.2	14.2 ± 3.1	4.6	0.7 ± 0.7	0.6	0.2 ± 0.11
18:4ω3	246	9.2 ± 4.9	25.9	14.5 ± 5.4	22.8	6.1 ± 2.4	ND	ND	0.4	0.1 ± 0.04
18:3ω6	39.2	23.1 ± 5.8	116	149 ± 12.8	43.0	34.2 ± 5.4	13.1	2.0 ± 0.8	0.4	0.7 ± 0.12
TFA	68422	70633 ± 8771	92504	105805 ± 8181	48751	56392 ± 5213	8481	6406 ± 294	673	635 ± 27
16DMA/16:0	0.024	0.107 ± 0.013	0.001	0.012 ± 0.002	0.003	0.114 ± 0.017	0.0003	0.08 ± 0.004	0.046	0.150 ± 0.004
18DMA/18:0	0.015	0.142 ± 0.016	0.005	0.023 ± 0.001	0.0009	0.102 ± 0.016	0.006	0.181 ± 0.011	0.025	0.101 ± 0.007
TP	770	5474 ± 1354	102	706 ± 42	46	2566 ± 328	12	371 ± 24	13	42 ± 1.2

ZS = Zellweger syndrome; TFA = total fatty acids; DMA = dimethyl acetals.
All the values are given on a molar basis: nmol/g of wet weight, for tissues; nmol/mg of protein, for fibroblasts; and nmol/ml of packed cells, for erythrocytes. The age of controls range from 0 to 5 postnatal months, and the figures represent the mean ± SE.

PUFA composition of the patient's tissues were discovered. The most important was an enormous decrease in the two products of Δ4 desaturation, 22:6ω3 and 22:5ω6, with a great decrease in the ratios 22:6ω3/22:5ω3 and 22:5ω6/22:4ω6. In the patient's kidney, the levels of 22:6ω3 fell below those of 26:0 and 26:1, causing an inversion of the normal situation, emphasized by a great increase in the index 26:0/22:6ω3 (and 26:1/22:6ω3) to over 150 times the

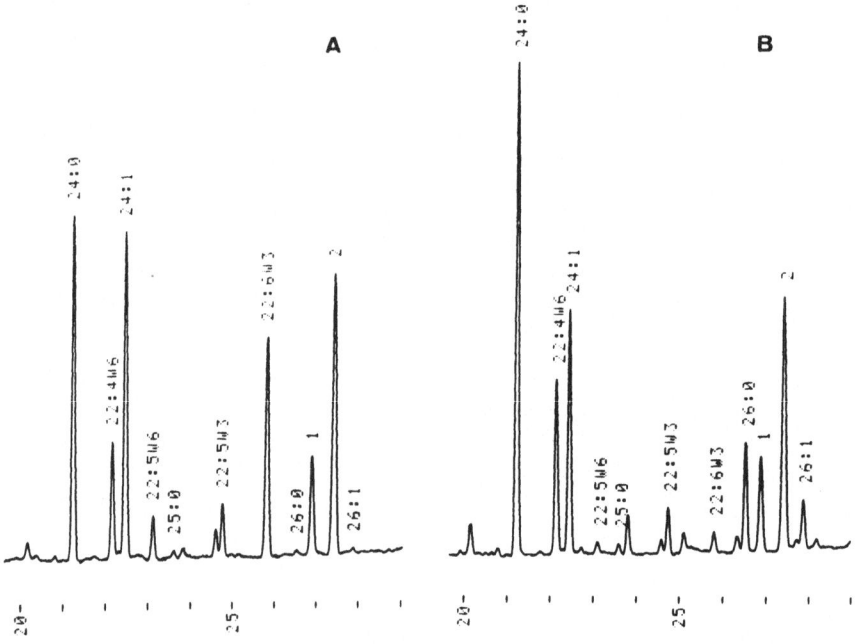

Fig. 1. Part of two gas chormatograms showing the long chain fatty acid methyl esters of renal tissue. A: Normal profile in a 3-month-old infant. B: Altered profile in the Zellweger patient. The peaks marked 1 and 2 are artifacts, totally resolved from 26:0 and 26:1. Note that 26:0 is bigger than 22:6 3 in the Zellweger kidney(B).

Table 2. Fatty acid composition of ethanolamine phosphoglycerides in Zellweger syndrome compared to controls.

	Cerebrum		Liver		Kidney	
	ZS	Controls (n = 5)	ZS	Controls (n = 5)	ZS	Controls (n = 5)
16:0	9.8	7.7 ± 0.5	23.6	22.6 ± 1.2	12.5	14.9 ± 1.6
16:1ω7	0.9	0.6 ± 0.2	1.1	1.0 ± 0.3	1.1	0.8 ± 0.1
18:0	29.5	28.5 ± 0.9	21.0	23.8 ± 0.6	25.9	24.7 ± 1.7
18:1ω9	13.0	7.6 ± 0.8	7.8	6.9 ± 1.0	17.7	15.2 ± 1.0
18:2ω6	1.2	0.4 ± 0.1	21.0	8.2 ± 2.5	10.8	5.7 ± 1.6
20:3ω6	4.4	2.1 ± 0.4	2.6	1.9 ± 0.1	2.7	2.2 ± 0.2
20:4ω6	20.9	16.3 ± 0.2	17.0	19.7 ± 2.3	22.2	25.1 ± 3.1
22:4ω6	5.5	13.3 ± 0.4	0.9	0.8 ± 0.1	2.1	2.0 ± 0.6
22:5ω6	0.4	5.2 ± 0.6	0.1	1.2 ± 0.2	0.1	1.0 ± 0.2
22:5ω3	0.2	0.4 ± 0.1	0.7	0.8 ± 0.3	0.4	0.3 ± 0.1
22:6ω3	2.8	12.9 ± 0.8	0.7	8.4 ± 1.8	0.4	3.1 ± 0.7

ZS = Zellweger syndrome. The values are molar per cent (mean ± S.E.).

normal values. There was an increase in the minor constituents 18:3ω6, 18:4ω3 and 20:4ω3 in some of the affected tissues, and the ratio 20:4ω6/20:3ω6 was significantly decreased in all the samples. Despite the striking changes in the fatty acid composition of the patient's tissues, it can be seen in Table 1 that the total amount of fatty acids was totally normal in all the samples analyzed.

Tables 2 and 3 show the most significant fatty acids in the EPG and CPG fractions of total cerebrum, liver and kidney, expressed as molar per cent. It can be seen that both phospholipid fractions were very poor in 22:6ω3 and 22:5ω6 in the Zellweger patient compared to controls. The parent fatty acid, 18:2ω6, was significantly increased in the two patient's phosphoglycerides, whereas 20:4ω6 was decreased in CPG, mainly in the kidney.

Table 3. Fatty acid composition of choline phosphoglycerides in Zellweger syndrome compared to controls.

	Cerebrum		Liver		Kidney	
	ZS	Controls (n = 5)	ZS	Controls (n = 5)	ZS	Controls (n = 5)
16:0	51.2	50.1 ± 1.3	39.2	36.2 ± 0.8	37.8	32.8 ± 2.0
16:1ω7	2.2	1.1 ± 0.2	2.0	2.7 ± 0.7	1.7	1.7 ± 0.3
18:0	6.3	11.1 ± 0.8	9.3	12.4 ± 0.7	11.2	11.2 ± 0.4
18:1ω9	21.4	19.6 ± 2.8	12.8	14.4 ± 1.6	20.6	19.3 ± 0.7
18:2ω6	1.4	0.5 ± 0.1	22.9	11.9 ± 3.0	12.6	8.6 ± 2.3
20:3ω6	1.6	0.9 ± 0.1	3.1	2.2 ± 0.2	2.6	1.8 ± 0.2
20:4ω6	3.5	5.0 ± 0.4	6.9	10.3 ± 1.6	5.7	14.6 ± 1.4
22:4ω6	0.3	1.2 ± 0.4	0.3	0.4 ± 0.1	0.6	1.2 ± 0.3
22:5ω6	tr.	0.5 ± 0.1	0.1	0.5 ± 0.1	0.1	0.3 ± 0.1
22:5ω3	tr.	tr.	0.3	0.4 ± 0.1	0.2	0.4 ± 0.1
22:6ω3	0.2	1.3 ± 0.3	0.1	3.1 ± 0.5	0.1	1.7 ± 0.4

The findings reported here clearly indicate that the reactions 22:5ω3 → 22:6ω3 and 22:4ω6 → 22:5ω6 were very poor in the patient's tissues, suggesting a defect of Δ 4 desaturase. To a lesser extent, other desaturation and elongation reactions may also be affected, as indicated by the increase in 18:2 ω6, 18:3ω6, 18:4ω3 and 20:4ω3 and by the decrease in 20:4ω6. It is worthwile noticing that fibroblasts, which were clearly altered in very long chain fatty acids and plasmalogen levels, were nearly normal in PUFA composition. This can be due to the fact of being cells cultured in an artificial medium. An alternative explanation could be that fibroblasts, in contrast to other tissues, may not be totally deficient in peroxisomes in Zellweger syndrome (6).

An enzyme defect of the long PUFA desaturase system in a peroxisomal disorder would favour the view that peroxisome derive from the endoplasmic reticulum (7). In fact, in the original description by Goldfischer at al (1) of peroxisomal defects in Zellweger patients, smooth endoplasmic reticulum was reported to be very scarce.

Although only one case has been studied so far, the significance of the findings is such that it urges investigators of peroxisomal disorders to study the desaturase system in these patients. The proposed desaturase defect could contribute to clarify the pathogenesis of Zellweger syndrome. So drastic a decrease in the main long PUFA 22:6ω3 (docosahexaenoic acid), a fatty acid mainly enriched in synaptic membranes and the retina, could explain the mental deterioration and visual impairment in Zellweger syndrome, especially when such a decrease cannot be compensated for by any increase in the other long PUFA 22:5ω6. Should this deficiency be confirmed in other cases of Zellweger syndrome it would be worthwile assaying diets rich in long polyunsaturated fatty acids in these patients.

References

1. S. Goldfischer, C.1. Moore, A.B. Johnson, A.J. Spiro, M.P. Valsamis, H.K. Wisniewski, R.H. Ritch, W.T. Norton, I. Rapin, and L.M. Gartner, Peroxisomal and mitochondrial defects in the cerebro-hepato-renal syndrome, Science 182: 62 (1973).
2. H.S.A. Heymans, R.B.H. Schutgens, R. Tan, H. van den Bosch, P. and P. Borst, Severe plasmalogen deficiency in tissues without peroxisomes (Zellweegr syndrome), Nature 306: 69 (1983).
3. F.R. Brown, A.J. McAdams, J.W. Cummins, R. Konkol, I. Singh, A.B. Moser, and H.W. Moser, Cerebro-hepato-renal (Zellweger) syndrome and neo-natal adrenoleukodystrophy: similarities in phemotype and accumula-tion of very long chain fatty acids, Johns Hopkins Med. J. 151: 344 (1982).
4. G. Lepage, and C.C. Roy, Direct transesterification of all classes of lipids in a one-step reaction, J. Lipid Res. 27: 114 (1986).
5. L.J. Macala, R.K. Yu, and S. Ando, Analysis of brain lipids by high performance thin-layer chromatography and densitometry, J. Lipid Res, 24: 1243 (1983).
6. J.A. Arias, A.B. Moser, and S.L. Goldfischer, Ultrastructural and cytochemical demonstration of peroxisomes in cultured fibroblasts from patients with peroxisomal deficiency disorders, J. Cell Biol. 100: 1789 (1985).
7. P.M. Novikoff, A.B. Novikoff, N. Quintana, and C. Davis, Studies on microperoxisomes. III. Observations on human and rat hepatocytes, J. Histochem. Cytochem. 21: 540 (1973).

SELECTIVE INCORPORATION OF EICOSAPENTAENOIC ACID IN POSITION 2 OF PLASMA TRIGLYCERIDES AFTER A DIETARY SUPPLY OF 2 EICOSAPENTAENOYL GLYCERIDE IN MAN

B. Descomps, S. El Boustani, L. Monnier, F. Mendy, and
A. Crastes de Paulet

INSERM U.58; Laboratoire Biochemie A; Service des Maladies
Metaboliques et Endocriniennes du CHR, Montpellier, France

The chemical form of polyunsaturated fatty acids (PUFA) in dietary sources is an essential determinant for their absorption and bio-availability. For example, we recently showed in man that enteral absorption of eicosapentaenoic acid (EPA) is higher and rapid when EPA is ingested as free fatty acid or triglyceride but is low when EPA is ingested as ethylester (1). This result was recently confirmed (2). Moreover, the ingestion of a synthetic triglyceride carrying EPA in position two, 1,3-dioctanoyl-2-eicosapentaenoyl glycerol (2EPA) is followed by a high incorporation of EPA in plasma triglycerides (1).

The present investigation was designed to check if in man, the ingestion of triglycerides carrying a PUFA in position 2 such as 2 EPA would preferentially result in insertion of this PUFA in position 2 of plasma triglycerides. For this purpose, three normal adult fasting subjects were given the equivalent of 1 g EPA in the form of the triglyceride : 2-EPA. The position of EPA in the plasma TG of the subjects was determined at the time of maximal incorporation into TG (4-6 h after ingestion) by combination of selective enzymatic hydrolysis with pancreatic lipase, thin layer chromatography and gas liquid chromatography.

Methods

The methods used for extraction of plasma lipids, separation of the plasma lipid fractions, transesterification of fatty and acid quantitation of fatty acids by gas liquid chromatography are described in ref. 3.

Reference TG: 1,3-dioctanoyl-2-eicosapentaenoyl glycerol (2-EPA) and 2,3-dioctanoyl-1-eicosapentanoyl glycerol (1-EPA) were pure samples from Roussel-Uclaf (Romainville, France) and were controlled by capillary column gas chromatography coupled with mass spectrometry.

Conditions for selective enzymatic hydrolysis of plasma TG and reference TG are the following: the TG (500 ug of plasma TG or 500 ug of the reference standards 2-EPA or 1-EPA) were submitted to enzymatic hydrolysis in 1 ml Tris HCl, pH 8 after addition of 100 ul of a $CaCl_2$ solution (22%) 250 ul of a taurocholate solution (0.1%) and 50-100 ug of porcine pancreatic lipase (Interchim, Sigma) all dissolved in bidistilled water. The mixture was maintained at $37^{\circ}C$ for 10 to 20 min. with moderate shaking. The incubation was stopped by addition of 0.6 N HCl (1 drop) and the lipids twice extracted with 3 ml ethyl ether. The extract was washed, dried, evaporated to dryness, dissolved in 100 ul of chloroform-methanol (1:1,v/v) and layered on Kieselgel GF254 for separation of the hydrolysis products by thin layer chromatography using hexane/ethyl ether/acetic acid (50/50/1,v/v/v) as developing solvent. After revelation by 2',7'-dichlorofluoresceine spraying the gel containing the free acid, monoglyceride, diglyceride and triglyceride fractions were scraped and the fatty acids of each fraction were transesterified after addition of C 19:0 and C 20:0 (20 ug each) as internal standards. The fatty acids were analyzed by GLC on a fused silica FFAP capillary column as described earlier (3).

Results

The experimental conditions for the selective enzymatic hydrolysis were checked in preliminary experiments in which were comapred the enzymatic hydrolysis of reference synthetic triglycerides : 1,3-di= octanoyl-2-eicosapentaenoyl glycerol and 2,3-dioctanoyl-1-eicosapenta= enoyl glycerol (1-EPA). Conditions were selected (50 ug of pancreatic lipase, 10 min. incubation) in which maximal enzymatic hydrolysis of the esters in position 1 and 3 was associated with minimal liberation of the fatty acid in position 2. In these conditions, the enzymatic hydrolysis of a TG sample containing EPA in a definite position results in the liberation of EPA as free fatty acid if it was esterifying position 1 or 3 of the initial TG, or to a 2 monoglyceride if the esterification was in position 2. Thus, the determination is reduced to the quantification of EPA present as a free acid or as a 2 mono- glyceride in the products of the selective enzymatic hydrolysis.

The plasma TG of the three subjects of the experiment were separated 4-6 hours after ingestion of the triglyceride 2-EPA and submitted to selective hydrolysis by pancreatic lipase in the conditions above described. The distribution of EPA between the products of hydrolysis, monoglycerides and other fractions is shown in the following table.

This result points out that 100% of the EPA recovered in the products of hydrolysis by pancreatic lipase remained as 2-monoglyceride. No EPA could be detected in the free fatty acid fraction.

Table 1

Subject	EPA monoglycerides (μg)	EPA in other fractions
1	1.46	Undetectable
2	0.96	Undetectable
3	0.36	Undetectable

In the same conditions of hydrolysis, the pancreatic lipase releases 75% of the EPA contained in the reference product 1-EPA (EPA in position 1), whereas more than 92% of EPA remains in the monoglycerides resulting from action of the enzyme on 2-EPA (EPA in position 2).

Since 100% of EPA recovered after selective enzymatic hydrolysis of the plasma TG of the subjects was found in the monoglycerides, it is concluded that this acid was in position 2 of the plasma TG. Then after ingestion of 2-EPA, the polyunsaturated fatty acid escapes to random distribution between the three positions of the triglyceride during the absorption process.

The insertion of PUFAs in position 2 of dietary TG appears as an interesting way for supplying directly and efficiently these acids in a critical position of the glycerol molecule in plasma lipids.

References

1. El Boustani S., C.Colette, L.Monnier, B.Descomps, A.Crastes de Paulet, F.Mendy (1987) Enteral absorption in man of eicosapentaenoic acid in different chemical forms, Lipids 22, 711
2. Lawson L.D. and B.G.Hugues (1988) Human absorption of fish oil fatty acids as triacylglycerols, free acids or ethyl esters, Biochem. Biophys.Res.Com. 152, 328
3. El Boustani S., B.Descomps, L.Monnier, J.Warnant, F.Mendy, A.Crastes de Paulet (1986) In vivo conversion of dihomogamma linolenic acid in man, Prog.Lipid Res. 25, 67 .

THE EFFECTS OF A FISH OIL PREPARATION (EPAGIS) ON SERUM LIPIDS, APOPROTEINS, CIRCULATING PLATELET AGGREGATES AND HEMORHEOLOGIC PARAMETERS IN HYPERLIPIDEMIC SUBJECTS

P. Green[1,4], J. Fuchs[2], P. Budowski[3], Y. Lurie[1,4], I. Beigel[1,4], J. Agmon[2], L. Leibovici[1], R. Mamet[5], and N. Schoenfeld[5]

1-Dept.Internal Medicine B,Beilinson Med.Ctr., Petah Tiqva, and Tel Aviv Univ. Sackler Sch.Med., Tel Aviv
2- Dept.Internal Med. A and Israel and Ione Massada Ctr.Heart Dis., Beilinson Med.Ctr., Petah Tiqva, and Tel Aviv Univ.Sackler Sch.Med., Tel Aviv
3- Fac.Agric., The Hebrew Univ. of Jerusalem, Rehovot
4- The Lipid Unit and "Salah" Lab.Lipid Res.,Beilinson Med.Ctr., Petah Tiqva
5- The Lab.Biochem.Pharmacol.,Beilinson Med.Ctr., Petah Tiqva, Israel

Fish oil as a means of preventing cardiovascular disease has recently attracted much attention (1). Many studies on the effects of fish oil ingestion on various cardiovascular risk factors were published, but only a few were well controlled and double-blind (2). The aims of our study were to examine the effects of a moderate intake of fish oil on several cardiovascular risk factors including serum lipids, apoproteins, some platelet functions and hemorheologic parameters in hyperlipidemic subjects, in a randomized, controlled, double-blind, cross-over fashion.

Subjects and methods

Twenty seven hyperlipidemic patients,15 Type IIB and 12 Type IV, participated in the study. Subjects ingested 15 g of an encapsulated fish oil preparation (EPAGIS) daily ("active"), providing 5.2 g w-3 polyunsaturated fatty acids (PUFAs), or an identical amount of encapsulated vegetable oil preparation, consisting in a mixture of olive and corn oil ("placebo"). Each supplement was taken for 8 weeks, with a 4 weeks wash-out period between the two trial periods. Serum total cholesterol, triglycerides and HDL-C (high density lipoprotein cholesterol) were determined at 2 weeks intervals, whereas determinations of apoproteins, the fatty acid content of serum, erythrocytes and platelets, the platelet function studies and the hemorheologic parameters were made at the beginning and the end of each period. Fatty acids were measured by gas liquid chromatography, apoproteins by an immunoturbidimetric method (3), cholesterol, triglycerides and HDL-C by commercial kits (Boehringer Mannheim), circulating platelet aggregates and mean platelet aggregate size as previously described (4,5), plasma viscosity by a capillary method using a Harkness Coulter viscometer, and deformability of red blood cells by a filtration method (6).

Statistical analysis: paired and non paired t tests for continous variables compared between two categories and analysis of variance and analysis of variance for repeated measures for comparisons among several categories.

Results

The changes in the fatty acid content of the serum, platelets and erythrocytes during the supplementation periods are shown in Figure 1.
The changes in serum total cholesterol, triglycerides and HDL-C are shown in Figure 2. No changes were observed in the apoprotein levels after the ingestion of either the fish oil or the vegetable oil (results not shown). The platelet function studies are summarized in Table 1 which also contains the normal values in our laboratory.
Hemorheologic changes are shown in Table 2.

Table 1

| | Fish oil (active) | | Veget. oil (placebo) | | Normal |
	Before	After	Before	After	values
Platelet count (* 10^3/ml)	226.8+65.9	217.0+80.6	240.1+69.7	242.0+60.5	
Percent of aggr. platelets	16.5+4.7	17.4+5.0	18.8+3.7	17.7+4.8	6.02+2.0
Percent of irrev. aggr. platelets	9.5+4.0	9.4+3.6	10.6+3.1	10.3+4.0	4.0+2.0
Percent of rev. aggr.platelets	6.8+5.1	8.0+5.8	7.7+3.1	7.0+4.1	2.0+1.0
Mean platelet aggregate size	2.2+0.2	2.2+0.1	3.3+3.9	2.8+3.3	2.2+0.4

* Significance of difference between after and before the oil feeding , p = 0.0001

Values are mean + SD

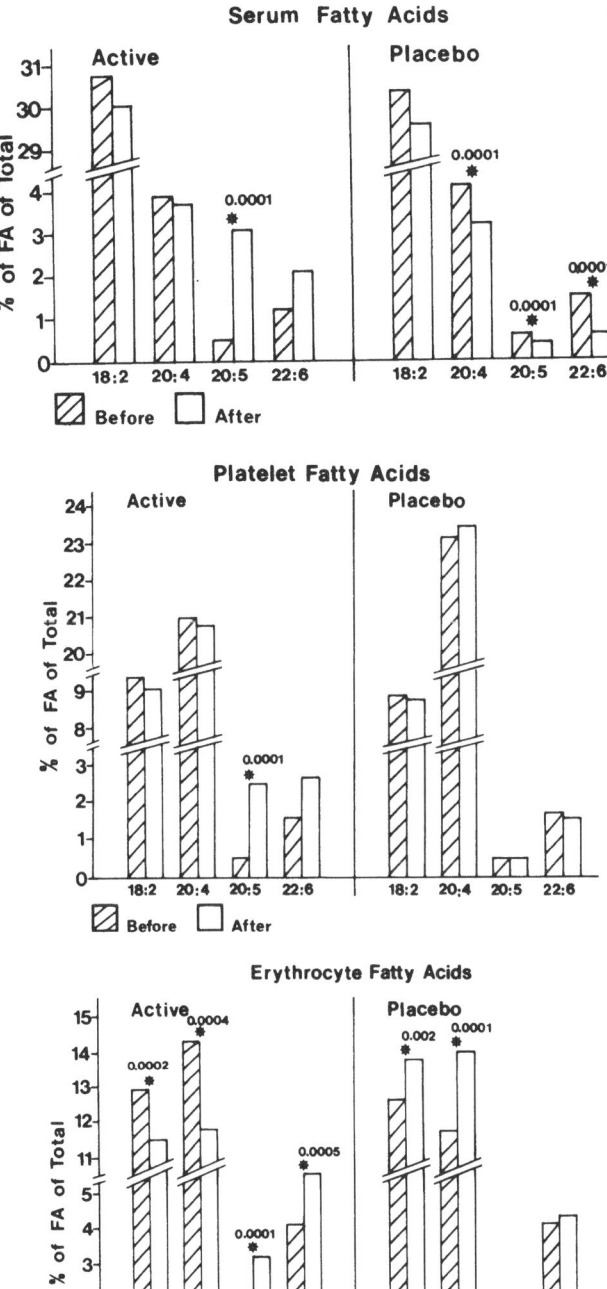

Figure 1. Serum, platelet and erythrocyte fatty acid composition changes during the feeding periods. Asterisks denote significant changes after the oil feeding as compared to before, and the numbers represent the significance of difference (p).

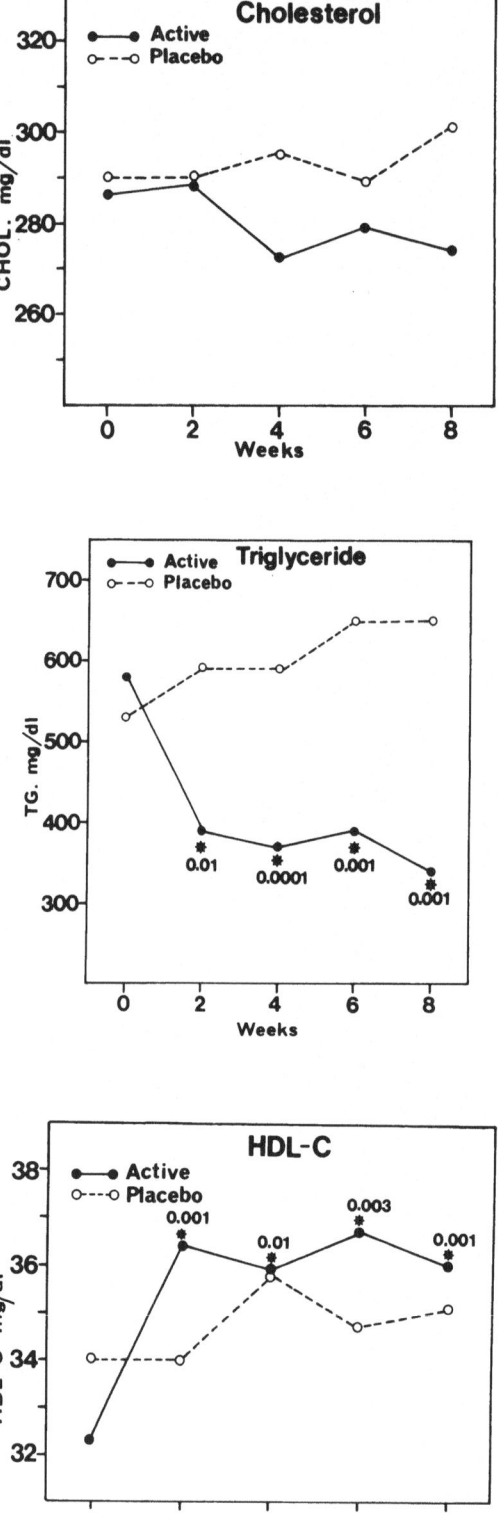

Figure 2. Serum total cholesterol, triglycerides and HDL-cholesterol
changes during the feeding periods. Asterisks as in Figure 1.

Discussion

The changes in the fatty acid content of the serum, platelets and erythrocytes were as expected, showing good compliance with the ingestion of the supplements (Fig.1). Concomitantly with the increase in the content of the w-3 PUFAs, a decrease in the w-6 PUFAs content of the erythrocytes was observed.

The well known (7) triglyceride lowering effect of fish oil ingestion was shown also in our study. This effect is more noteworthy if we take into account the high linoleic acid content of the food (8)

Table 2

	Fish oil (active)		Vegetable oil (placebo)	
	Before	After	Before	After
Plasma viscosity (cp.)	1.4000+ 0.08552	1.3495+ 0.088647*	1.3969+ 0.07707	1.3757+ 0.07933
Filtrability of RBC (sec./ml)	33.6+5.4	34.0+8.3	35.6+6.3	34.0+8.1

Values are means $\pm$ SD

and the adipose tissue (9) of the population in Israel. Furthermore patients in whom a significant triglyceride lowering effect was achieved with the fish oil capsules stated that they prefered this form of treatment over their previous treatment with drugs, mostly a fibric acid derivative. No significant changes were observed in the total serum cholesterol levels following the fish oil ingestion, as was found in several other investigations (10-12). HDL cholesterol levels were significantly higher after the fish oil ingestion, a finding which was observed by some investigators (12), but not by others (10,11).

No changes were observed in the apoprotein levels of our subjects following ingestion of either fish oil providing 5.2 g/d w-3 PUFAs, or vegetable oil. Similar results were reported by investigators who gave fish oil providing 3 and 7.5 g/d w-3 PUFAs to hypertriglyceridemic subjects (12,13). Increases in apoprotein B levels were observed in some clinical trials in which hypertriglyceridemic subjects were fed fish oil in amounts providing 4.6 g/d (14) or 6-7 g/d (15) w-3 PUFAs. In contrast, significant decreases in apoprotein B levels were noted after feeding large amounts of fish oil providing 20-30 g/d w-3 PUFAs (16).

The percentage of circulating aggregated platelets was found elevated in our study population, most of them being irreversibly aggregated. Treatment with fish oil or vegetable oil did not affect their number.

Similar distribution of aggregated platelets was found in patients with stable angina (4) and it may represent a chronic state of platelet activation.

Although w-3 PUFAs were incorporated into erythrocytes, no change in their deformability was observed after the ingestion of fish oil. This finding is in contrast to that of Tamura et al.(17) who found increased erythrocyte deformability in hyperlipidemic subjects following ingestion of ethyl-ester of EPA. We observed a significant decrease in the plasma viscosity of the patients after the fish oil feeding but not after the supplementation of the vegetable oil. The mechanism of this decrease is not clear, since no changes in the plasma proteins were observed and there was no correlation between the decrease in the viscosity and the changes in the lipid levels.

The results of this study suggest that fish oil ingestion in a moderate dose is a feasible treatment for hypertriglyceridemia. Furthermore, the improvement in the lipid profile observed in our patients together with the decrease in the plasma viscosity point toward a potential beneficial effect of fish oil ingestion on the atherosclerotic process.

References

1. Leaf A. and P.C.Weber,(1988) Cardiovascular effects of n-3 fatty acids. N.Eng.J.Med. 318, 549

2. Herold P.M. and J.E.Kinsella (1986) Fish oil consumption and decreased risk of cardiovascular disease: a comparison of findings from animal and human trials. Am.J.Clin.Nutr. 43, 566

3. Slutzky et al.(1987) Quantitative determination of apolipoproteins A-1 and B with a rapid immunoturbidimetric assay. Clin.Chem. 33 897

4. Fuchs J., I.Weinberger, Z.Rotenberg et al.(1987) Circulating aggregated platelets in coronary artery disease. Am.J.Cardiol. 60, 534

5. Weinberger I., J.Fuchs, Z.Rotenberg et al.(1986) Circulating platelet aggregate size in ischemic heart disease. Angiology 37, 676

6. Reid H.J., A.J.Barner, P.J.Lock et al.(1976) A simple method for measuring erythrocyte deformability. J.Clin.Pathol. 29, 855

7. Connor W.E., Hypolipidemic effects of dietary omega-3 fatty acids in normal and hyperlipidemic humans: effectiveness and mechanisms. In: Health effects of polyunsaturated fatty acids in seafoods, A.P.Simopoulos, R.R.Kifer and R.E.Martin eds.,Academic Press,Inc., 1986

8. Central Bureau of Statistics, Jerusalem: Food Balance Sheets, 1985/86

9. Blondheim S.H., T.Horne, R.Davidovich et al. (1976) Unsaturated fatty acids in adipose tissue of Israeli Jews. Isr.J.Med.Sci. 12, 658

10. Sanders T.A.B., D.R.Sullivan, J.Reeve et al. (1985) Triglyceride lowering effect of marine polyunsaturates in patients with hypertriglyceridemia. Arteriosclerosis 5, 459

11. Simons L.A., J.B.Hickie, S.Balasubramaniam (1985) On the effects of dietary n-3 fatty acids (MaxEPA) on plasma lipids and lipoproteins in patients with hyperlipidemia. Atherosclerosis 54, 75

12. Boberg M., B.Vessby, I.Selinus (1986) Effects of dietary supplementation with n-6 and n-3 long chain polyunsaturated fatty acids on serum lipoproteins and platelet function in hypertriglyceridemic patients. Acta Med.Scand. 220, 153

13. Schectman G., S.Kaul, A.H.Kissebah (1987) Effect of fish oil concentrate on HDL-cholesterol and apo A-1 levels in hypertriglyceridemic subjects. Arteriosclerosis 7, 508a (abstract)

14. Sullivan D.R., T.A.B.Sanders, I.M.Trayner et al.(1986) Paradoxical elevation of LDL apoprotein B levels in hypertriglyceridemic patients and normal subjects ingesting fish oil. Atherosclerosis 61, 129

15. Harris W.S., C.A.Dujovne, M.L.Zucker et al.(1987) Fish oil supplements raise low density lipoprotein cholesterol levels in hypertriglyceridemic patients. Arteriosclerosis 7, 509a (abstract)

16. Phillipson B.E., D.W.Rothrock, W.E.Connor et al.(1985) Reduction of plasma lipids, lipoproteins, and apoproteins by dietary fish oils in patients with hypertriglyceridemia. N.Eng.J.Med. 312, 1210

17. Tamura Y., A.Hirai, T.Terano et al.(1986) Clinical and epidemiological studies of eicosapentaenoic acid (EPA) in Japan. Prog.Lipid Res. 25, 461 .

DIFFERENTIAL MOBILIZATION OF ESSENTIAL FATTY ACIDS INTO THE SERUM FREE FATTY ACID POOL IN RESPONSE TO GLUCOSE INGESTION

Stephen C. Cunnane, T.M.S. Wolever, J.K. Armstrong, and D.J.A. Jenkins

Department of Nutritional Sciences, Faculty of Medicine
University of Toronto, Toronto, Canada

Introduction

Although it is well known that the serum free fatty acid (FFA) pool is reduced during carbohydrate loading primarily in response to insulin secretion, it has not been established whether all or specific FFA are affected. This is of particular interest in view of the fact that the rate of carbohydrate absorption has a dramatically different effect on the FFA response. Rapid carbohydrate loading in a short period of time induces a rapid change in blood glucose and FFA depression followed by FFA rebound after 2h. Slow carbohydrate loading, on the other hand, results in a sustained depression of FFA over at least 4h. We have therefore compared the effects of rapid carbohydrate loading (glucose bolus) with slow carbohydrate loading (glucose sipping) on the response of the total FFA pool over 4h and have determined the changes in individual FFA, especially essential fatty acids, in healthy male volunteers.

Methods

Following a 12h fast, 8 healthy male volunteers consumed 50 g of glucose in 750ml water over a 5' period (BOLUS) or over a 210' period (SIP). Each participant consumed the glucose by both BOLUS and SIP in randomized order with at least 6 days between tests. Venous blood samples were drawn from an anticubital forearm vein at 0', 30', 60', 90', 120', 180' and 240'. Blood glucose was determined colorimetrically and serum·insulin was assayed by radioimmunoassay. Total FFA were measured using a spectrophotometric assay (WAKO, Japan) on aliquots of frozen serum.

A second aliquot of serum was used for determination of the composition of the individual fatty acids in the FFA pool. Total lipids in these aliquots were extracted with chloroform:methanol (2:1) containing 0.02% butylated hydroxytoluene as antioxidant. Total FFA were separated by thin layer chromatography, converted to methyl esters using 14% boron trifluoride in methanol and analysed by gas liquid chromatography on a 30m capillary column and a Hewlett-Packard 5890 gas liquid chromatograph with automated sample injection.

Results

Blood glucose rose from 4.7 to 7.9 mmol/L at 30' after the glucose BOLUS and returned to baseline levels by 120'. After the glucose SIP, blood glucose peaked at 5.6 mmol/L between 60' and 180' and returned to baseline by 240'. Serum insulin duplicated the time course of blood glucose, rising 6X and returning to baseline after 120' after the glucose BOLUS or doubling between 60' and 180' and returning to baseline by 240' after SIPPING the glucose.

Serum FFA averaged 575 umol/L at entry (0') and dropped to 250-300 umol/L by 120' after both the glucose SIP or BOLUS. After the glucose SIP, the FFA remained between 200-300 umol/L during the 120-240' period but after the glucose BOLUS, they rebounded to 700 umol/L at 240'. 240' after the glucose SIP, all the measured FFA (myristic acid- 14:0 to docosahexaenoic acid- 22:6n-3) were decreased 20-40%. Stearic acid (18:0) was decreased the least (20%) and myristic acid (14:0) and palmitoleic acid (16:1n-7) were decreased the most (both 40%). After the glucose BOLUS, the maximal depression of FFA was at 120' and all the fatty acids were decreased 50-80% (18:0 the least at 52% and palmitoleic acid -16:1n-7 - the most at 85%). The rebound of FFA between 120' and 240' after the glucose BOLUS was primarily due to an increase of 18:1n-9 (79 umol/L; +33%) and linoleic acid (18:2n-6)(18 umol/L; +14%). Palmitic acid (16:0), 18:0 and arachidonic acid (20:4n-6) did not contribute significantly to the FFA rebound after the glucose BOLUS ($<$ 1 umol/L each). During the rebound period (120' - 240'), 22:6n-3 showed the greatest percent increase (700%) and 18:0 and 20:4n-6 the least (30%).

Discussion

These results confirm that the serum FFA pool is differentially responsive to carbohydrate loading depending on the rate of carbohydrate absorption. Furthermore, individual fatty acids within the serum FFA pool make markedly different contributions to the FFA rebound that occurs 120-240' after rapid glucose ingestion (BOLUS). On a mass basis, 18:1n-9 and 18:2n-6 comprised the majority of the total FFA rebound. Nevertheless, from their individual levels at 120' (nadir), essential fatty acids such as 22:6n-3 and 18:3n-3 (alpha-linolenic acid) also increased markedly after the BOLUS glucose ingestion. 18:0 and 20:4n-6 responded the least both in decreasing after the glucose BOLUS. The suppression of individual serum FFA during carbohydrate loading by either SIP or BOLUS appears to have been relatively uniform but there was a differential rebound of individual FFA after

Table 1

Serum free fatty acid (FFA) response at 240' to the ingestion of 50 g glucose as a BOLUS (consumed within 5') or sipped over 210' (SIP).

	Baseline (0') umol/L	BOLUS (240' versus 0')		SIP (240' versus 0')	
		umol/L	% change	umol/L	% change
Total FFA	575	+ 107	+ 19	- 190	- 33
Individual FFA					
14:0	5	+ 2	+ 40	- 2	- 40
16:0	95	< 1	< 1	- 24	- 25
18:0	44	< 1	< 1	- 9	- 20
16:1n-7	20	+ 3	+ 15	- 8	- 40
18:1n-9	240	+ 79	+ 33	- 92	- 38
18:2n-6	127	+ 18	+ 14	- 42	- 33
20:4n-6	22	- 2	- 9	- 8	- 36
18:3n-3	13	+ 3	+ 23	- 3	- 23
22:6n-3	6	+ 4	+ 67	- 2	- 33

the glucose BOLUS. The timing of this differential FFA rebound corresponded to the phase of the counter-regulatory response after glucose loading and the rise in serum growth hormone and glucagon. It appears to be related to the rate of oxidation of the FFA involved; those FFA which are readily oxidized (14:0, 18:1n-9, 18:3n-3 and 22:6n-3) were those which had the highest rebound 120-240' after the glucose BOLUS whereas those which are not readily oxidized (18:0 and 20:4n-6 in particular) showed minimal rebound (< 1% above baseline (0') for 18:0 and 9% below baseline for 20:4n-6).

Whether differences in the fatty acid composition of the adipose tissue (from which the serum FFA are derived) alter the relative FFA response to glucose ingestion was not established.

Acknowledgement

We thank the Natural Sciences and Engineering Research Council of Canada for financial support of this research.

FLAX AS A SOURCE OF ALPHA-LINOLENIC ACID

Paul A. Stitt

Essential Nutrient Research Co., 4300 Country
Road CR, Manitowoc, WI 54220, U.S.A.

Flax is a very rich source of alpha-linolenic acid. The exact alpha-linolenic acid and other fatty acid content depends on species and growing conditions. The fatty acids contained are shown in Table 1.

Table 1. Fatty Acid Content of Flax.

16:0	Palmitic	1.6 to 2.5
18:0	Stearic	0.9 to 2.0
18:1	Oleic	6.8 to 8.0
18:2	Linoleic	4.7 to 6.5
18:3	Linolenic	16.7 to 22.0

Flax is also a rich source of nutritionally complete protein, minerals and fiber; both soluble and insoluble.

One of the drawbacks to flax is the low zinc content and the anti-B-6-factor. This situation can cause problems in feeding studies, because if not corrected, it can affect the utilization of the essential fatty acids. Adding 100 ppm of vitamin B-6 and 200 ppm of Zinc Sulfate will correct these deficiencies. Several feeding studies are underway using flax fortified at these levels, and good growth rates are being observed in chickens and pigs. The nutritional analysis is shown in Table 2.

Table 2. Nutritional Content of Flax

Moisture	7.20%	Potassium	700 mg/100g
Protein	19.80%	Phosphate	550 mg/100g
Fiber (crude)	14.40%	Calcium	250 mg/100g
Fiber (soluble)	17.21%	Magnesium	350 mg/100 g
Fat	38.70%	Boron	1.5 mg/100 g
Ash	2.69%	Iron	10.2 mg/100 g
		Manganese	6.0 mg/100 g
		Zinc	2.0 mg/100 g

We have observed very small increases over time in the peroxide level with the flax that is fortified with Zinc and B-6. Other have not found this and we are studying the effect of cultivar, growing, storing, and grinding conditions on the stability of flax after it is ground. We have not observed any palatability problems with humans and animals. Our results are shown in the following chart.

Flax seed was finely ground and fortified with zinc and vitamin B-6. Then samples were taken for analysis. Other samples were stored in sealed foil bags for various periods of time. Air was not excluded before sealing. The results are shown above.

In summary, we have found fortified flax to be an excellent source of stable alpha-linolenic acid.

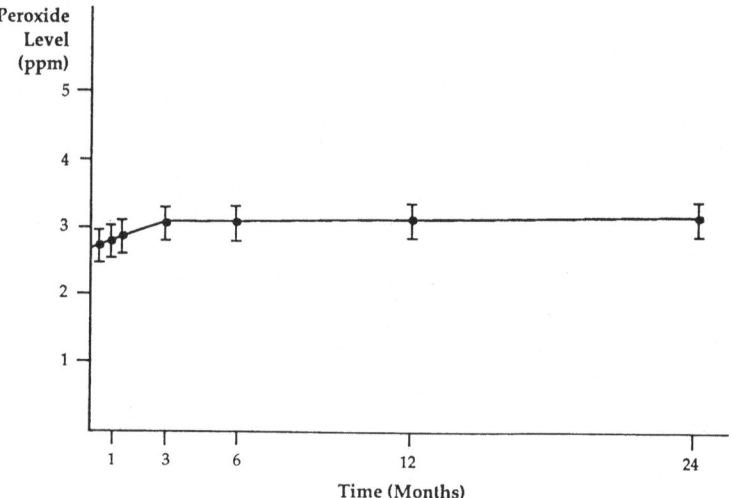

Figure 1. Stability of fortified flax seed.

EXECUTIVE SUMMARY

Artemis P. Simopoulos [1]

Division of Nutritional Sciences
International Life Sciences Institute Research Foundation
1126 Sixteenth Street, N.W.
Washington, D.C. 20036

The executive summary consists of highlights of papers presented at
the workshop and at the poster session, including the discussions at the
round table. The summary is presented under the following headings:
Dietary Sources of Omega-3 and Omega-6 Fatty Acids; Chemistry,
Biosynthesis and Interactions of Omega-3 and Omega-6 Fatty Acids; The
Role of Omega-3 and Omega-6 Fatty Acids in Development; Biological
Effects of Omega-3 and Omega-6 Fatty Acids on Cell Activation Processes;
and The Role of Omega-3 and Omega-6 Fatty Acids in Human Diseases. A
summary of the conclusions completes the executive summary.

DIETARY SOURCES OF OMEGA-3 AND OMEGA-6 FATTY ACIDS

It was most appropriate to open the workshop with a presentation
given by Dr. Konrad Bloch who reviewed some early studies using alga
euglena gracilis. The studies indicated that with increasing
photosynthetic efficiency there was an increase in the production of
alpha-linolenic acid from linoleic acid by chloroplasts.

A splendid review of the evolutionary aspects of the omega-3 and
omega-6 fatty acids in the food chain was given by Dr. Michael Crawford.
He emphasized the differences that exist between the marine and the
terrestrial food chain. The former is rich in long chain fatty acids
[eicosapentaenoic (EPA) and docosahexaenoic (DHA)] whereas the
terrestrial food chain is rich in the short chain fatty acids, namely
linoleic and alpha-linolenic acid. Since DHA is required for brain
function Dr. Crawford suggested that man probably evolved near a marine
rather than in a land environment. Dr. Joyce L. Beare-Rogers reviewed
the omega-3 fatty acid content of various foods and presented her study
on the erythrocyte fatty acid content of cynomolgus monkeys fed three
different diets. The diets had a similar distribution of saturated,

---

[1]Present address is The Center for Genetics, Nutrition and Health,
American Association for World Health, 2001 S Street, N.W., Suite 530,
Washington, D.C. 20009.

monounsaturated and polyunsaturated fatty acids, but differed in the amount of linoleic, linolenic, EPA and DHA. When the diet contained similar amounts of linoleic and linolenic acid the ratio of omega-6 to omega-3 fatty acids in the erythrocytes remained constant at approximately 2.4. When linseed oil was fed, DHA was the predominant fatty acid whereas EPA predominated when fish oil was fed. The data indicated a close regulation of the conversion of essential fatty acids to their long chain derivatives and their incorporation into the cellular membranes.

The need to know precisely the food composition of omega-3 and omega-6 fatty acid was extensively discussed. Most scientists agree that the determination of an omega-3 fatty acid requirement depends on the quantity or the ratio of ω6/ω3. The data on food composition are not accurate but an attempt was made by Drs. Olaf Adams and Edward J. Hunter to estimate the quantities of omega-3 and omega-6 fatty acids in the Western diet based on world production data, and per capita availability. Of interest is the fact that both speakers determined the ratio of ω6/ω3 to be 10-11/1. The methods used at present for the production of fish oils for dietary use were discussed by Dr. H.J. Willie.

CHEMISTRY, BIOSYNTHESIS AND INTERACTIONS OF OMEGA-3 AND OMEGA-6 FATTY ACIDS

The different distributions of omega-3 and omega-6 fatty acids and their long chain derivatives EPA, DHA and arachidonic acid (AA) in the various tissues - adipose tissue, plasma phospholipids, cell membrane, brain, and retina - have been of intense interest over many years. Yet, despite extensive studies, it is not known why linoleic and arachidonic acids are found in membrane lipids whereas dietary alpha-linolenic acid is not incorporated into membrane lipids. This point was extensively discussed by Dr. Howard Sprecher who summarized his findings as follows. "When fish oil is added to the diet there are increased levels of EPA and DHA in heart, kidney and liver lipids. Conversely platelets and neutrophils do not incorporate large amounts of 22:5ω3 and 22:6ω3 into their lipids. The apparent discrimination against long chain (omega-3) acids in vivo is not consistent with ex vivo studies which show that these acids are incorporated into platelet and neutrophil lipids. Enzymatic studies with macrophages show that 22:6ω3 is transferred to the sn-2 position of lysoplatelet activating factor. In addition, when fish oils are fed to rats the 20:5ω3 is incorporated into various lipid classes of neutrophils to different extents. Molecular species analysis shows that 20:5ω3 pairs with the same saturated acids, and in the same ratio, as does arachidonate. When these cells are incubated with lyso-platelet activating factor both 20:5ω3 and 20:4ω6 are transferred to the sn-2 position. These findings are consistent with the concept that 1-O-alkyl-2-acyl-GPC is the source of both 20:4ω6 and 20:5ω3 for LTB$_4$ and LTB$_5$ synthesis as well as for the platelet activating factor synthesis."

Evidence continues to accumulate that the relative proportions of omega-3 versus the omega-6 polyunsaturated fatty acids (PUFAs) in the phospholipids of circulating human platelets can greatly influence the amount and type of eicosanoids which are formed in both normal and pathophysiological states.

Dr. Bruce J. Holub spoke on the relative uptake and utilization of EPA and AA in membrane phospholipids of resting and stimulated human platelets. The relative _in vitro_ entry of radio-labelled AA and EPA into the individual phospholipids of human platelets has been studied using intact platelets or via the corresponding membrane lysophospholipid acyltransferases. These studies have exhibited moderate selectivities between AA/EPA entry at the level of individual phospholipid type. However, the dramatic discrimination against EPA accumulation in phosphatidylinositol as observed upon fish oil consumption _in vivo_ was not accounted for. Interestingly, mass analyses have revealed the ether-containing ethanolamine phospholipid (PE), 1-alkenyl 2-EPA PE, to be a major reservoir of EPA in the platelets of human subjects consuming a fish oil concentrate. Intervention studies with fish oil have revealed that the alterations in selected omega-3 and omega-6 fatty acid compositions as seen in the 1-alkenyl 2-acyl PE are most pronounced relative to other human platelet phospholipids. Feeding human subjects a canola oil enriched diet for 25 days resulted in a threefold increase in EPA in platelet phospholipids of 1-alkenyl-2-acyl- phosphatidyl ethanolamine, compared to feeding a sunflower oil enriched diet. Previous studies reporting conversion of dietary alpha-linolenic acid to EPA in platelet phospholipids have focused only on total platelet phospholipids rather than on specific phospholipid classes.

The importance of DHA as an inhibitor of platelet aggregation was emphasized by Dr. Michel Lagarde who showed that DHA is more potent than EPA in inhibiting platelet aggregation in vitro. The metabolism of these fatty acids appears to be entirely different whereas EPA may compete with AA metabolism at several steps. DHA might inhibit platelets at the membrane level as a structural component of phospholipids. Dr. Pierre Budowski reviewed the important role of alpha-linolenic acid in the control and as a modulator of AA metabolism and consequently in reducing the tendency to platelet aggregation. Using autoradiography, Becker reported his studies on comparative uptake in rats and man of omega-3 and omega-6 fatty acids. A most remarkable finding was an increase of radioactivity in the adrenal cortex from EPA, indicating some specific function of EPA in the adrenal.

THE ROLE OF OMEGA-3 AND OMEGA-6 FATTY ACIDS IN DEVELOPMENT

In this session, the role of omega-3 fatty acids, particularly DHA, and their relationship to omega-6 fatty acids during fetal life, infancy and lactation were extensively discussed.

The effects of PUFA deficiency on the developing brain has been widely documented in the experimental animal whereas information from humans is scarce. However, recent work by Drs. Manuela Martinez, Sheila Innis, Susan E. Carlson, Martha Neuringer and Jean-Marie Bourre, all of whom presented papers at this session and at the poster session, have added considerably to our knowledge that was pioneered by Drs. Lampty, Walker, Crawford and Clandinin. It is now well recognized that nutrition during the first weeks of life can have a decisive influence on brain development. Since fatty acid patterns of all organs change during development, it is necessary to know the normal profiles during the various stages of development in order to understand the role of nutritional influences.

Dr. Martinez described her studies on the composition of omega-3 and omega-6 fatty acids in brain, liver and retina in human fetuses during the last trimester of pregnancy. After 30 weeks of gestation there is a preferential desaturation of the long chain omega-3 fatty acids in the brain. The liver shows a similar profile. In both tissues 22:6ω3 increases in a quadratic way and 20:4ω6 and 18:1ω9 decrease linearly in PE. In the retina, as in the forebrain and the liver, the proportion of omega-3 fatty acids increases whereas that of omega-6 fatty acids decreases throughout development. These changes can be clearly illustrated by using the ratio of 22:6ω3 to 20:4ω6. The value of this ratio in the human retina doubles between 24 weeks of gestation and term, and continues to increase with age. These findings should be the standards or guiding posts for the feeding of prematurely born infants.

Dr. Martinez described investigations in the liver and forebrain of infants receiving total parenteral nutrition (TPN) consisting of high doses of linoleate (Intralipid) for 4-12 days. At autopsy a significant decrease in the proportion of 22:6ω3 was found in liver phosphoglycerides compared to control values. There were a number of other changes in long chain PUFAs and a significant increase in 18:2ω6, not consistent with the values noted in normal fetal development.

At the poster session, Dr. Martinez showed data obtained from the retinas of two postnally malnourished infants. Interestingly the findings were similar to those described in the liver of children receiving high doses of intravenous 18:2ω6. One of the malnourished children had mucoviscidosis. Both children had an increase in 22:5ω6 in retina phosphatidlyethanolamine (PE) and phosphatidylcholine (PC) as a sign of PUFA deficiency. The other infant, a premature (25 weeks gestation), received commercial milk formulas with ω6/ω3 ranging between 18:1-66:1 during 4 months of life. The retina of this premature was very deficient in 22:6ω3.

It can be concluded that diets with a high ratio of ω6 to ω3 can be considered unbalanced relative to human breast milk and that these diets are damaging to the PUFA composition of the human developing CNS. Dr. Martinez stated that "when high doses of 18:2ω6 are given intravenously, the inhibiting effect on the omega-3 series is very strong, even with a theoretically correct ω6/ω3 ratio, probably because substrate inhibition adds to competition between families of fatty acids for the desaturase systems. This should serve as a warning against manufacturing such unphysiological fatty acid mixtures for use in pediatric nutrition."

Dr. Innis in her presentation "Fats from Terrestrial and Marine Animals and Milk and Tissue Fatty Acids in Humans" documented the passage of omega-3 fatty acids through the food chain and emphasized the compositional differences among marine mammal, human and fish lipid. The lipids of marine mammal contained higher levels of 22:5ω3 and 22:6ω3 relative to 20:5w3 than fish lipid. The breast milk of the Inuit contained three to fourfold more 20:5ω3 and twofold more 22:6ω3 than comparable mature breast milk collected in Vancouver. The Inuit of course have a diet that is high in omega-3 fatty acids from eating marine mammals, caribou and fish, whereas the women in Vancouver consumed a mixed Western type diet high in omega-6 fatty acids. This difference in the food content of omega-3 fatty acids was reflected in the red blood cell PC and PE as well as in their breast milk.

Dr. Hornstra at the poster session presented preliminary data of DHA plasma phospholipid concentrations indicating that DHA levels are decreased in pregnant women suffering with toxemia, and in the elderly. Thus it appears that the number of diseases and conditions that are associated with abnormal levels or metabolites of ω3 fatty acids continue to increase.

The role of "PUFA in Infant Nutrition" was discussed by Dr. Carlson who presented a series of studies indicating the differences in the composition of human breast milk and formula. The latter does not contain any long chain fatty acids of either the omega-3 or omega-6 series. The content of the omega-3 fatty acids in human milk is reflected in the red blood cell membrane phospholipid. Analyses of cord blood and blood from preterm infants fed human milk or formula confirmed that membrane 22:6ω3 was highest at birth, declined with time in formula-fed infants, and that fish oil could be used as a source of 22:6ω3. A month after delivery, preterm infants had lower DHA plasma phospholipid concentrations similar to the DHA concentrations found in monkeys that had been fed safflower oil. These low levels of DHA found in premature infants are analogous to those at which demonstrable deficits in visual acuity occur in the monkey infant.

Dr. Neuringer spoke on "The Essentiality of ω3 Fatty Acids in the Development and Function of the Retina and Brain." Dr. Neuringer described in detail her classic studies in which monkeys made DHA deficient pre- and postnatally showed that the biochemical abnormalities were also associated with abnormalities in visual function and learning.

Whether or not the biochemical abnormalities of 22:6ω3 deficiency noted in the red cell membranes of the human premature infant are associated with functional abnormalities is under investigation. The hypothesis is that membrane 22:6ω3 in unsupplemented infants will remain low throughout infancy and that visual acuity and habituation learning will be related to 22:6ω3 status as have been noted in the rat and monkey. Continuing with this theme, Dr. Bourre reviewed PUFAs of the omega-3 series and their role in nervous system development in the rat animal model. Feeding animals with oils that have a low omega-3 fatty acid content results in serious anomalies in the composition of brain membranes. In all brain cells and organelles, reduced amount of 22:6ω3 is compensated for by an increase in 22:5ω6. The speed at which recovery occurred from these anomalies is extremely slow for brain cells, organelles and microvessels, in contrast with other organs. Dr. Bourre stated that in the rat, during the period of cerebral development "there is a linear relationship between the omega-3 fatty acid content of the brain and that of food until linolenic acid represents approximately 200 mg per 100 gm of food (for 1100 mg linoleic acid). Beyond that point there is a plateau in the brain." He determined the dietary requirements during brain development to be 0.4 percent of calories for 18:3ω3 and 2.2 percent of calories for 18:2ω6. Dr. Bourre concluded that "The level of 22:6ω3 in membranes is not affected much by the dietary quantity of 18:2ω6 if 0.4 percent of calories, at least, come from 18:3ω3. A decrease in longer chain fatty acids of the linolenic series in the membrane results in a 40% reduction of Na-K-ATPase in nerve terminals and a 20% reduction in 5[1]-nucleotidase in whole brain homogenate. A diet low in linolenic acid leads to anomalies in the electroretinogram which disappear partially with age. The presence of linolenic acid in the diet confers a greater resistance to certain neurotoxic agents (triethyltin, i.e.). Deficiency in linolenic acid has little effect on motor activity but it seriously affects learning tasks."

BIOLOGICAL EFFECTS OF OMEGA-3 AND OMEGA-6 FATTY ACIDS ON CELL ACTIVATION PROCESSES

Human studies suggest that dietary EPA supplementation will not reduce PGI_2 formation and in addition will lead to substantial PGI_3 formation. This finding has added further impetus for the use of dietary EPA supplements in the prevention of coronary heart disease. However, Dr. Spector indicated that PGI_2 release by umbilical endothelial cell cultures is reduced when omega-3 PUFAs are added. EPA produces the largest reductions whereas linolenic acid ($18:3\omega3$) has little or no inhibitory action. On the other hand, DHA ($22:6\omega3$) has an intermediate effect which is more pronounced at long incubation times. Studies with (U^{14}-C) $22:6\omega3$ indicate that retroconversion of $22:6\omega3$ to EPA occurs under these conditions, suggesting that the inhibitory effect on PGI_2 might be produced by the EPA that slowly accumulates. This difference between the human studies and the endothelial cell cultures on PGI_2 production needs to be resolved. Another unexplained finding with umbilical endothelial cell cultures is that high concentrations of linoleic acid ($18:2\omega6$) also reduce PGI_2 formation since very little $18:2\omega6$ is converted to $20:4\omega6$ in these cultures. The mechanism by which $18:2\omega6$ can reduce PGI_2 formation is not understood. In human umbilical endothelial cell cultures Dr. Spector has identified the formation of a polar product whose formation is reduced by indomethacin or acetylsalicylic, and this finding suggests that the polar product is formed by the cycloxygenase reaction. Further studies demonstrated that the structure of this compound is 18:2 (9-OH). It is not known if 18:2 (9-OH) has any functional effects. The production of 20:2 and the formation of 18:2 (9-OH) in endothelial cultures indicate that human umbilical endothelial cell cultures have a more complex metabolism of the main dietary essential fatty acid, linoleic acid, than previously recognized.

Dr. C. Galli reported on studies carried out on male rabbits on three different semi-synthetic diets rich in omega-9, omega-6, and omega-3 on the following parameters: a)fatty acid composition, absolute levels and distribution of PUFAs among the different plasma lipid and lipoprotein classes and platelet glycerophospholipids; b) generation of thromboxane B_2 by stimulated platelet rich plasma (PRP); and c) generation of inositol phosphates by inositol labelled washed platelets. As expected, the diets resulted in significant modifications of the fatty acid profiles of plasma and platelet phospholipids and in marked changes in the activation of inositol phosphate generation in platelets after thrombin stimulation. The data indicate that complex interactions and displacements of PUFA of omega-3 and omega-6 series take place in plasma and cellular lipids after dietary manipulations and that early steps of cell activation, such as generation of inositol phosphates, are induced by dietary fatty acids. The effects of dietary fatty acids on the inositol phosphate pathway indicate that dietary induced modifications of PUFA at the cellular level affect the activity of the enzymes responsible for the generation of lipid mediators, in addition to the formation of products (eicosanoids) directly derived from their fatty acid precursors. This shows that the impact of fats in the diet affects key processes in cell function.

Dr. Leat attempted to answer the questions 1) does linoleic acid have any function that cannot be fulfilled by linolenic acid, and its corollary; 2) does linolenic acid have any function that cannot be fulfilled by linoleic acid? He carried out studies in male and female rats. The data indicate that in both male and female rats there was no difference in growth rate between these two groups of rats. The female rats fed linolenate conceived without difficulty and gestation was uneventful but parturition was prolonged and difficult. This impaired parturition could be prevented by replacing the linolenate with a supplement of linoleate ($18:2\omega6$) from day 18 of pregnancy, indicating that linoleate ($18:2\omega6$) was essential for the process of parturition. Male rats fed linolenate ($18:3\omega3$) as a sole source of EFA became sterile, with complete degeneration of the germinal cells, although the Leydig cells appear normal. In subsequent studies it was shown that of the omega-6 fatty acids tested, $20:4\omega6$ had a greater effect on testis development than $22:4\omega6$, suggesting that prostaglandins of the two series are involved in spermatogenesis.

Studies with guinea pigs fed high $18:2\omega6$/low $18:3\omega3$ diets resulted in virtual depletion of linolenate from the retina by the second generation. However, the retinae of these animals were still responsive to light and their electroretinograms were not statistically different from those of controls. It should be pointed out that the results of these studies on retinae differ from previous studies carried out in rats and the studies by Neuringer and Connor in monkeys where abnormalities in the electroretinogram were noted in monkeys deficient in dietary $18:3w3$.

Dr. Bazan spoke on "The Supply of Omega-3 Polyunsaturated Fatty Acids to Photoreceptors and Synapses." He reviewed previous work on this subject that has shown that during $18:3\omega3$ dietary deprivation, DHA is replaced by $20:5\omega6$ in the retina and brain of animals. This replacement with $20:5\omega6$ - the fatty acid that most closely resembles DHA - suggests activation of a cellular compensatory mechanism. Dr. Bazan summarized recent studies mainly from his laboratory using the retina and photoreceptor cells as models to study the composition, metabolism, supply, alterations in pathological conditions, and possible physiological significance of omega-3 PUFAs.

Previous studies have shown that the plasma phospholipid content of AA and DHA is decreased in patients with inherited retinal degeneration. In the inherited retinal degeneration model, the rd mouse, developing photoreceptor cells fail to differentiate outer segments and they degenerate before reaching maturity.

THE ROLE OF OMEGA-3 AND OMEGA-6 FATTY ACIDS IN HUMAN DISEASES

There were two sessions held on this subject and a total of eleven papers were presented. Dr. K.S. Bjerve spoke on the role of "$\omega3$ and $\omega6$ Fatty Acids in Serum Lipids and Their Relationship to Human Disease". Dr. Bjerve reviewed his previous studies and presented additional data on adults and one child with omega-3 fatty acid deficiency who were fed orally for several years by gastric tube with diets containing very low amounts of omega-3 fatty acids. Studies based on clinical findings and determinations of plasma and red cell lipids following supplementation of patients with soya and cod-liver oil strongly suggest that the patients had omega-3 fatty acid deficiency. The results indicate that omega-3 fatty acids are essential for normal growth and cell function in man in a similar way as they are essential in several animal species.

Assuming linear relationships between dietary intake of omega-3 fatty acids and the measured concentrations of omega-3 fatty acids in plasma and erythrocyte lipids, the optimal intake of 18:3ω3 has been estimated to be 800-1100 mg/day, while the optimal intake of very long chain omega-3 fatty acids was estimated to be 300-400 mg daily. In addition Dr. Bjerve presented data on patients with Type II diabetes. The patients with diabetes had lower serum phospholipid levels of 22:5ω3 and 22:6ω3. No differences were noted between diabetics and controls in the omega-6 fatty acid series. The workshop participants suggested that the dietary requirements of omega-3 as well as omega-6 fatty acids should be stated in mg or grams per day and not only as a per cent of calories.

Much more is known about the effects of omega-3 fatty acids on cardiovascular disease than on any other clinical entity. Dr. T.A.B. Sanders discussed the relationship of omega-3 fatty acids to lipoprotein metabolism. Total plasma cholesterol concentrations in some instances are reduced by fish consumption but most of the reduction occurs in the very low density lipoprotein (VLDL) or low density lipoprotein (LDL) fractions. Very high intakes of fish oil (90-120 g/day) do lower the concentrations of both LDL cholesterol and LDL apoB by decreasing the rate of LDL synthesis. Such high intakes also prevent the rise in plasma cholesterol obtained with dietary cholesterol. However, the reduction in LDL pool size is smaller than would be predicted from the decrease in synthesis.

With lower intakes of fish oils (15 g/day) or oily fish providing 3-5 g omega-3 fatty acids there is a tendency for LDL cholesterol and LDL apoB concentrations to rise. In patients with Type V hyperlipoprotein-emia, LDL cholesterol levels rise even at high intakes – (this occurs with most forms of triglyceride lowering therapy). A likely explanation is that a moderate intake of fish oil decreases hepatic triglyceride synthesis, so that smaller than normal VLDL particles are secreted. These small particles are known to be more readily converted to LDL than the larger triglyceride rich ones. Dr. Sanders concluded that moderate intakes of fish oil concentrates are not useful for the treatment of hypercholesterolemia but offer a safe and effective means of treating patients with hypertriglyceridemia resulting from excessive VLDL synthesis especially in patients with Type IV and Type V hyperlipo-proteinemias. Others however have reported lowering of serum cholesterol levels in normal volunteers and in patients with cardiovascular disease even with moderate amounts of fish oils. Further studies are needed that distinguish the effects of fish oils in the various types of hypercholesterolemias.

Preliminary data presented at the poster session by Dr. Beitz indicate that omega-3 fatty acids in the form of cod liver oil decrease Lp(a) which is known to be an atherogenic lipoprotein that is genetically determined. Previous studies had shown that diet did not affect Lp(a) levels. This is the first evidence of a nutrient - omega-3 fatty acid from cod liver oil - lowering Lp(a), and that this lowering effect is potentiated by exercise (swimming). The effects of fish oils on gene expression is a very promising area for future research.

Dr. Kromhout reviewed the evidence on the relationship of fish (oil) consumption to coronary heart disease in epidemiologic studies. It appears that in population studies different mechanisms may explain the inverse relationship between fish consumption and coronary heart disease at the low and very high levels of marine food consumption. Dr. B.J. Horrobin presented studies suggesting possible beneficial effects of gamma-linolenic, dihomo-gamma-linolenic, AA, EPA and DHA in certain disease states, even though the amount of dietary intake of linoleic and alpha-linolenic acids appear to be satisfactory. These studies need to be confirmed using appropriate test materials and controls.

Since the late 1950s, linoleic acid has been exclusively used to lower cholesterol levels in the United States. Recently interest has been expressed in the fact that stearic acid, unlike other saturated fatty acids, does not raise serum cholesterol levels. So much emphasis has been put on lowering serum cholesterol levels through diet and drugs that the importance of thrombosis in coronary heart disease has not been properly discussed. Dr. Renaud pointed out that the risk of thrombosis increases with increasing content of saturated fat in the diet, and particularly with an increase in stearic acid intake. Therefore while stearic acid lowers serum cholesterol, at the same time it increases the tendency to thrombosis. In human trials conducted by Renaud and colleagues, the principal effect of dietary alpha-linolenic acid has been to reduce platelet aggregation and possibly decrease the risk from coronary artery disease by reducing clotting and thrombus formation. It was felt that, generally, western diets should contain more alpha-linolenic acid and less saturated fat.

Dr. Spielmann presented studies using black currant seed oil. Black currant seed oil has a particular composition: 47 percent of $18:2\omega6$; 17 percent of $18:3\omega6$; 13 percent of $18:3\omega3$; and 3.5 percent of $18:4\omega3$ which has an $\omega6/\omega3$ ratio of 5/1. Therefore this oil has been used in studies employing rats, guinea pigs and humans and has been found to lower LDL cholesterol while increasing HDL cholesterol. This is the first time that a vegetable oil has been shown to increase HDL. Further research is needed to determine which HDL fraction is increased by black currant seed oil.

Studies with fish oils on the control of blood pressure indicate that MaxEPA® lowers both systolic and diastolic blood pressure. Dr. Howard Knapp presented his latest study on "Dietary PUFAs, Blood Pressure and Prostanoid Metabolism." A decrease in both systolic and diastolic blood pressure and an increase in PGI_3 excretion in the urine were noted with MaxEPA® administration, whereas comparable amounts of safflower oil had no effect on blood pressure.

Linoleic acid deficiency was recognized by Burr and Burr more than 50 years ago. In the skin linoleic acid is a constituent of stratum corneum ceramides which may play an important role in preventing transepidermal water loss. Dr. H.S. Hansen indicated that dietary arachidonic acid can prevent the transepidermal water loss seen in EFA-deficient rats, but it is linoleic acid which is found in the epidermal ceramides of the arachidonic acid- supplemented rats, indicating that dietary arachidonic acid can be retroconverted to linoleate by EFA-deficient rats. This suggests that at least in rats, linoleic acid has an essential function of its own without being converted to arachidonic acid. In addition Dr. Hansen stated that in rats, when water is given ad libitum, alpha-linolenic acid can promote growth.

Dr. J. Kremer reviewed his previous work and presented additional data on omega-3 fatty acid supplementation in rheumatoid arthritis. The data indicate clinical improvements were dose dependent, and consistent with alterations in LTB_4 and interleukin metabolism, in a group of 49 rheumatoid arthritis patients given a supplement of EPA and DHA for 24 weeks. The levels of LTB_4 decreased and LTB_5 increased with fatty acid treatment. During the discussion it became evident, that fish oil supplements have an advantage over drugs currently available for arthritis in that the drugs often produce toxic effects which have not been seen today with fish oils. In addition fatty acids work through the lipoxygenase and cycloxygenase pathways whereas aspirin works at the cycloxygenase pathway only.

Dr. Weber presented an update on the "Modification of the Arachidonic Acid Cascade by Long-Chain W-3 Fatty Acids" that included a review of eicosanoids as modulators of cell function; nutrition and the eicosansoid system; dietary modification of cell membrane fatty acid composition; dietary omega-3 fatty acids and the modification of the eicosanoid system; and functional effects of dietary omega-3 fatty acids related to modified eicosanoid formation. Dr. Weber said that future research should include both basic research at the cellular level and clinical investigations, "These include the study of the effects of dietary eicosanoid precursors on eicosanoid formation from various long- versus short-chain w-6 and w-3 fatty acids; the reabsorption, organ distribution, cellular and phospholipid subclass incorporation of eicosanoid precursor fatty acids and their metabolism during cell formation and differentiation; and the large scale controlled evaluation of their potential role in the modification of the proliferative, prothrombotic and inflammatory response as it occurs clinically after cell injury and during cell repair, such as after percutaneous transluminal coronary angioplasty."

Dr. Karmali gave a thorough review of the state of the art on the effects of dietary omega-3 and omega-6 fatty acids in cancer. Animal studies in various tumor models in general indicate that omega-3 fatty acids have anti-tumor effects and omega-6 fatty acids potentiate the tumorigenic effect. On the other hand, recent reports suggest that gamma-linolenic acid and dihomogamma-linolenic acid have anti-tumor effects compared to linoleic acid. Preliminary data suggest that the mechanism of action involves down-regulation of arachidonic acid metabolism.

In conclusion, the participants felt that the knowledge base on the role of omega-3 fatty acids in growth and development and in health and disease has expanded over the past 3 years. Western diet contains much more omega-6 fatty acids and much less of omega-3 fatty acids than what it should be based on man's evolution and genetic development and adaptation. This "deficient state" is more marked in infant feeding practices, particularly in the feeding of premature infants, because the premature infant is born with a deficient amount of DHA in the brain and liver. Yet infant formula today does not contain any of the longer chain fatty acids of either the omega-6 or the omega-3 fatty acid series. Questions remain if this biochemical deficiency is associated with functional deficiencies. Studies are in progress to answer this very important question.

It was generally agreed that DHA is essential for the premature infant and possibly the full-term as well. Although a DHA requirement has not been determined, most of the participants agreed that the amount of DHA in infant formula should be similar to that found in human milk.

In terms of the overall western diet, the current amount of saturated fat was considered excessive. Therefore a recommendation was made to decrease saturated fat and increase the omega-3 fatty acid content of the overall diet. Despite the evidence that omega-6 fatty acids are also excessive in relation to omega-3 fatty acids the participants could not agree either on the amount of omega-3 fatty acids as a percent of calories or on the ratio of $\omega 6/\omega 3$ in the diet. The current estimate of this ratio in the western diet is 10-11/1. Evidence based on estimates from paleolithic nutrition and from terrestrial animals (mammals) in the wild indicate a ratio of $\omega 6$ to $\omega 3$ to be 1 to 1 in the diet. To prevent deficiency in adults alpha-linolenic acid is estimated to be 800-1100 mg/day and for the longer chain omega-3 fatty acids (EPA and DHA) to be 300-400 mg/day.

Finally, particular attention should be given that omega-3 fatty acids are included in the proper amounts in intravenous and intraoral feedings and in total parenteral nutrition throughout the life cycle.

For diseases and conditions, despite extensive research, the dose response is not accurately known yet, for either prevention, or treatment. However, the participants agreed that omega-3 fatty acids lower serum triglycerides and at high doses lower cholesterol, increase PGI_3, decrease the tendency to thrombosis and platelet aggregation, have anti-inflammatory properties, lower LTB_4 and increase LTB_5 in patients with rheumatoid arthritis and therefore ought to be beneficial in the management of patients with cardiovascular disease, hypertension and arthritis. The role of omega-3 fatty acids in diabetic patients with hyperlipidemia requires careful investigation because of some reports that suggest increases in blood glucose levels following high doses of fish oils.

Further work is needed to define the relationships of omega-6 and omega-3 fatty acids and their various metabolites at the cellular and clinical level. Alpha-linolenic acid is metabolized to EPA and DHA in human beings. The conversion rate is not as slow as was originally suggested. The availability of pure fatty acid test materials should help define the conversion rate for various stages of the life cycle and in diabetics and hypertensives. Preliminary data indicate that in diabetics and hypertensives the delta-6-desaturase is limited. It is quite possible that further research may reveal other diseases where limitations occur at various points of elongation and desaturation of alpha-linolenic acid.

Clinical trials are indicated in patients with cardiovascular disease, thrombosis and atherosclerosis, hypertension, diabetes, arthritis and other autoimmune disorders in order to specifically define the effective dose of omega-3 fatty acids, type of omega-3 fatty acid and the mechanisms involved. Preliminary studies already indicate the beneficial roles of omega-3 fatty acids in many of these disorders and their essentiality in normal growth and development.

ACKNOWLEDGEMENT

This work was supported by the Howard Heinz Endowment.

GENERAL RECOMMENDATIONS ON DIETARY FATS FOR HUMAN CONSUMPTION

A. Introduction and General Rationale

1. The total fat intake should be < 30% of daily calories.

2. The saturated fat intake should be < 10% of daily calories.

3. The total polyunsaturated fat intake should represent 6-7% of total calories (with 10% being a maximum).

4. The polyunsaturated fat should be a mixture of n-6 plus n-3 polyunsaturates.

5. Research has suggested that the minimal nutritional requirements for the polyunsaturated fatty acids be as follows :

   (i) Linoleic acid (18:2n-6) = 3.0% of calories

   (ii) Linolenic acid (18:3n-3) = 0.25 to 0.54% of calories

   (iii) Eicosapentaenoic acid (20:5n-3)
        plus docosahexaenoic acid (22:6n-3) = 0.15% of calories.

B. Current Dietary Intakes (North American Adults)

   (Assuming 2600 kcal of energy/day)

Saturated fatty acids (S) = 15% of calories (43 g/d)

Monounsaturated fatty acids (M) = 18% of calories (52 g/d)

Polyunsaturated fatty acids (P) = 7% of calories (20 g/d)

   18:3n-3 = 0.7% of calories (2 g/d)
   20:5n-3 = 0.03% of calories (0.1 g/d)
   22:6n-3 = 0.03% of calories (0.1 g/d)

Current w6/w3 (n-6/n-3) ratio = 9/1

C. Recommendations for Dietary Fat/Fatty Acid Intake

   (Assuming 2600 kcal/day)

   (For normal healthy adults free of genetic disorders, medical
    disorders, etc.)

   (For prevention of cardiovascular disease, etc.)

1. Polyunsaturated Fatty Acids

   18:2n-6 = 14 g/d (4.8% of calories)

   18:3n-3 = 3 g/d (1.0% of calories)

   (EPA) 20:5n-3
     plus 22:6n-3 (DHA) $\quad$ = 0.8 g/d (0.27% of calories)

   Total PUFA = 18 g/d (6-7% of calories)

   (n-6/n-3) ratio = (4/1)

   (n-3 as 18:3/n-3 as EPA plus DHA ratio = (4/1)

2. Saturated Fatty Acids

   Total = 18 g/d (6-7% of calories)

     (P/S) ratio = (1/1)

3. Monounsaturated Fatty Acids

   Total = 36 g/d (12-14% of calories)(mainly as oleic acid)

4. Total Fat = 6-7% as PUFA

           6-7% as Saturated Fat

          <u>12-14% as Monounsaturated Fat</u>

   Fat = 24-28% of dietary calories.

   (P/S/M) ratio = (1/1/2) .

A DOUBLE-BLIND CONTROLLED TRIAL OF OILS CONTAINING LONG-CHAIN-n-3-POLYUNSATURATED FATTY ACIDS IN THE TREATMENT OF MULTIPLE SCLEROSIS

D. Bates, J. French, S. Nightindale, and
D. Shaw (Newcastel upon Tyne)

S. Hawkins and H. Millar (Belfast)

M. Sidey, A. Smith, R. Thompson, and K. Zilka (London)

M. Gale and H. Sinclair (Oxford), U.K.; with others

A trial of n-3 polyunsaturated fatty acids in the treatment of multiple sclerosis has been conducted over a 5-year period. 312 ambulant patients with acute remitting disease were randomly allocated to treatment and placebo controlled groups. Both were given dietary advice to increase the intake of n-6 polyunsaturated fatty acids and the treatment group in addition received capsules containing n-3 polyunsaturated fatty acids. The results showed no significant difference at the usual 95% confidence level, but there was a trend in favour of the n-3 treated group in all parameters examined including rate of deterioration, frequency and severity of attacks and number of patients improving or remaining unchanged during the course of the trial.

REGULATION OF HEPATIC AND INTESTINAL LIPID METABOLISM BY DIETARY n-3 AND n-6 FATTY ACIDS

J.E. Bauer, P. Schenck, and C.H. Beauchamp

University of Florida, College of Veterinary Medicine
Gainesville, Florida

The regulation of hepatic and intestinal 3-hydroxy-3-methyl-glutaryl-CoA (HMG-CoA) reductase, acyl-CoA:cholesterol acyltransferase (ACAT) and serum arylesterase activities by dietary fish or vegetable oils was examined in the rabbit. Diets containig menhaden (MHO), safflower (SAF), cocoa butter (COB), or olive (OLV) oil (14% w/w) were used in conjuction with a known atherogenic casein/wheat starch basal diet. Serum free and esterified cholesterol and phospholipid concentrations were elevated in the MHO diet fed rabbits with a decrease in free fatty acid concentration. Animals fed the other diets were similarly affected with the SAF diet resulting in the least hyperlipemic effect. Serum arylesterase activities were unchanged. Marked elevations of lipoprotein cholesterol was found in all the MHO group low density fractions, while animals fed the OLV diet had more lipoprotein cholesterol in the very low density fraction. In the liver, microsomal ACAT activities were unchanged, and HMG-CoA reductase activities were decreased by the ingestion of MHO and SAF diets. In the intestine, microsomal ACAT activities were increased in the MHO diet fed rabbits, while HMG-CoA reductase activities were similar in all groups. Lipid class analysis of hepatic and intestinal microsomes revealed no significant changes except an increase in the intestinal esterified cholesterol concentration of MHO diet fed rabbits. Fatty acid compositional analysis of microsomes demonstrated an enrichment of n-3 polyunsaturated fatty acids (PUFA) in the MHO group, and n-6 PUFA in the SAF group. The COB and OLV diets resulted in nearly identical microsomal fatty acid patterns in spite of dietary fat source, most likely the result of desaturase and elongase activities.

Bile analysis revealed increased free cholesterol concentrations with MHO diet feeding. No other changes were observed including bile acid content. The changes seen with fish oil feeding indicate a possible exacerbation of casein-induced entero-hepatic recirculation of biliary cholesterol. This possibility is supported by increased intestinal ACAT activities and the subsequent assembly of cholesterol rich post-

prandial lipoproteins with their presentation to the liver.
Decreased hepatic microsomal ACAT activities also support a net increased flux of biliary cholesterol and hypercholesterolemia when fish oil casein diet are fed.

A RAPID STIMULATION OF LIVER PALMITOYL-CoA SYNTHETASE, CARNITINE PALMITOYLTRANSFERASE AND GLYCEROPHOSPHATE ACYLTRANSFERASE COMPARED TO PEROXISOMAL beta-OXIDATION AND PALMITOYL-CoA HYDROLASE IN RATS FED HIGH FAT DIETS

R.K. Berge and A. Aarsland
Laboratory of Clinical Biochemistry, University
of Bergen, Bergen, Norway

Key enzymes involved in oxidation and esterification of long-chain fatty acids were investigated in male rats fed different types and amounts of oil in the diet. A diet with 20% (w/w) fish oil, partially hydrogenated fish oil and partially hydrogenated soybean oil was shown to stimulate the mitochondrial and microsomal palmitoyl-CoA synthetase activity compared to soybean oil-fed animals after 1 week of feeding. Rapeseed oil had no effect.

Partially hydrogenated oil in the diet resulted in significant higher levels of mitochondrial glycerophosphate acyltransferase compared to unhydrogenated oils. Rats fed 20% (w/w) rapeseed oil had a decreased activity of this mitochondrial enzyme whereas the microsomal glycero-phosphate acyltransferase activity was stimulated to a comparable extent with 20% (w/w) rapeseed oil, fish oil or partially hydrogenated fish oil in the diet.

Increasing the amount of partially hydrogenated fish oil (from 5% to 25% (w/w)) in the diet for 3 days led to increased mitochondrial- and microsomal palmitoyl-CoA synthetase activities and microsomal glycerophosphate acyltransferase activity with 5% of this oil in the diet. The mitochondrial glycerophosphate acyltransferase was only marginal affected with increasing oil doses.

Administration of 20% (w/w) partially hydrogenated fish oil, the mitochondrial- and microsomal palmitoyl-CoA synthetase activities, carnitine palmitoyltransferase activity and microsomal glycerophosphate acyltransferase activity were rapidly increased almost to their maximum value within 36 h. In contrast, the glycerophosphate acyltransferase and palmitoyl-CoA hydrolase of the mitochondrial fraction and the peroxisomal beta-oxidation reached their maximum activities after administration of the dietary oil for 6 1/2 days.

This sequence of enzyme changes a) are in accordance with the proposals that increased cellular level of long-chain acyl-CoA act as metabolic message for induction of peroxisomal beta-oxidation and palmitoyl-CoA hydrolase i.e. regulated by a substrate-induced mechanism and b) indicate that with partially hydrogenated fish oil, a greater part of the activated fatty acids are directed from triacyl-glycerol esterification and hydrolyzation toward oxidation in the mitochondria. It is also conceivable that the mitochondrial beta-oxidation is proceeding before enhancement of peroxisomal beta-oxidation.

THERAPEUTIC EFFECTS OF DIETARY FISH AND FISH OIL SUPPLEMENTATION IN HYPERLIPIDAEMIA

R. Clarke, A. Tobin, C. O'Morain, and I. Graham
Adelaide Hospital, Dublin, Ireland

Endothelial damage by neutrophil superoxide release has been implicated in the pathogenesis of atherosclerosis. Neutrophils from hyperlipidaemic subjects exhibit increased oxidative metabolism and superoxide generation as measured by chemiluminescence, though whether this is causally related to elevated serum lipids in unknown.

Twenty hyperlipidaemic patients (cholesterol 7 mmol/L on three consecutive occasions) were randomised to receive daily supplements of fish oil containing 6 g of n-3 fatty acids or a placebo containig olive oil, and ten hyperlipidaemic patients were invited to take a dietary regime of four fish meals per week. All patients were maintained on a standard lipid lowering diet throughout the study period of six months. Serum lipids were measured at monthly intervals; neutrophil superoxide release was determined in a luminal-dependent assay at entry and on completion of the study.

Fish oil supplementation was associated with a 23% reduction in triglycerides and a 21% reduction in neutrophil chemiluminescence (P .05). There was no significant change in lipids or neutrophil function on placebo, while the fish diet was associated with a 14% reduction in chemiluminescence.

The lipid lowering effect of fish oil supplements observed in this study is consistent with previous reports. The parallel reduction in neutrophil superoxide release may reflect fish oil induced changes in membrane lipid composition which inhibit their formation. Reduction of cardiovascular risk by fish oil supplements may be partly due to reduced superoxide mediated endothelial injury.

DIFFERENTIAL MOBILIZATION OF ESSENTIAL FATTY ACIDS INTO THE SERUM FREE FATTY ACID POOL IN RESPONSE TO GLUCOSE INGESTION

S.C. Cunnane, T.M.S. Wolever, J.K. Armstrong, and D.J.A. Jenkins
Department of Nutritional Sciences, University of Toronto
Toronto, Canada

Ingestion of 50 g glucose as a bolus (within 5') induces a 60% fall (p $\leq$ 0.01) in the serum total free fatty acid (tFFA) pool over 60-120'. Between 120-240' after the glucose ingestion, tFFA rebound, overshooting the baseline (0') level by 20% (p $<$ 0.01). In comparison with the bolus ingestion of glucose, sipping 50 g glucose over the 240' results in a decrease in serum tFFA to 50-60% of baseline (p $<$ 0.01). We have determined the fatty acid composition of the tFFA pool during the time course of the tFFA response to both methods of glucose ingestion and have found that although the n-3 and n-6 EFA are qualitatively minor components of the serum tFFA pool, some EFA make a quantitatively significant response to bolus glucose ingestion. The net increase in serum tFFA 240' after the glucose bolus was 110 umol/L. The fatty acids responsible for this difference were (umol/L) 18:1n-9 (79), 18:2n-6 (18), 22:6n-3 (4), 16:1n-7 (3), 18:3n-3 (3), 14:0 (2) and others (<1). Compared to their baseline values (0'), these fatty acids increased: 22:6n-3 (55%), 18:1n-9 (45%), 14:0 (28%), 18:3n-3 (24%), 18:2n-6 (21%), 16:1n-7 (17%), 20:4n-6 (- 11%), others ($\leq$ 1%). After the maximal fall in tFFA (120' post-glucose bolus), the quantitative increase in individual FFA (at 240'; relative to their lowest mean value) was: 22:6n-3 (7.1X), 16:1n-7 (6.5X), 18:2n-6 (3.8X), 18:1n-9 (3.6X), 18:3n-3 (2.6X), 16:0 (2.4X), 14:0 (2.3X), 20:4n-6 (2.2X) and 18:0 (1.8X). These data indicate that the serum FFA response to the bolus ingestion of glucose is provided mainly by 18:1n-9 (on a mass basis) but also included a significant response by EFA, especially 18:2n-6, 18:3n-3 and 22:6n-3. The net response by 16:0 and 18:0 was 0 and by 20:4n-6 was - 11%. In contrast, the net change (0'-240') of individual FFA during glucose sipping was less variable (decrease of 29-52%) for all the fatty acids measured. Whether a change in the dietary n-6/n-3 ratio affects the net and differential FFA response to glucose ingestion has not been determined. There is considerable agreement between the whole body rates of oxidation of long chain fatty acids by the rat and the ranking shown here of fatty acids mobilized into the serum tFFA pool (14:0 > 18:1n-9 > 18:2n-6 > 18:3n-3 > 16:0 > 18:0 > 20:4n-6). Whether the EFA mobilized into the tFFA pool during glucose ingestion are in fact oxidized or re-esterified at another site, e.g. liver, remains to be established.

ACKNOWLEDGEMENT

Supported by the Natural Sciences and Engineering Research Council of Canada.

COMPARED ABSORPTION of MARINE FATTY ACIDS in HUMANS AFTER TWO WEEKS

ORAL INTAKE of a TRIGLYCERIDE or an ETHYLESTER CONCENTRATE

Knut H. Dahl, G. ØI, and H.E. Krokan

Norsk Hydro Research Centre, Porsgrunn, Norway

The marine fatty acids Eicosapentaenoic acid (EPA) and Docosa-hexaenoic acid (DHA) are known to influence human health, and in particular, a preventive effect against cardiovascular diseases has been suggested. One clinical study has indicated a less effective absorption of EPA in the form of a ethylester compared to a tri-glyceride. In order to further elucidate this question, an EPA/DHA-ethylester concentrate (K85, containing 55% EPA and 30% DHA) or an EPA/DHA-triglyceride concentrate (TG30, containing 18% EPA and 12% DHA) was given to healthy male volunteers for two weeks and the fatty acid profile of the total serum lipids were analyzed on GC before and after the supplementation period. Two groups of 8 volunteers took 12 or 24 g/day of triglyceride TG30 corresponding to 2.2 or 4.4. g EPA/day. Three groups of 8 volunteers took 4, 8 or 14 g/day of ethylester K85 corresponding to 2.2, 4.4 or 7.7 g EPA/day. The supplementation produced a dose-dependent increase in the relative content of EPA and DHA in the serum lipids. After the ethylester (K85) supplementation the increases in EPA were 4.6, 8.2 and 8.9% respectively while the increases after supplementation with triglyceride (TG30) were 5.1 and 7.5%.
The serum lipid fatty acid profile changes were the same whether EPA was taken as ethylester or as triglyceride. This shows that a EPA-ethylester is absorbed with the same efficacy as an EPA-tri=glyceride.

THE EFFECT OF DIETARY n3-FATTY ACIDS UPON THE BIOSYNTHESIS OF UNSATURATED FATTY ACIDS, ENERGY METABOLISM, LIPID COMPOSITION AND FUNCTION of MITOCHONDRIA

R. De Schrijver

Laboratory of Nutrition, Catholic University of Leuven
Leuven, Belgium

For a period of 10 weeks, five groups of rats (60-65 g) received diets with 10% fat and 6 energy % linoleic acid. The control diet was practically free from n3-fatty acids; four diets contained 0.8% and 1.6% linolenic acid (as the ethylester) and 0.8% and 1.6% long chain-polyunsaturated n3-fatty acids (as menhaden oil) respectively. Higher incorporation of long chain n3-fatty acids in the liver microsomes and mitochondria were found in the animals fed menhaden oil, as compared to the groups which had to synthesize these n3-fatty acids from linolenic acid. The incorporation of long chain n3-fatty acids in the phospholipids occurred at the expense of higher n6-fatty acids, while there was an increase of the linoleic acid level. The increase of the amounts of linoleic acid in the phospholipid fractions of liver microsomes and mitochondria was due to the inhibition of its conversion into higher n6-homologues. This was shown by the inhibition of the "in vitro" activities of the 5- and 6- desaturases in the groups fed 1.6% linolenic acid, 0.8% and 1.6% long chain-polyunsaturated n3-fatty acids. In the experimental conditions used, the inhibition of the biosynthesis of arachidonic acid had no deleterious effects upon the animals as could be concluded from their growth rates, feed conversions and basal metabolic rates. However it must be considered that dietary n3-fatty acids may raise the minimum dietary requirements of linoleic acid, especially when diets with low contents of linoleic acid are fed. Feeding n3-fatty acids had no significant effect upon the contents of triglycerides, cholesterol esters, free cholesterol and phospholipid classes in the liver mitochondria. Feeding long chain n3-fatty acids probably activated the phospholipase A_2 and increased the free fatty acid concentration in the liver mitochondria.

The liver mitochondria of the 1.6% linolenic acid- and both menhaden-groups, showed stimulated "in vitro" respiration rates and reduced ADP/O ratios. Such mitochondria had three characteristics in common:

1) elevated levels of free fatty acids; 2) the desaturation-indexes of the total phospholipids were higher than 200; 3) high linoleic acid contents of diphosphatidylglycerol. However, it was not clear what lipid-factor was precisely involved in the impairment of oxidative phosphorylation. Possibly, an inhibitory effect of free fatty acids upon mitochondrial energy conversion was enhanced by the more fluid state of the lipid bilayer in the mitochondrial inner membrane, induced by the high unsaturation of the phospholipids. Also the linoleic acid-rich diphosphatidylglycerol, having a possible function as co-factor for the ATP-synthetase system, may have brought about a too fluid state of the lipid domain in which the enzyme operates and may have induced an unfavourable configuration of the enzyme.

SELECTIVE INCORPORATION OF EICOSAPENTAENOIC ACID IN POSITION 2 OF PLASMA TRIGLYCERIDES AFTER A DIETARY SUPPLY OF 2 EICOSAPENTAENOYL GLYCERIDE IN MAN

B. Descomps, S. El Boustani, L. Monnier, F. Mendy, and
A. Crastes de Paulet

INSERM U.58, Montpellier; Laboratoire de Biochimie A
et Service des Maladies Metaboliques et Endocriniennes
du CHR de Montpellier, Montpellier , France

The chemical form of polyunsaturated fatty acids (PUFA) in dietary sources is an essential determinant for their absorption and bio-availability. For example, we recently showed that enteral absorption in man of eicosapentaenoic acid (EPA) is strongly dependent on its chemical form. The ingestion of the triglyceride 1,3-dioctanoyl-2-eicosapentaenoyl glycerol (2-EPA) is followed by earlier and higher EPA incorporation into plasma triglycerides (TG) than after ingestion of the ethylester (1). The present investigation was designed to check if in man, the ingestion of triglycerides carrying a PUFA in position 2 such as 2-EPA would preferentially result in insertion of this PUFA in position 2 of plasma triglycerides.

For this purpose, three normal adult fasting subjects were given the equivalent of 1 g EPA in the form of the triglyceride 2-EPA. The position of EPA in the plasma TG of the subjects was determined at the time of maximal incorporation into TG (4-6 h after ingestion) by combination of selective enzymatic hydrolisis with pancreatic lipase, thin layer and gas liquid chromatography. The experimental conditions for the selective enzymatic hydrolysis were checked in preliminary experiments in which were compared the enzymatic hydrolysis of reference synthetic triglycerides: 1,3 dioctanoyl-2-eicosapentaenoyl glycerol and 2,3 dioctanoyl-1 eicosapentaenoyl glycerol (1-EPA). Conditions were selected in which maximal enzymatic hydrolysis of the esters in position 1 and 3 was associated with minimal liberation of the fatty acid in position 2. In these conditions, the enzymatic hydrolysis of a TG sample containing EPA in an unknown position results in the liberation of EPA as free fatty acid if it was esterifying position 1 or 3 of the initial TG or to a 2 monoglyceride if the esterification was in position 2. Thus, the determination is reduced to the quantification of EPA present as a free fatty acid or as a 2 monoglyceride in the products of the selective enzymatic hydrolysis.

The overall procedure can be summarized as follows: after Folch's extraction the plasma TG were separated by thin layer chromatography, extracted from the gel and submitted to hydrolysis by pancreatic lipase (50 ug for a 500 ug TG sample, 10-20 min. at $37^{o}C$). After extraction the products of hydrolysis were separated by thin layer chromatography, transesterified and the EPA content of each fraction determined by gas liquid chromatography on a FFAP capillary column.

The analysis of the plasma TG of the subjects after ingestion of the triglyceride 2-EPA pointed out that 100% of the EPA recovered in the hydrolysis products of pancreatic lipase remained as 2-monoglyceride. No EPA could be detected in the free fatty acid fraction.

This result demonstrates that after ingestion of 2-EPA, this polyunsaturated fatty acid escapes to random distribution between the three positions of the triglyceride during the absorption process. The insertion of PUFAs in position 2 of dietary TG appears as an interesting way for supplying directly and efficiently these acids in a critical position of the glycerol molecule in plasma lipids.

REFERENCES

(1) B. Descomps, S.El Boustani, C.Colette, L.Monnier, F.Mendy, A.Crastes de Paulet, Lipids 10, 711, 1987.

THE EFFECTS OF A FISH OIL PREPARATION (EPAGIS) ON PLASMA LIPIDS, APOPROTEINS, CIRCULATING AGGREGATED PLATELETS AND HEMORHEOLOGIC PARAMETERS IN HYPERLIPIDEMIC SUBJECTS

P. Green, J. Fuchs, P. Budowski, Y. Lurie, I. Beigel,
J. Agmon, L. Leibovici, R. Mamet, and N. Schoenfeld

Beilinson Medical Center, Petah Tiqva, Israel

Twenty-seven hyperlipidemic subjects- 15 type IIB and 12 typeIV participated in a randomized double-blind, crossover clinical trial which compared effects of a fish oil preparation (EPAGIS) providing 4.8 g n-3 PUFA per day, to that of a placebo containing vegetable oil on the following laboratory parameters: total serum cholesterol, HDL-cholesterol, serum triglycerides, apoproteins A1 and B, circulating aggregated platelets, erythrocyte filtrability and plasma viscosity. Fatty acids were analysed in the serum and in the membrane phospholipids of erythrocytes and thrombocytes.

There was no significant change in the serum total cholesterol and in the HDL-cholesterol in the fish oil period compared to that of placebo, although there was a trend of decrease in the former and a small trend of increase in the latter. There was, however, a significant decrease in the triglyceride level: pretreatment levels of 576.2 $\pm$ 59.5 (mg/dl, mean $\pm$ SEM) versus post-treatment levels of 343.8 $\pm$ 31.2, $P = 0.001$. There was no significant change in the apoprotein levels. There was a significant decrease in plasma viscosity: pretreatment levels of 1.400 $\pm$ 0.016 (cp. mean $\pm$ SEM) versus post-treatment levels of 1.349 $\pm$ 0.017, $P = 0.001$, but no significant difference in erythrocyte filtrability.

These findings are discussed with special emphasis on their relevance to prevention of atherosclerosis.

EFA-STATUS of NEONATES BORN AFTER NORMAL AND COMPLICATED PREGNANCIES INFLUENCE OF A FISH-ENRICHED DIET

G. Hornstra[1], R. van Houwelingen[2], J. Gerrard[3]
and H. Huisjes[4]

Department of Biochemistry(1) and Human Biology(2)
Limburg University, Maastricht, The Netherlands;
Manitoba Institute of Cell Biology, Winnipeg, Canada(3)
Department of Gynaecology and Obstetrics, Academic Hospital
Groningen, The Netherlands(4).

To assess the EFA status of Dutch neonates born after uncomplicated pregnancies, fatty acid compositions were determined, isolated from umbilical veins and arteries, reflecting foetal EFA supply and return, respecyively. The fatty acid profiles were characterized by the absence of timnodonic acid (20:5(n-3) eicosapentaenoic acid), a low (2-3%) content of linoleic acid (18:2(n-6)) and reasonable amounts of arachidonic acid (20:4(n-6), 10-15%) and cervonic acid (CA, 22:6(n-3) docosahexa-enoic acid, 3-5%). In each cord, the efferent blood vessels (arteries) contained significantly less fatty acids of the linoleic (n-6) and linolenic (n-3) families and more fatty acids of the oleic (n-9) family than the afferent blood vessel (the vein). In the arteries considerable amounts (2-4%) of mead acid (20:3(n-9)) and its elongation product 22:3(n-9) were found. These observations indicate that the EFA-requirement of peripheral foetal tissue is not adequately covered.

In contrast to all other (n-6) fatty acids, 22:5(n-6) was significantly higher in arteries compared to veins. Since a shortage of cervonic acid is known to stimulate the compensatory synthesis of 22:5(n-6), this observation indicates that the CA status of Dutch newborn is not optimal. This condition may have negative implications for the developing nervous system. In Dutch neonates, born after severe toxaemia, this putative CA 'deficiency' seems more pronounced since in these cases, the arterio-venous difference of the 'marker fatty acid' 22:5(n-6) is 5-10 times higher than in neonates born after un-complicated pregnancies ($P_2 < 0.001$).

To investigate the possible role of dietary fish, comparative studies were done among meat or fish eating Inuit women in Northern Canada. Umbilical cords obtained after normal pregnancies were analyzed as described above. Moreover, blood pressure records of 70 meat-eating

423

and 43 fish-eating pregnant Inuit women were checked retrospectively, to investigate the pregnancy-associated changes in blood pressure. In the cords from fish-eating Inuit mothers the arteriovenous difference for 22:5(n-6) was negative and significantly different from zero, similar to that for the other (n-6) fatty acids. In cords from meat-eating Inuit mothers the 22:5(n-6) content did not change significantly from artery to vein.

This observation suggests that dietary enrichment with fish improves the CA status in such a way that the efferent blood still contains sufficient CA to prevent a CA-shortage in the 'downstream' foetal tissues.

Blood pressure after 20 weeks of pregnancy did not differ significantly between fish-eating and meat-eating Inuit women. The same holds for pregnancy durations of 30 and 35 weeks. The mean blood pressure for the last 6 hours before delivery, however, was higher in the meat-eating women ($P_2 < 0.10$ and < 0.005 for systolic and diastolic pressures, respectively). This suggests that a fish-enriched diet may reduce the stress-related increase in blood pressure. No clear cut cases of toxaemia were observed in this pilot, which will be repeated and extended this year.

DIETARY n-9, n-6 AND n-3 FATTY ACIDS AFFECT INOSITOL PHOSPHATE BUT NOT THROMBOXANE FORMATION BY PLATELETS, IN THE RABBIT

L. Medini, S. Colli, E. Tremoli, and C. Galli
Institute of Pharmacological Sciences and Grossi-Paoletti
Center for the Study of Hyperlipidemias, University of Milan
Milan, Italy

Semisynthetic diets containing 5% (w/w) of either olive oil (OO), corn oil (CO) or fish oil (FO, as MaxEPA) were fed to male rabbits (8 animals/group) for a period of 5 weeks. At the end of the feeding period, blood was drawn from incannulated arteries under anaesthesia and both platelet-rich-plasma and washed platelets were separately prepared. Plasma lipids and platelet phospholipids were analyzed as reported by Mosconi et al.. Platelet thromboxane formation by PRP was assessed by measuring levels (RIA) at 5 min of incubation with 5 U thrombin. Washed platelets were labelled with ^{3}H-myo-inositol and subsequently stimulated with 1 U thrombin for 10 and 90 s..
Lipids and water soluble products were extracted and partitioned. Inositol phosphates were separated by ion exchange chromatography and radioactivity was measured in lipid extracts and in the inositol phosphate fractions. Thromboxane formation was not appreciabily affected by dietary treatments and this observation may be related to the minimal differences of AA contents in PC and PI in platelets from the three groups.   Remarkable differences were, instead, observed in phospho-inositide and inositolphosphate labelling. Lipid labelling was quite lower in both non stimulated and stimulated platelets from the FO group than in the other groups. Labelling of the water soluble products was greatly increased after stimulation and the maximal increment occurred, at both 10 and 90 s, in the IP_2 fraction, in all groups. Basal levels of IP_3 and IP_2 were highest in the CO group and lowest in the FO, but the percent increments after stimulation were highest in the OO and lowest in the FO group. The data indicate that inositol phosphate production in platelets is markedly affected by dietary manipulations, even in conditions in which the AA cascade is not appreciably modified.

DIFFERENTIAL ACCUMULATION OF EPA AND ARACHIDONIC ACID IN PLASMA AND PLATELET LIPIDS IN MaxEPA FED RABBITS

C. Mosconi, M. Blasevich, M Gianfranceschi, and C. Galli

Institute of Pharmacological Sciences, University of Milan
Milan, Italy

Three groups of male rabbits (8 animals/group) were fed for a period of 5 weeks semisynthetic diets containing 5% (w/w) of either olive oil (OO), corn oil (CO) or fish oil (FO, as MaxEPA). At the end of the feeding periods blood was drawn from incannulated carotid arteries under anaesthesia, plsma was obtained and washed platelets prepared. The fatty acid compositions of plasma phospholipids (PL), triglycerides (TG) and cholesterol esters (CE) and of total lipids and individual PL classes of platelets were analyzed by combined TLC and GLC chromatography. In addition, absolute levels of linoleic (LA), arachidonic (AA) and eicosapentaenoic acids (EPA) were measured by quantitative GLC in individual lipid classes. The fatty acid composition of plasma lipids generally reflected that of dietary fats, but considerably differences were detected especially in the distribution of the three measured PUFA in FO vs OO and CO fed animals. In plasma of the FO group, EPA was found mainly in TG and to lower levels, in PL and CE. In addition, the distribution of AA among the different plasma lipid classes was significantly different from that in the other animal groups, since levels were reduced in PL and increased in CE and TG.

The administration of FO vs the other oil supplements resulted also in considerable changes in platelet fatty acids. All the major glycero PL (PS,PC,PI and PE) were depleted of LA, the largest decrement occurring in PE. Significant reduction of total AA was also observed, and the decrement in PE accounted for more than 90% of the total AA loss in platelets. Small decline of AA occurred in PI and some elevation in PS. EPA accumulated mainly in PE, but significant increments as percentage of total fatty acids occurred also in PC and PS with a minimal increment in PI. The unsaturation indexes of individual PL were very similar in the OO and CO groups, but they were markedly higher in the FO group, due to replacement by EPA mainly of LA rather than of AA.

The data indicate that exogenously administered n-3 PUFA follow a pattern of distribution in plasma and cell lipids quite different from that of the endogenous n-6. This should be considered in studies on the biological effects of n-3 fatty acids and in the interpretation of their mechanisms of action.

MARINE MAMMALS : ANIMAL MODELS FOR STUDYING the DIGESTION AND TRANSPORT OF DIETARY FATS ENRICHED IN OMEGA-3 FATTY ACIDS

D.L. Puppione, R.J. Jandacek, S.T. Kunitake, and D.P. Costa

University of California at Santa Cruz, Los Angeles and San Francisco; The Proctor and Gamble Co., Miami Valley Laboratories, U.S.A.

In contrast to seed or plant oils which contain triacylglycerol molecules with the relative percent distribution at the three positions of the glycerol backbone being essentially the same for each of the major fatty acids, the distribution of fatty acids in animal fats are asymmetrically distributed. Fats of marine animals are not exception. However, the fatty acid distributions on the triacylglycerol molecules in the blubber of marine mammals and in fish oils differ. In marine mammalian blubber, the polyenoic omega-3 fatty acids are located on the outer positions of the molecule, i.e. the 1 and 3 positions. In fish oils, the polyenoic omega-3 fatty acids are located almost exclusively in the 2 position.

Because most marine mammals eat either fish or plankton (killer whales, polar bears and leopard seals being exceptions), we were interested in determining if the fat of marine mammalian milk contained triacylglycerol molecules with the polyenoic fatty acids distributed as in fish oil or as in the blubber of the lactating female. In our studies, we have obtained data on the milk fat of two species of pinnipeds the northern elephant seal (Mirounga angustirostris) and the northern fur seal (Callorhinus ursinus). These data demonstrate that the polyenoic omega-3 fatty acid together with the long chain monoenoic acids are located on the 1 and 3 positions. We intend to use newborn pinnipeds to determine if differences in the ingestion and absorption of fat which would result following the ingestion of milk as compared with fish, would demonstrably alter the physicochemical properties and metabolism of chylomicra as well as the other lipoprotein classes.

NEW HEALTHY EFFECTS OF w3 FATTY ACIDS

THEIR ROLE AS DIETARY ANTIMUTAGENS

H.W. Renner

Federal Research Centre of Nutrition, Karlsruhe, F.R.G.

While looking for possible antimutagens in foodstuffs, the esters of unsaturated fatty acids were tested for their potential to reduce chromosomal damage induced by a test mutagen.

Busulfan (50 mg/kg), a mutagen acting without metabolic activation, was given to Chinese hamsters orally, along with unsaturated fatty acid esters dosed 10-500 mg/kg. After 30 h sampling time, bone marrow cells were prepared for a chromosome aberration test.

Fatty acids with 12 to 18 C-atoms in the acid seem to disclose antimutagenic effects by reducing the range of induced chromosomal damage from about 10% to about 3%. The threshold for the observed effect of the unsaturated fatty acids lies at about 10 mg/kg and thus corresponds to amounts of PUFA taken in the normal daily diet.

Further studies applying the same experimental design and using the essential fatty acids linoleic (w6), linolenic (w3) and their derivatives, revealed for the mother compounds once more equal anti-mutagenic effects, but the derivatives of both fatty acids produced clearly different effects: linoleic acid derivatives (e.g. arachidonic acid) exhibited no antimutagenic properties while all the linolenic acid derivatives gave a distinct antimutagenic response. Relations between w3 fatty acids and inhibitory effects on tumor development in the mammalian organism are discussed.

MARINE LIPID CONCENTRATE AND ATHEROSCLEROSIS IN THE RABBIT MODEL

D.S. Sgoutas, J.A. Hearn, K.A. Robinson, S.B. King, and G.S. Roubin

Andreas Gruentzig Cardiovascular Center, Clinical Research Facility, and Department of Pathology and Laboratory Medicine, Emory University School of Medicine, Atlanta U.S.A.

Twenty-seven New Zealand white rabbits underwent balloon de-endothelialization of the aorta and iliac arteries while consuming a 2% cholesterol, 10% peanut oil rabbit chow. Ten of these rabbits were fed one milliliter of concentrated marine fish lipid (MaxEpa^{Tm}) daily. Six weeks after de-endothelialization, angiography of the treated arteries was performed and histologic cross sections of the terminal aorta were measured with a planimeter. Iliac artery luminal diameters were also measured at consecutive 3mm divisions from the aortic bifurcation and found to have a mean lumen diameter of 1.60 ± 0.08 mm in the marine supplemented group (M) and 1.38 ± 0.12 mm in the control group (C) ($p < 0.001$). Analysis of variance on individual segmental diameters confirmed this difference. However, neither the angiographic diameters nor histologic, cross-sectional, luminal areas of the terminal aorta were different between groups. Instead, the mean cross-sectional area of the terminal aortic wall was significantly greater in the marine lipid fed group (4.4 ± 1.2 mm^2 in M and 3.1 ± 0.6 mm^2 in C, $p < 0.01$). In addition, the vessel wall area showed a positive correlation with RBC incorporation of docosahexaenoic acid ($r = 0.82$, $p < 0.005$) in both groups. In the M group, RBC eicosapentaenoic acid and docosahexaenoic acids increased 100% and 650%, respectively, over baseline.

We concluded that marine fish lipids 1) are incorporated into rabbit RBC cell membranes during feeding, 2) increase vessel wall thickness of the distal aorta in proportion to the docosahexaenoic acid present, without compromising the luminal area, and 3) provide mild sparing of the luminal diameters of more distal arteries in this model of atherosclerosis.

FLAX AS A SOURCE OF OMEGA-3

Paul A. Stitt

ENRECO/Essential Nutrient Research Company, Manitowoc U.S.A.

Flax seed has been found to contain from 18 to 22% alpha-linolenic acid, the vegetarian form of omega-3. The ground whole seed is easily emulsified in water and has a sweet, nutty flavour. No deleterious effects have been found in short or long-term feeding experiments in animals with nutritionally balanced diets containing high levels of flax.

Feeding flax that contains 10 grams of alpha-linolenic acid per day to humans produced reductions of serum triglycerides in hypertriglyceridemics of 49% in three weeks. Flax is a good source of omega-3 fatty acids for humans and animals.

INHIBITION OF MONOCYTE SUPEROXIDE PRODUCTION AND CHEMILUMINESCENCE BY DIETARY N-3 FATTY ACID SUPPLEMENTATION

B.H. Weiner

Cardiovascular Medicine, University of Massachusetts
Medical School, Worcester, MA

Increased dietary N-3 fatty acid consumption has been associated with reduced atherogenesis in animal models and reduced cardiovascular morbidity and mortality in human populations. The mechanism of this inhibitory effect remains uncertain. Since tissue macrophages, derived from circulating monocytes, are important contributors to the atherogenic process, we assessed the effect of dietary N-3 fatty acid supplementation on monocyte inflammatory potential in 9 healthy volunteers. Superoxide production, as measured by cytochrome c reduction, and chemiluminescence were measured in stimulated monocytes. The measurements were made before and following 6 weeks of daily dietary supplementation with 3.6 gms of eicosapentanoic acid (EPA) and 2.4 gms docosahexanoic acid (DHA). Compliance with dietary supplementation was determined by monitoring fatty acid changes in lymphocyte composition by gas chromatography. Sufficient monocytes were not available to perform fatty acid determination to be made in these cells. Superoxide declined from 10.9 ± 6.1 NMoles/30 min/1×10^6 monocytes (mean $\pm$ S.D.) at baseline to 4.5 ± 3.2 after N-3 fatty acids were added ($P < .01$).

Chemiluminescence decreased from 941 ± 141 counts $\times 10^{-3}$/5 min/monocyte at baseline to 506 ± 156 after N-3 fatty acid supplementation ($P < .01$). Superoxide and chemiluminescence measurements were lower in every individual after taking the N-3 fatty acids. Lymphocyte arachidonic levels declined from $22.2 \pm 1.6\%$ (total fatty acids) to $14.2 \pm 5.4\%$ ($P < 0.1$). EPA increased from 0.8 ± 0.1 to $2.7\% \pm 1.4\%$, while DHA increased from $0.7 \pm 0.9\%$ to $2.1 \pm 1.1\%$ ($P < .01$).

These data demonstrated that dietary N-3 fatty acid supplementation can inhibit monocyte free radical production in association with substantial changes of fatty acid content. This finding may be important in elucidating the mechanisms of beneficial effects observed with these compounds.

CONTRIBUTORS